STATISTICAL INFERENCE

Classical and Bayesian Inference

STATISTICAL INFERENCE

Classical and Bayesian Inference

Dr. S.R. Patel
Retired Professor of Statistics
Sardar Patel University
Vallabh Vidyanagar
Gujarat, India

TECHSAR™
NEW DELHI

Published and Printed by

Techsar Pvt. Ltd.
4435-36/7, Ansari Road, Daryaganj
New Delhi–110 002 (India)
E-mail: tech_info@techsarworld.com
Website: www.techsarworld.com

ISBN: 978-93-90620-06-7

Reprint 2025

Preface

Nowadays statistical techniques are widely used in all faces of the society and scientific research. So in this book mathematical theory of statistical inference: estimation and testing of hypotheses have been developed. The purpose of this book is to give up-to-date theory of statistical inference with less rigorous mathematical treatment, giving realistic illustrations and simple proofs of the theorems so that the reader may easily go through it. My thirty-three years teaching experience to postgraduate classes and refreshers courses will surely help the students and teachers with the help of this book.

There are fourteen chapters out of which five chapters are on theory of estimation and nine chapters are on testing of hypotheses in this proposed book.

Sufficiency and completeness have been introduced in chapter 1. Having defined sufficiency with some illustrations, the Neyman-Fisher factorization theorem is proved in very simple way. Many examples have been worked out by using factorization theorem. Concept of minimal sufficient statistic is given. After establishing the need of completeness and defining completeness it has been shown that completeness is a much stronger property than minimality and hence completeness is widely used in estimation theory and testing of hypotheses. Both properties of sufficiency and completeness are established for the wide class of exponential family of distributions.

The second chapter on unbiasedness deals with Rao-Blackwell and Lehman-Scheffe theorems. Using Basu's and Mitra's results along with Rao-Blackwell and Lehman-Scheffe theorems, many examples have been worked out. Special types of research problems have been considered for truncated families (non-regular families) which may be utilized by research workers.

The efficiency of estimator is the subject matter of the third chapter in which the lower bound of the variance of the unbiased estimator is obtained and after defining minimum variance bound (MVB) estimator many results have been given for it. It has been noted that the bound is independent of the unbiased estimator used. Chapman Robbins inequality is established without using Rao-Cramer (R-C) regularity conditions. For sharpening the R-C lower bound Bhattacharya bounds have been established. After obtaining Rao-Cramer lower bound matrix for several parameters many examples have been worked out for Bhattacharya bounds and R-C lower bound matrices.

Chapter 4 is devoted for different large sample criteria for estimation and some methods of estimation even though these methods do not guarantee that they satisfy desirable properties. After getting the estimator one has to show separately whether the property is satisfied by the estimator in which one is interested.

One of the important large sample properties of an estimator is the consistency. After defining it some theorems have been proved to check the consistency of proposed estimator. Another large sample properties of an estimator is the extension of consistency, known as consistent asymptotically normal (CAN) estimator. Some examples are worked out for illustrating CAN estimator. Further, it has been extended to multivariate normal distribution.

Some methods of estimation are examined for their large sample properties. Method of moments is very old and since many large samples desirable properties are not satisfied so it is rarely used. A widely used method of estimation is maximum likelihood estimation (MLE). Many illustrations are given for bad estimation or no estimation while using MLE for small samples. For large samples and with some mild conditions, it is shown that MLE is not only consistent but it is a CAN estimator. Uniqueness of MLE is also established for large sample. Minimum χ^2, modified minimum χ^2, some distances : Halinder, Haldane, Kullback Leibler are mentioned.

Bayes and Minimax estimation as a core part of decision theory has been discussed extensively in chapter 5. Explaining the concepts of Minimax and Bayes estimators, some theorems on Bayes estimators are proved. Obtaining Bayes estimator of normal distribution, multivariate extension has been done. To reduce the difficulty for the choice of prior distribution, some methods describing prior distribution have been discussed, especially non-informative prior, conjugate prior, etc. Along with capture-recapture problem and some other sampling problems are given as applications of Bayes estimator. Lastly, utility function is used as the complementary to loss function in order to get loss function, when it is difficult to get it. Some real-life illustrations are given for utility function.

The second part of Statistical Inference is on testing of hypotheses. Basic principles of testing of hypotheses are given for two types of errors, power function, Neyman theory for the choice of the best test.

Many times choice of the null hypothesis is very difficult. In this case many real-life illustrations are given for setting of null hypothesis. Another concept of the test function $\phi(x)$ is given and it is illustrated by examples. Usage of $\phi(x)$ include randomized and non-randomized tests. The problem of testing of hypothesis can be looked upon as a linear programming problem and hence the problem of testing hypotheses can be solved by using linear programming technique. All these concepts along with real-life illustrations is the subject matter of chapter 6.

Chapter 7 deals with Neyman-Pearson theory in which Neyman-Pearson lemma is proved for non-randomized version rather differently from the proofs given by the existing books and the proof given here for randomized version is lucid and simple. For composite alternatives the proof has been extended by restricting the monotone class of distributions. Many examples have been worked out. A section is devoted for locally most powerful test. Considering a least favourable distribution testing a composite hypothesis versus simple hypothesis test is carried out in the last section. Defining unbiased test and giving the form of generalized Neyman-Pearson lemma some illustrations have been given for unbiased test. Three theorems for the following three hypotheses have been proved by using generalized Neyman-Pearson Lemma and taking exponential family of distributions.

Hypothesis H vs Hypothesis K

(1) $H_1 : \theta \leq \theta_1 \text{ or } \theta \geq \theta_2 K_1 : \theta_1 < \theta < \theta_2$

(2) $H_2 : \theta_1 < \theta < \theta_2 K_2 : \theta \leq \theta_1 \text{ or } \theta \geq \theta_2$

(3) $H_3 : \theta = \theta_0 K_3 : \theta \neq \theta_0$

By using these theorems some examples are given. Moreover, introducing locally most powerful unbiased test, it is illustrated by examples in this chapter 8.

Introducing similar tests with illustration and connecting them with unbiased tests and Neyman structure tests, applications are made with exponential family of distributions for testing a parameter when other parameters are nuisance in chapter 9. Also comparisons of two Poisson and two Binomial distributions are made. Applications to normal parameters are done and confidence bounds have been discussed.

Sequential probability ratio test (SPRT) is the subject matter of chapter 10. After establishing approximate boundaries, OC and ASN functions have been obtained and are illustrated for some distributions. Wald's fundamental identity has been derived and it has been applied to get OC and ASN functions.

After introducing some results of order statistics, non-parametric tests, sign test, median test, run test, Kolmogorov-Smirnov test, Mann-Whiteney test, Wilcoxon signed rank test, etc. are discussed in chapter 11. Asymptotic results for the above tests have been derived in detail for those who are doing research in this area.

In chapter 12, UMP tests for testing two-sided hypothesis in the class of truncated one-parameter family of distributions have been derived and applications have been made to some specific distributions. Asymptotic tests for several exponential families of distributions have been obtained and applied to some particular distributions in chapter 13 whereas asymptotic tests for several truncated one parameter family of distributions have been discussed and applications have been carried out for some particular cases in chapter 14.

In short, the book is designed for the students who are offering major statistical courses in postgraduate classes. Some of the topics in estimation and some in testing are useful for those who are doing research work in this area.

S.R. Patel

Contents

1

CHAPTER

Sufficiency and Completeness

1.1 INTRODUCTION

Statistics stems from collection of data with their analysis and interpretation. For processing data one has to transform them into most suitable form so as to make decision easier. This requires the reduction of data to a smaller number of statistics. However, it is desirable that while reducing the data no information is lost contained in the distribution under study. Such type of statistics defined on the sample space is called sufficient statistic. Sufficient statistic itself does not necessarily provide good estimator in estimation theory and the best test in testing of hypothesis but it may help in arriving at the decision. Thus, the idea of sufficient statistic is to obtain a function defined on a random sample (X_1, X_2, ..., X_n) of a random variable (X) so as to reduce the sample space generated by a random sample without losing any information contained in a sample. For example, in some distributions, $T = \sum_{i=1}^{n} X_i$, contains all information relevant to the parameter (or distribution of X).

If the probability density function (p.d.f.) or probability mass function (p.m.f.) of T involves the parameter θ then T contains information about θ, otherwise it is independent of θ and in this case T is called sufficient statistic or ancillary statistic. Sufficient statistic and factorization theorem are dealt with in section 1.2 whereas minimal sufficient statistic is defined and obtained in section 1.3. Completeness is discussed in section 1.4. Exponential family of distributions is defined and its properties are given in section 1.5.

1.2 SUFFICIENCY

We can base our inference about θ by using T which is considerably simpler than X. Thus, we must define a function, $u(X) = u(X_1, X_2, ...,$

X_n) on $X = (X_1, X_2, \ldots, X_n)$ such that given the value of $u(X)$ no more information is provided by further observations on X. We define sufficient statistics as under.

Definition 1.1

Let $\mathbb{P} = \{P_\theta(x), \theta \in \Omega\}$ be the family of distributions of an observable random sample $X = (X_1, X_2, \ldots, X_n)$ having realized value $x = (x_1, x_2, \ldots, x_n)$ of X in R^n space. Let $T = u(X)$ be a function of X defined on the sample space $X = \{(x_1, x_2, \ldots, x_n)\}$, all possible observable values of X. Then T is said to be sufficient statistics for $\mathbb{P}$ or for θ if the conditional distribution of X given $T = t$ is independent of θ for all values of t. Note that the independence of θ means neither the probability function of the conditional distribution, $P(X/T = t)$ nor the domain contains θ.

In this definition how sufficient statistics contains all the information about θ? Whether the knowledge of the values of T only will provide the original observations? One can generate values of X, say X' by random mechanism so that the conditional distribution of X given $T = t$ is independent of θ.

If X' is another observation whose conditional distribution given $T = t$ is independent of θ where T is sufficient for θ or the family $\mathbb{P} = \{P_\theta, \theta \in \Omega\}$ of X then unconditional distribution is same as that of X. Let us prove this for discrete random sample, $X = (X_1, X_2, \ldots, X_n)$

$$
\begin{aligned}
P_\theta(X = x) &= \sum_t P_\theta(X = x, T = t) \\
&= \sum_t P_\theta(T = t)P(X = x/T = t), \ T \text{ being sufficient} \\
&= \sum_t P_\theta(T = t)P(X' = x/T = t) \\
&= \sum_t P_\theta(X' = x, T = t) \\
&= P_\theta(X' = x)
\end{aligned}
$$

Thus, knowledge of T alone X' gives information on θ random mechanism which provides same information on θ as X provides.

Example 1.1

Let $X = (X_1, X_2, \ldots, X_n)$ be a random sample from Bernoulli distribution with probability of success as θ. Examine whether $T = \sum_{i=1}^{n} X_i$ is sufficient for θ.

$$P_\theta(X = x/T = t) = \frac{P_\theta(X = x, T = t)}{P_\theta(T = t)}$$

$$= \frac{P_\theta(X_1 = x_1, \ldots, X_{n-1} = x_{n-1}), X_n = t - \sum_{i=1}^{n} x_i}{P_\theta\left(\sum_{i=1}^{n} X_i = t\right)}$$

$$= \frac{\prod_{i=1}^{n-1}(\theta^{x_i}(1-\theta)^{1-x_i})\theta^{t-\sum_{i=1}^{n-1} x_i}(1-\theta)^{1-t+\sum_{i=1}^{n-1} x_i}}{\binom{n}{t}\theta^t(1-\theta)^{n-t}}$$

$$= \frac{1}{\binom{n}{t}}$$

which being independent of θ, T is sufficient for θ.

Given the value of $T = 1$ for $n = 4$, the conditional probability of X given $T = 1$ is $\frac{1}{4}$. By random mechanism we can construct the original observations as 1000,0100,0010,0001 where 1 denotes success and 0 denotes failure. This does not require the knowledge of X.

Notes

(1) θ and T could be vectors.
(2) If there is one-to-one transformation $(X_1, X_2, \ldots, X_n) \to (T_1, T_2, \ldots, T_k, U_1, U_2, \ldots, U_{n-k})$ then $(T_1, T_2, \ldots, T_k)$ is sufficient for the parameters if conditional distribution of $(U_1, U_2, \ldots, U_{n-k})$ given $(T_1, T_2, \ldots, T_k)$ is independent of the parameters.

Example 1.2

If $X = (X_1, X_2, \ldots, X_n)$ is a random sample from uniform distribution $u(0, \theta)$ then examine whether $T = \max(X_1, X_2, \ldots, X_n)$ is sufficient for θ. As T has probability density function (p.d.f.)

$$f_T(t) = \frac{nt^{n-1}}{\theta^n}, 0 < t < \theta$$

and as joint probability function of order statistics, $X_{(1)}, X_{(2)}, \ldots, X_{(n)}$, is

$$f_{(X_{(1)}, X_{(2)}, \ldots, X_{(n)})}(x_{(1)}, x_{(2)}, \ldots, x_{(n)}) = \frac{n!}{\theta^n},$$

the conditional distribution of $X_{(1)}, X_{(2)}, \ldots, X_{(n)}$ given $T = t$ is

$$f_{X/T}(x/t) = \frac{n!}{\theta^n} \bigg/ \frac{nt^{n-1}}{\theta^n} = (n-1)!/t^{n-1}$$

which being independent of θ, T is sufficient for θ.

Example 1.3

Let $X_1, X_2, \ldots, X_n$ be a random sample from $N(\theta, 1)$ distribution. Examine whether $T = \sum_{i=1}^{n} X_i$ is sufficient for θ.

Making one–to–one transformation $(X_1, X_2, \ldots, X_n) \to (T, Y_2, \ldots, Y_n)$ where $Y_j = X_j - X_1$; $j = 2, 3, 4, \ldots, n$.

We have to find the conditional distribution of $Y_1, Y_2, \ldots, Y_n$ given $T = t$.

As $\quad \text{Cov}(T, Y_j) = \text{Cov}(T, X_j) - \text{Cov}(T, X_1)$ for $j = 2, 3, 4, \ldots, n$
$= 1 - 1 = 0$

T and $(Y_2, \ldots, Y_n)$ are independently distributed. So the conditional distribution of $Y_2, \ldots, Y_n$ given $T = t$ is unconditional distribution of $Y_2, \ldots, Y_n$, that is,

$$f_{Y_2,\ldots,Y_n/T}(y_2,\ldots,y_n/T = t) = f_{Y_2,\ldots,Y_n}(y_2,\ldots,y_n).$$

As $\text{Var}(Y_j) = 2$ for $j = 2, 3, \ldots, n$
$\text{Cov}(Y_i, Y_j) = 1$ for $i \neq j$, $i, j = 2, 3, \ldots, n$.

The variance covariance matrix Σ of $Y_2, \ldots, Y_n$ is

$$\Sigma = \begin{pmatrix} 2 & 1 & \ldots 1 \\ 1 & 2 & \ldots 1 \\ \vdots & \vdots & \vdots\ \vdots \\ 1 & 1 & \ldots 2 \end{pmatrix} \quad \text{and} \quad E(Y_j) = 0 \text{ for } j = 2, 3, \ldots, n,$$

$(Y_2, \ldots, Y_n)$ follows $(n-1)$ variate normal distribution $N_{n-1}(0, \Sigma)$. This being independent of θ, T is sufficient for θ.

Example 1.4

Let $X_1, X_2, \ldots, X_n$ be a random sample from the distribution having p.d.f. as

$$f_X(x;\alpha,\beta)=\beta^{-1}e^{-\frac{(x-\alpha)}{\beta}},x\geq\alpha,\quad\alpha,\beta>0$$

and let $Y_1, Y_2, \ldots, Y_n$ be ordered sample of $X_1, X_2, \ldots, X_n$. Then show that $\left(Y_1,\sum_{i=2}^{n}Y_i\right)$ is sufficient for (α, β). Making the transformation

$$Z_i=(n-i+1)(Y_i-Y_1),\ i=1,2,\ldots,n,\ y_0=0$$

the joint p.d.f. of $Y_1, Y_2, \ldots, Y_n$ is

$$f_{(Y_1,Y_2,\ldots,Y_n)}(y_1,y_2,\ldots,y_n)=n!\,\beta^{-n}e^{-\frac{\Sigma Y_i}{\beta}+\frac{n\alpha}{\beta}},\ Y_1<Y_2<\ldots<Y_n$$

and using $\Sigma Z_i = \Sigma Y_i$,

$$f_{(Z_1,Z_2,\ldots,Z_n)}(z_1,z_2,\ldots,z_n)=n!\,\beta^{-n}e^{-\frac{\Sigma z_i}{\beta}+\frac{n\alpha}{\beta}},$$

again making the transformation

$$Z_1=Z_1,U_1=Z_2,U_2=Z_2+Z_3,\ldots,U_{n-1}=Z_2+Z_3+\ldots+Z_n$$

the joint distribution of $U_1, U_2, \ldots, U_{n-1}$ will be

$$f_{(U_1,U_2,\ldots,U_{n-1},Z_1)}(u_1,u_2,\ldots,u_{n-1},z_1)=n!\,\beta^{-n}e^{-\frac{u_{n-1}}{\beta}+\frac{n\alpha}{\beta}-\frac{z_1}{\beta}},$$

$$0<u_1<u_2<\ldots<u_n,\ z_1\geq n\alpha,$$

integrating out $u_1, u_2, \ldots, u_{n-2}$ we have

$$f_{(U_{n-1},Z_1)}(u_{n-1},z_1)=\beta^{-1}e^{-\frac{z_1-n\alpha}{\beta}}\frac{e^{-\frac{u_{n-1}}{\beta}}u_{n-1}^{n-2}}{\beta^{n-1}(n-2)!}$$

which implies Z_1 has negative exponential distribution with parameters (n, α, β) and U_{n-1} has gamma distribution with parameters $\left(\frac{1}{\beta},n-1\right)$. Both Z_1 and U_{n-1} are independently distributed. So the conditional distribution of $U_1, U_2, \ldots, U_{n-2}$, given (Z_1, U_{n-1}), is

$$f_{U_1,U_2,\ldots,U_{n-2}/Z_1,U_{n-1}}(u_1, u_2, \ldots, u_{n-2}/z_1, u_{n-1}) = \frac{(n-2)!}{u_{n-1}^{n-2}}$$

which being independent of (α, β), (Z_1, U_{n-1}) is sufficient for (α, β),

that is, $\left(Y_1, \sum_{i=2}^{n} Y_i\right)$ is sufficient for (α, β).

Although definition (1.1) is easy to understand but it is unable to provide the sufficient statistic and secondly it is usually difficult to verify that the assumed statistic is sufficient due to conditional distribution. The following theorem overcomes both the difficulties.

Theorem 1.1: Neyman - Fisher factorization criterion

A necessary and sufficient condition for a statistic T to be sufficient for a family $\mathbb{P} = \{P_\theta(x), \theta \in \Omega)$ of $X = X_1, X_2, \ldots, X_n$ is that there exist non-negative functions $g_\theta(t)$, a function of t only and $h(x)$, a function of x only such that the probability function $p_\theta(x)$ of $P_\theta(x)$ satisfies

$$p_\theta(x) = g_\theta[t(x)]h(x) \qquad \ldots(1.2.1)$$

Case I Let $X = X_1, X_2, \ldots, X_n$ have discrete distribution, $p_\theta(x) = P_\theta(X = x)$

Necessary Condition

As T is given to be sufficient for θ, $P(X = x/T = t)$ is independent of θ and

$$P_\theta(X = x) = \sum_t P_\theta(X = x, T = t)$$

$$= P_\theta(T = t)P(X = x / T = t)$$

The necessary condition follows by defining

$$g_\theta(t) = P_\theta(T = t) \text{ and } h(x) = P(X = x/T = t)$$

Sufficient Condition

$$P_\theta(X = x/T = t) = \begin{cases} \dfrac{P_\theta(X = x, T = t_0)}{P_\theta(T = t_0)}, & \text{when } t = t_0 \\ 0, & \text{otherwise} \end{cases}$$

$$= \begin{cases} \dfrac{P_\theta(X = x)}{P_\theta(T = t_0)}, & \text{when } t = t_0 \\ 0 \quad , & \text{otherwise} \end{cases}$$

$$= \begin{cases} \dfrac{P_\theta(X = x)}{\sum_{x \in A(t_0)} P(X = x)}, & \text{when } t = t_0 \\ 0 & , \text{otherwise} \end{cases}, \text{ where } A(t_0) = \{x : t = t_0\}$$

$$= \begin{cases} \dfrac{g_\theta(t_0)h(x)}{g_\theta(t_0)\sum_{x \in A(t_0)} h(x)}, & \text{when } t = t_0 \\ 0 & , \text{otherwise} \end{cases}$$

which being independent of θ, T is sufficient for θ.

Case II Let X have absolutely continuous distribution.

Necessary Condition

Let $T = (T_1, T_2, \ldots, T_r)$, $r < n$ and $U = (U_{r+1}, U_{r+2}, \ldots, U_n)$. Then making one-to-one transformation $X \to (T, U)$

$$f_X(x; \theta) = f_{T,U}(t, u; \theta)\left|\frac{\partial(t,u)}{\partial x}\right|$$

$$= f_T(t; \theta) f_{U/T}(u/t; \theta)\left|\frac{\partial(t,u)}{\partial x}\right|$$

$$= g_\theta(t)h(x)$$

Sufficient Condition

Consider the inverse transformation $(U, T) \to X$,

$$f_{T,U}(t, u; \theta) = f_X(x(t, u); \theta)\left|\frac{\partial x}{\partial(t, u)}\right|$$

$$= g_\theta(t)h(x)\left|\frac{\partial x}{\partial(t,u)}\right|,$$

so that
$$f_T(t) = g_\theta(t) \int h(x(t,u)) \left| \frac{\partial x}{\partial(t,u)} \right| du.$$

Hence,

$$f_{U/T}(u/t) = \frac{f_{U,T}(t,u;\theta)}{f_T(t)} = \frac{h(x(t,u)) \left| \frac{\partial x}{\partial(t,u)} \right|}{\int h(x(t,u)) \left| \frac{\partial x}{\partial(t,u)} \right| du}$$

which being independent of θ, T is sufficient for θ.

Remarks

(i) If T is sufficient for θ then every one-to-one function of T is also sufficient for θ because if $U = k(T)$ is one-to-one function of T then

$$T = k^{-1}(U) \quad \text{and} \quad p_\theta(x) = g_\theta(t)h(x)$$
$$= g_\theta(k^{-1}(U))h(x)$$
$$= g^*_\theta(u)h(x)$$

(ii) If vector T is sufficient for θ then it does not follow that every component of T is sufficient for the corresponding component of θ. For example, if $X_1, X_2, \ldots, X_n$ is random sample from $N(\mu, \sigma^2)$ then

$$p_\theta(x) = \sigma^{-n}(2\pi)^{-\frac{n}{2}} e^{-\frac{\Sigma x_i^2}{2\sigma^2} + \frac{\mu \Sigma x_i}{\sigma^2} - \frac{n\mu^2}{2\sigma^2}}$$

$\Rightarrow (\Sigma x_i, \Sigma x_i^2)$ is sufficient for (μ, σ^2) by using factorization theorem, but if σ^2 is unknown, Σx_i is not sufficient for μ.

(iii) If T is sufficient for θ then it is sufficient for any function of θ.

(iv) It is not necessary that only one-to-one function of sufficient statistic is sufficient for the parameter, because it $X \sim N(0, \sigma^2)$, $f_X(x; \sigma^2) = (\sqrt{2\pi}\sigma)^{-1} e^{-\frac{x^2}{2\sigma^2}}$, by factorization theorem not only x^2 but any one of the functions: $|X|$, e^{X^2}, X^4 is sufficient for σ^2.

(v) Sometimes more than one statistic is sufficient for a single parameter. Consider a random sample $X_1, X_2, \ldots, X_n$ from uniform distribution, $u(\theta, \theta + 1)$.

Here, $$p_\theta(x) = \begin{cases} 1 & \text{if } \theta < \min_i x_i < \max_i x_i < \theta + 1 \\ 0 & \text{otherwise} \end{cases}$$

$$= H\left(\theta, \min_i x_i\right) H\left(\max_i x_i, \theta + 1\right),$$

where $$H(a, b) = \begin{cases} 1 & \text{if } a < b \\ 0 & \text{if } a \geq b \end{cases}$$

$$= g_\theta\left(\min_i x_i, \max_i x_i\right) h(x),\ h(x) \equiv 1$$

which implies $\left(\min_i x_i, \max_i x_i\right)$ is sufficient for θ.

(vi) If $X_{(1)}, X_{(2)}, \ldots, X_{(n)}$ is an ordered random sample of $X_1, X_2, \ldots, X_n$ then $X_{(1)}, X_{(2)}, \ldots, X_{(n)}$ is always sufficient for the distribution because

$$f_{X_1, X_2, \ldots, X_n / X_{(1)}, X_{(2)}, \ldots, X_{(n)}}(x_1, x_2, \ldots, x_n / x_{(1)}, x_{(2)}, \ldots, x_{(n)}) = \frac{1}{n!}$$

1.3 MINIMAL SUFFICIENT STATISTIC

If more than one sufficient statistics are available for the parameter then one should select one of them which may achieve the largest reduction in data. Let us consider the following example to understand the largest reduction with minimum number of statistics.

Let $X_1, X_2, \ldots, X_n$ be a random sample having Bernoulli distribution with probability of success as θ, that is,

$$P(X_j = x_j) = \theta^{x_j}(1-\theta)^{1-x_j},\ x_j = 0, 1;\ j = 1, 2, \ldots, n,\ 0 < \theta < 1$$

that is,

$$P(X_1 = x_1, X_2 = x_2, \ldots, X_n = x_n) = \theta^{\sum_{j=1}^{n} x_j}(1-\theta)^{n - \sum_{j=1}^{n} x_j}$$

Using factorization theorem with $h(x) \equiv 1$, we have the following sufficient statistics for θ

1. $T_1 = (X_1, X_2, \ldots, X_n)$
2. $T_2 = (X_1 + X_2, X_3, X_4, \ldots, X_n)$

3. $T_3 = (X_1 + X_2 + X_3, X_4, X_5, \ldots, X_n)$

$$\vdots$$

4. $T_n = (X_1 + X_2 + X_3 + \ldots + X_n) = \sum_{j=1}^{n} X_j$

Out of these n sufficient statistics $T_1, T_2, \ldots, T_n$ for θ, T_n has the largest reduction with minimum number of components because T_2 is a function of T_1, T_3 is a function of T_1 and $T_2, \ldots, T_n$ is a function of T_1, $T_2, \ldots, T_{n-1}$. Thus, for achieving the largest reduction of data without losing any information on the parameter, θ, one must choose that sufficient statistic which is a function of all other sufficient statistics for that parameter.

Definition 1.2

A sufficient statistic $T(x)$ for θ is minimal sufficient statistic if it is a function of all other sufficient statistics for θ. [$v(x)$ is a function of $u(x)$ if for any $x \neq y \in D$, domain, $u(x) = u(y) \Rightarrow v(x) = v(y)$]

Theorem 1.2: A sufficient statistic $T(x)$ for θ is minimal sufficient statistic if for any two points $x \neq y \in D$, $T(x) = T(y)$.

Proof: Let $S(x)$ be any other sufficient statistic for θ. Then for any two points $x \neq y \in D$

$$S(x) = S(y) \;\Rightarrow\; P_\theta(x) = g_\theta[S(x)]h(x)$$

$$= g_\theta[S(y)]h(x)$$

$$= g_\theta[S(y)]h(y)\frac{h(x)}{h(y)}$$

$$= k(x, y)P_\theta(y), \text{ where } k(x, y) = \frac{h(x)}{h(y)}$$

$\Rightarrow h(x)g_\theta[t(x)] = k(x, y)g_\theta[t(y)]h(y)$, as T is sufficient

$\Rightarrow T(x) = T(y)$

$\Rightarrow T(x)$ is a function of $S(X)$,

which shows that $T(x)$ is minimal sufficient statistic. Thus for showing that a sufficient statistic is minimal, one has to show that $p_\theta(x)/p_\theta(y)$ is independent of θ.

Example 1.5

If $X_1, X_2, \ldots, X_n$ is a random sample from uniform distribution, $u(0, \theta)$ then show that $T(x) = \max_i X_i$ is minimal sufficient statistic.

$$\text{As} \quad p_\theta(x) = \frac{1}{\theta^n} H\left(\max_i X_i, \theta\right), \text{ where } H(a,b) = \begin{cases} 1 & \text{if } a < b \\ 0 & \text{if } a \geq b \end{cases}$$

and $p_\theta(y) = \dfrac{1}{\theta^n} H\left(\max_i Y_i, \theta\right)$, $p_\theta(x)/p_\theta(y)$ is independent of θ if $\max_i X_i = \max_i Y_i$ that is $T(x) = Y(y)$. Thus, $T(x) = \max_i X_i$ is minimal sufficient statistic.

Example 1.6

Let $X_1, X_2, \ldots, X_n$ be a random sample from $N(\mu, \sigma^2)$ distribution, μ and σ^2 are unknown. Obtain minimal sufficient statistic for $\theta = (\mu, \sigma^2)$.

$$\text{Here} \qquad \frac{p_\theta(x)}{p_\theta(y)} = e^{-\frac{1}{2\sigma^2}\left(\Sigma x_i^2 - \Sigma y_i^2\right)} e^{-\frac{\mu}{\sigma^2}(\Sigma x_i - \Sigma y_i)},$$

which is independent of θ if and only if $\Sigma x_i = \Sigma y_i$ and $\Sigma x_i^2 = \Sigma y_i^2$. Hence, $(\Sigma X_i, \Sigma X_i^2)$ is minimal sufficient statistic for (μ, σ^2).

Example 1.7

If $X_1, X_2, \ldots, X_n$ is a random sample from the distribution having p.d.f. as

$$f_X(x; \alpha, \beta, \gamma) = \begin{cases} \dfrac{1}{\beta(1 - e^{-(\gamma - \alpha)})} e^{-\frac{x-\alpha}{\beta}} & , \alpha < x < \gamma, \alpha, \beta, \gamma > 0, \\ 0 & , \text{otherwise} \end{cases}$$

then obtain minimal sufficient statistic for (α, β, γ).

The joint p.d.f. of $X_1, X_2, \ldots, X_n$ is

$$f_X(x_1, x_2, \ldots, x_n; \alpha, \beta, \gamma) = \left[\frac{1}{\beta(1 - e^{-(\gamma - \alpha)})}\right]^n e^{-\sum_{i=1}^{n}\left(\frac{x_i - \alpha}{\beta}\right)},$$

$$\alpha < x_1, x_2, \ldots, x_n < \gamma$$

$$= \left[\frac{1}{\beta(1-e^{-(\gamma-\alpha)})}\right]^n e^{-\frac{\Sigma x_j}{\beta}} e^{\frac{n\alpha}{\beta}}, \alpha < \min_i x_i < \max_i x_i < \gamma$$

$$= \left[\frac{1}{\beta(1-e^{-(\gamma-\alpha)})}\right]^n e^{-\frac{\Sigma x_i}{\beta}} e^{\frac{n\alpha}{\beta}}, \alpha < x_{(1)} < x_{(n)} < \gamma,$$

where $x_{(1)} = \min(x_1, x_2, ..., x_n)$ and $x_{(n)} = \max(x_1, x_2, ..., x_n)$

$$= \left[\frac{e^{\frac{\alpha}{\beta}}}{\beta(1-e^{-(\gamma-\alpha)})}\right]^n e^{-\frac{\Sigma x_i}{\beta}} H(\alpha, x_{(1)})H(x_{(n)}, \gamma)$$

where $$H(a, b) = \begin{cases} 1 & \text{if } a < b \\ 0 & \text{if } a \geq b \end{cases}$$

so that for two distinct points $x = (x_1, x_2, ..., x_n)$ and $y = (y_1, y_2, ..., y_n)$, we have

$$\frac{f(x; \alpha, \beta, \gamma)}{f(y; \alpha, \beta, \gamma)} = e^{-\frac{1}{\beta}(\Sigma x_i - \Sigma y_i)} \frac{H(\alpha, x_{(1)})H(x_{(n)}, \gamma)}{H(\alpha, y_{(1)})H(y_{(n)}, \gamma)}$$

which is independent of (α, β, γ) if and only if

$$\Sigma x_i = \Sigma y_i, \; x_{(1)} = y_{(1)}, \; x_{(n)} = y_{(n)}$$

Hence, $T = (X_{(1)}, \Sigma X_i, X_{(n)})$ is minimal sufficient statistic for (α, β, γ).

1.4 COMPLETENESS

In section 1.3 it was shown that minimum number of statistics achieves the largest reduction of the data but the degree of reduction achieved through sufficiency is extremely variable and hence one has to deal with situation where substantial degree of reduction must be achieved and the criterion suggested must be stronger than that of minimal sufficiency. Such type of criterion known as completeness is suggested by Lehmann.

Definition 1.3

A family of distributions $[P_\theta, \theta \in \Omega]$ of X is said to be complete if for every function $u(x)$, not involving θ

$$Eu(x) = 0, \forall \theta$$

$\Rightarrow \quad u(x) = 0$ for a.e. x.

Definition 1.4

A statistic T is said to be complete if for every function $g(T)$, not involving parameter of the distribution of T

$$Eg(T) = 0, \forall \theta$$

$\Rightarrow \quad g(t) = 0$ for a.e. t

If a statistic T is complete as well as sufficient for θ then T is called complete sufficient statistic for θ. A complete sufficient statistic plays an important role in estimation theory and testing of hypotheses theory.

In the following theorem it is shown that completeness is a stronger property than minimality.

Theorem 1.3: Completeness always implies minimality but the converge may not be true.

Proof: Let $T(X)$ be a complete sufficient statistic for θ in the family of distribution $\{P_\theta, \theta \in \Omega\}$ of $X = (X_1, X_2, \ldots, X_n)$ and let $S(X)$ be any other sufficient statistic for θ such that $S(X)$ is a function of $T(X)$. Let

$$g(T) = T - E(T/S)$$

Then $g(T)$ is a function of T only as S is a function of $T(x)$.

Now $$Eg(T) = E(T) - EE(T/S) = ET - ET$$

$= 0$, for a.e. $\theta \in \Omega$.
Since T is complete,

$$Eg(T) = 0, \forall \theta \;\Rightarrow\; g(T) = 0,$$

for a.e. t, that is, $T - E(T/S) = 0$ for (a.e.) t, that is, $T = E(T/S) =$ a function of s, which shows that T is minimal sufficient. Converse part is proved by a counter example.

If $X_1, X_2, \ldots, X_n$ is a random sample from $N(\theta, \theta^2)$ distribution then $T = (\Sigma X_i, \Sigma X_i^2)$ is minimal sufficient statistic for θ but it is not complete for θ, because $E(\Sigma X_i^2) = 2n\theta^2$ and

$$E(\Sigma X_i)^2 = E(\Sigma X_i^2) + E\Sigma X_i X_j;\ i \neq j$$

$$= 2n\theta^2 + n(n-1)\theta^2$$

$$= n(n+1)\theta^2$$

so that, $E[(n+1)\Sigma X_i^2 - 2(\Sigma X_i)^2] = 0,\ \forall\ \theta$

$$\not\Rightarrow \quad (n+1)\,\Sigma X_i^2 = 2(\Sigma X_i)^2 \text{ for a.e. } x.$$

Basu (1955) has shown that any ancilliary statistic is independent of complete sufficient statistic.

Definition 1.5

A statistic $u(x)$ is said to be ancilliary if its distribution does not depend on the parameter, θ and it is the first order ancilliary if $Eu(X)$ is a constant.

Note that as ancillairy statistic does not contain any information about θ whereas sufficient statistic contains all information about θ, independence of sufficient statistic and ancilliary statistic arises which needs further consideration of completeness. A theorem by Basu throws light on this aspect.

Theorem 1.4: Basu's theorem

If T is complete sufficient statistic for the family of distributions $\mathbb{P} = \{P_\theta, \theta \in \Omega\}$ of $X = (X_1, X_2, \ldots, X_n)$ then any ancilliary statistic $V = V(X_1, X_2, \ldots, X_n)$ is independent of T.

Proof: Let X have discrete distribution. Fix any value (v) of V and let

$$g(t) = P[V = v/T = t]$$

Then,

$$E_\theta g(T) = \Sigma_t P[V = v/T = t]\ P(T = t)$$

$$= \sum_t P(V = v, T = t)$$

$$= P(V = v)$$

so that

$$E_\theta g(T) - P(V = v) = 0 \ \forall\ \theta$$

$$\Rightarrow \qquad g(t) = P[V = v/T = t] = P(V = v) \text{ for a.e. } t,$$

due to completeness of T.

Thus V and T are independent. For continuous distribution Σ is replaced by integration in the above proof.

Example 1.8

Let $X = (X_1, X_2, \ldots, X_n)$ be identically independently distributed (i.i.d) random variables with probability function

$$P(X_i = x_i) = \theta^{x_i}(1-\theta)^{1-x_i}, 0 < \theta < 1, x_i = 0, 1, i = 1, 2, \ldots, n$$

Then show that the family is complete, and $T = \sum_{i=1}^{n} X_i$ is also complete.

Let $E_\theta g(X) = 0$, $\forall\ \theta$, that is,

$$g(0) + \theta[g(1) - g(0)] = 0, \ \forall\ \theta$$

$$\Rightarrow \qquad g(0) = g(1) = 0$$

that is, $g(x) = 0$ for $x = 0, 1$.

Now if $n = 2$, $P(0, 1) = P(X = 0) + P(X = 1) = 1$ and if

$$E = (0, 1)^2 = [(0, 0), (0, 1), (1, 0), (1, 1)]$$

$$\begin{aligned} P(E) = P(0, 1)^2 &= P(0, 0) + P(0, 1) + P(1, 0) + P(1, 1) \\ &= 2 - 2\theta - (1 + 1) + 2\theta \\ &= 0, \text{ for all } \theta \end{aligned}$$

Thus, E is μ – null set for all θ, so that

$$x \notin E \Rightarrow Eg(X) = 0, \ \forall\ \theta$$

$\Rightarrow \ g(x) = 0$ for a.e. x

If $T = \sum_{i=1}^{n} X_i$, then T is sufficient and its distribution is binomial, $b(n, \theta)$. Suppose $Eg(T) = 0$, $\forall\ \theta$

i.e.,
$$\sum_{t=0}^{n} g(t)\binom{n}{t}\left(\frac{\theta}{1-\theta}\right)^t = 0, \ \forall\ \theta$$

$$\Rightarrow g(t)\binom{n}{t} = 0, \ \forall \ t, \text{ being positive coefficient of polynomial}$$

$\Rightarrow g(t) = 0$, for a.e. t

which shows that T is complete.

Example 1.9

Let probability function of X be defined as

$$P(X=0)=\frac{\theta}{3},\ P(X=1)=\frac{2\theta}{3},\ P(X=2) = 1 - \theta, 0 < \theta < 1.$$

Show that the family is not complete.
Suppose $Eg(X) = 0, \forall \ \theta$

i.e.,
$$g(0)\frac{\theta}{3}+g(1)\frac{2\theta}{3}+g(2)(1-\theta)=0 \ \ \forall \ \theta$$

i.e.,
$$g(2)+\theta\left[\frac{1}{3}g(0)+\frac{2}{3}g(1)-g(2)\right]=0 \ \ \forall \ \theta$$

$$g(2)=0 \text{ and } \frac{1}{3}g(0)+\frac{2}{3}g(1)=0$$

$\not\Rightarrow \ g(x) = 0$ for a.e. x,

which can also be shown by taking $g(x) = (x - 2)(2x - 1)$,

$$Eg(x)=2\frac{\theta}{3}+\frac{2\theta}{3}(-1)=0 \text{ but } g(x) \neq 0 \text{ for a.e. } x \text{ as}$$

$$g(0) = 2, g(1) = -1$$

Example 1.10

If $X_1, X_2, \ldots, X_n$ is a random sample from uniform distribution, $u(0, \theta)$ then show that $T=\max\limits_i X_i$ is complete for θ.

Let
$$E_\theta g(T) = 0, \forall \ \theta$$

i.e.,
$$\int_0^\theta g(t)\frac{nt^{n-1}}{\theta^n}dt=0, \forall \ \theta$$

i.e., $$\int_0^\theta g(t)t^{n-1}dt = 0, \forall\ \theta$$

i.e., $$\frac{d}{dt}\int_0^\theta g(t)t^{n-1}dt = 0, \forall\ \theta$$

$\Rightarrow$ $$g(\theta)\theta^{n-1} = 0, \forall\ \theta$$

$\Rightarrow$ $$g(\theta) = 0, \forall\ \theta$$

i.e., $$g(t) = 0, \forall\ t$$

which proves that T is complete for θ. Note that we have used Libnitz rule of derivative under integration such as [Libnitz rule: if $f(x, \theta)$ is integrable and differentiable function of both x and θ then

$$\frac{d}{d\theta}\int_{a(\theta)}^{b(\theta)} f(x,\theta)dx = f(b(\theta),\theta)\frac{db(\theta)}{d\theta} - f(a(\theta),\theta)\frac{da(\theta)}{d\theta} + \int_{a(\theta)}^{b(\theta)} \frac{df(x,\theta)}{d\theta}dx$$

Example 1.11

Let $X = (X_1, X_2, \ldots X_n)$ be a random sample from uniform distribution, $u\left(\theta - \frac{1}{2}, \theta + \frac{1}{2}\right)$ and let $T_1 = \min\limits_i X_i$ and $T_2 = \max\limits_i X_i$, $T = (T_1, T_2)$. Then show that T is minimal sufficient but not complete.

Since $$P_\theta(x) = \begin{cases} 1, & \theta - \frac{1}{2} < \min\limits_i X_i < \max\limits_i X_i < \theta + \frac{1}{2} \\ 0, & \text{otherwise} \end{cases}$$

$$= H\left(\theta, T_1 + \frac{1}{2}\right)H\left(T_2 - \frac{1}{2}, \theta\right)$$

where $$H(a,b) = \begin{cases} 1, & a < b \\ 0, & a \geq b \end{cases}$$

which implies $T = (T_1, T_2)$ is sufficient and it can be shown that it is minimal. Making the transformation $R = T_2 - T_1$, $U = T_1$ in the joint distribution of T_1, T_2

$$f_{T_1,T_2}(t_1,t_2)=n(n-1)(t_2-t_1)^{n-2},\ \theta-\frac{1}{2}<t_1<t_2<\theta+\frac{1}{2}$$

the joint distribution of R and U will be

$$f_{R,U}(r,u)=n(n-1)r^{n-2},\quad \theta-\frac{1}{2}<u<\theta+\frac{1}{2}<r,$$

and hence

$$f_R(r)=n(n-1)r^{n-2}(1-r),\ 0<r<1,$$

so that $\quad E\left(R-\frac{n-1}{n+1}\right)=0,\ \forall\ \theta \not\Rightarrow r-\frac{n-1}{n+1}=0$ for a.e. r

i.e $\not\Rightarrow t_2-t_1-\frac{n-1}{n+1}=0$ for a.e. t_2, t_1, which shows that $T=(T_1, T_2)$ is not complete.

1.5 EXPONENTIAL FAMILY OF DISTRIBUTIONS

A family $\mathbb{P}=\{P_\theta,\ \theta\in\Omega\}$ of distributions of $X=(X_1, X_2, \ldots, X_n)$ is said to form a k-parameter exponential family of distributions if the probability function is of the form

$$p_\theta(x)=c(\theta)h(x)\exp\left(\sum_{i=1}^{k}\prod\nolimits_i(\theta)t_i(x)\right) \qquad \ldots(1.5.1)$$

with some measure μ. $\prod_i$ and c are real valued functions of parameters and t_i are real valued statistics. Binomial, Poisson, Normal, Negative Binomial, Exponential, Gamma are some of the distributions having exponential family with $k=1$ or 2.

It is more convenient to use $\prod_i$ as the parameters and write P_θ in the canonical form as

$$p(x,\Pi)=c(\Pi)h(x)\exp\left(\sum_{i=1}^{k}\prod\nolimits_i t_i(x)\right) \qquad \ldots(1.5.2)$$

where $\prod=\prod_1, \prod_2, \ldots, \prod_k)$. This form is not unique as multiplying $\prod_i$ by a and dividing t_i by a, it remains unchanged. The probability function defined in (1.5.2) defines probability distribution iff the constant $c(\prod)$ exists such that

$$\int[\exp(\Sigma\Pi_i t_i(x))]h(x)dx<\infty \text{ for continuous random variable } X$$

or

$$\sum_x \left[\exp\left(\sum \Pi_i t_i(x)\right)\right] h(x) < \infty \text{ for discrete random variable } X$$

The class Π of points $\Pi = (\Pi_1, \Pi_2, \ldots, \Pi_k)$ is called the natural parameter space of family (1.5.2). Π is convex. We assume that there is no linear relation between $t_1, t_2, \ldots, t_k$ and also there is no linear relation between $\Pi_1, \Pi_2, \ldots, \Pi_k$ and if there exists linear relationship it reduces dimensionality of E_k. Following theorem is given without proof.

Theorem 1.5: For any integrable function g and Π in the interior of Π the integral

$$\int g(x) c(\Pi) \exp\left[\sum_{j=1}^{k} \Pi_j t_j(x)\right] h(x) dx$$

is continuous and has derivatives with Π's of all orders under integral sign is valid.

As an application of the above theorem, differentiating

$$c^{-1}(\Pi) = \int \exp\left[\sum_{j=1}^{k} \Pi_j t_j(x)\right] h(x) dx$$

w.r.t. Π_j and then w.r.t. Π_i, one gets

$$E_\Pi(T_j) = c(\Pi) \frac{\partial}{\partial \Pi_j} [c^{-1}(\Pi)]$$

$$E_\Pi(T_j T_i) = c(\Pi) \frac{\partial^2}{\partial \Pi_j \, \partial \Pi_i} [c^{-1}(\Pi)].$$

Example 1.12

Let n independent trials be made with $(k + 1)$ possible outcomes and let p_i be the probability of ith outcome in each trial. Let X_i be number of trials resulting i^{th} outcome. Then show that $X_1, X_2, \ldots, X_{k+1}$ has exponential family and obtain variance-covariance matrix of X_i, $i = 1, 2, \ldots, k + 1$ by using Theorem 1.5

$$P_\theta(x) = P(X_1 = x_1, X_2 = x_2, \ldots, X_{k+1} = x_{k+1})$$

$$= \frac{n!}{x_1! x_2! \ldots x_{k+1}!} p_1^{x_1} p_2^{x_2} \ldots p_{k+1}^{x_{k+1}}$$

where $\sum_{i=1}^{k+1} p_i = 1, \sum_{i=1}^{k+1} x_i = n$

This can be expressed as

$$P_\theta(x) = h(x)\left[\exp\left(n\log p_0 + x_1 \log\frac{p_1}{p_0} + x_2 \log\frac{p_2}{p_0} + \ldots + x_k \log\frac{p_k}{p_0}\right)\right]$$

where $h(x) = \dfrac{n!}{x_1!x_2!\ldots x_{k+1}!}, p_0 = 1 - \sum_{i=1}^{k} p_i,$

which is k-parameter exponential family with

$$\Pi_i = \log\frac{p_i}{p_0}, t_i(x) = x_i, \quad i = 1, 2, \ldots, k$$

$$c(\Pi) = \left(1 - \sum_{i=1}^{k} p_i\right)^n = p_0{}^n = \left(1 + \sum_{i=1}^{k} e^{\Pi_i}\right)^{-n}$$

Thus, $P_\theta(x) = c(\Pi)h(x)\exp\left(\sum_{i=1}^{k} \Pi_i x_i\right)$

Using these results in Theorem 1.5, we have

$$E(X_j) = c(\Pi)\frac{\partial}{\partial \Pi_j}(c(\Pi))^{-1} = \frac{ne^{\Pi_j}}{1 + \sum_{i=1}^{k} e^{\Pi_i}} = np_j, j = 1, 2, \ldots, k$$

As

$$\frac{\partial^2}{\partial \theta_j^2}(c(\Pi))^{-1} = n\left(1 + \sum_{i=1}^{k} e^{\Pi_i}\right)^{n-1} + n(n-1)\left(1 + \sum_{i=1}^{k} e^{\Pi_i}\right)^{n-2} e^{2\Pi_j}$$

$$\frac{\partial^2}{\partial \Pi_i \, \partial \Pi_j}(c(\Pi))^{-1} = n(n-1)\left(1 + \sum_{i=1}^{k} e^{\Pi_i}\right)^{n-2} e^{\Pi_i} e^{\Pi_j}$$

$$E(X_j^2) = \frac{ne^{\Pi_j}}{1+\sum_{i=1}^{k} e^{\Pi_i}} + \frac{n(n-1)e^{2\Pi_j}}{\left(1+\sum_{i=1}^{k} e^{\Pi_i}\right)^2} = np_j + n(n-1)p_j^2$$

$$E(X_i X_j) = \frac{n(n-1)e^{\Pi_i} e^{\Pi_j}}{\left(1+\sum_{i=1}^{k} e^{\Pi_i}\right)^2} = n(n-1)p_i p_j,\ i \neq j, i, j = 1, 2, \ldots, k$$

so that $\quad \text{Var}(X_j) = np_j(1-p_j), j = 1, 2, \ldots, k$ and

$$\text{cov}(X_i X_j) = -np_i p_j,\ i \neq j,\ i, j = 1, 2, \ldots, k$$

The variance covariance matrix Σ of $X_1, X_2, \ldots, X_k$ is

$$\Sigma = \begin{pmatrix} np_1(1-p_1) & -np_1p_2 & \cdots & -np_1p_k \\ -np_2p_1 & np_2(1-p_2) & \cdots & -np_2p_k \\ \vdots & \vdots & \vdots & \vdots \\ -np_1p_k & -np_kp_2 & \cdots & np_k(1-p_k) \end{pmatrix}$$

Note that $p_\theta(x)$ is the probability function of multinomial distribution with parameter $(n, p_1, p_2, \ldots, p_k)$.

Theorem 1.6: If $X = (X_1, X_2, \ldots, X_n)$ is distributed according to an exponential family of distributions with probability function given in (1.4.1) then $T = (T_1, T_2, \ldots, T_k)$, $T_i = \sum_{j=1}^{n} t_i(X_j), i = 1, 2, \ldots, k$ is distributed according to exponential family of distributions with probability function as $[c(\theta)]^n h_0(t) e^{\sum_{i=1}^{k} \Pi_i t_i}$.

Proof: Considering discrete case, the joint probability function of $X_1, X_2, \ldots, X_n$.

$$P(X_1 = x_1, X_2 = x_2, \ldots, X_n = x_n) = [c(\theta)]^n \left[\prod_{j=1}^{n} h(x_j)\right] \exp\left[\sum_{i=1}^{k} \Pi_i(\theta) \sum_{j=1}^{n} t_i(x_j)\right]$$

transforms to probability function of $T_1, T_2, \ldots, T_k$ by the transformation

$$T_i = \sum_{j=1}^{n} t_i(x_j),\ i = 1, 2, \ldots, k \text{ as}$$

$$P(T_1 = t_1, T_2 = t_2, \ldots, T_k = t_k)$$

$$= [c(\theta)]^n h_0(t) \exp\left(\sum_{i=1}^{k} \Pi_i(\theta) t_i\right) \qquad \ldots(1.5.3)$$

where $h_0(t) = \prod_{j=1}^{n} h(x_j)$ over all $(x_1, x_2, \ldots, x_n)$ for which

$$\sum_{j=1}^{n} t_i(x_j) = t_i,\ i = 1, 2, \ldots, k.$$

This proves that T has exponential family of distributions with probability function as (1.5.1). Hence for natural parameterization the probability function of T will be

$$f_T(t) = c(\Pi) h_0(t) \exp\left(\sum_{i=1}^{k} \Pi_i t_i\right) \qquad \ldots(1.5.4)$$

The following theorem is concerned with the sufficiency and completeness for the exponential family of distributions.

Theorem 1.7: If $X = (X_1, X_2, \ldots, X_n)$ is a random sample from (1.5.2) and if $n \geq k$ and if the natural parameter space Π contains open set in E_k then

$$T = \left(\sum_{j=1}^{n} t_1(x_j), \sum_{j=1}^{n} t_2(x_j), \ldots, \sum_{j=1}^{n} t_k(x_j)\right)$$

is complete sufficient statistic for $\Pi = (\Pi_1, \Pi_2, \ldots, \Pi_k)$.

Proof: Since the probability function of T given in (1.5.2) can be rewritten as

$$f_T(t) = g(\Pi, t) h_0(t), \quad \text{where} \quad g(\Pi, t) = c(\Pi) \exp\left(\sum_{i=1}^{k} \Pi_i t_i\right),$$

T is sufficient for Π by the use of factorization theorem.

For proving completeness of T let $g(T)$ be a function of T, such that $Eg(T) = 0, \forall\ \theta$

that is, $$\int g(t)h_0(t)\exp\left(\sum_{i=1}^{k}\Pi_i t_i\right)dt = 0,\ \forall\ \theta,$$

$\Rightarrow$ $g(t)h_0(t) = 0$ for a.e. t, by virtue of Laplace theorem.

$\Rightarrow$ $g(t) = 0$ for a.e. t, if $h_0(t) \neq 0$,

which shows that T is complete [Laplace theorem: If $\int_{-\infty}^{\infty} f(x)e^{-\theta x}dx = 0$ then $f(x) = 0$ for a.e. x]. Thus, if the given distribution belongs to exponential family of distributions, the complete sufficient statistic is easily available.

Remarks

(vii) If $f(x;\ \theta)$ is density of X and if T is sufficient for θ and if $\{x : f(x;\ \theta) = 0\}$ does not depend on θ with certain mild regularity conditions, the distribution $f(x;\ \theta)$ has exponential family. This is called regular family and if $\{x : f(x) = 0\}$ depends on θ it is called non-regular family, such as

$$f(x;\theta) = c(\theta)h(x),\ \ a(\theta) < x < b(\theta).$$

If $X_1, X_2, \ldots, X_n$ is a random sample from such irregular family then it can be shown that $\left(\min_i X_i, \max_i X_i\right)$ is sufficient for θ but completeness is to be shown separately. Uniform distribution defined on $(\theta_1,\ \theta_2)$ is a special case of non-regular family of distributions.

(viii) If T is complete for θ and if $E\phi_1(T) = E\phi_2(T) = \Psi(\theta)$ for all θ then $\phi_1(t) = \phi_2(t)$ for a.e. t.

(ix) If T is complete for θ then any function of T is complete for any function of θ.

PROBLEMS

1.1 Using basic definition of sufficient statistic show that ΣX_i is sufficient for λ when a random sample $X_1, X_2, \ldots, X_n$ is available from Poisson distribution with parameter λ. Also show that ΣX_i is minimal sufficient statistic for λ.

1.2 By using definition show that ΣX_i is sufficient for p if X_1, X_2, ..., X_n is a random sample from geometric distribution with probability function

$$P(X=x)=pq^{x-1}, x=1,2,\ldots.$$

Further show that it is minimal sufficient.

1.3 Given a random sample of size n from discrete uniform distribution with probability function.

$$P_\theta(X=x)=\frac{1}{\theta}, x=1,2,\ldots,\theta$$

obtain minimal sufficient statistic for θ.

1.4 If X_1, X_2, ..., X_n is a random sample from $N(0, \sigma^2)$ distribution then obtain minimal sufficient statistic for σ^2.

1.5 Given a random sample X_1, X_2, ..., X_n from (i) $N(\theta, \theta)$ and (ii) $N(\theta, \theta^2)$ distributions obtain (a) Sufficient statistic for θ in each case (b) Minimal sufficient statistic for θ in each case.

1.6 Let X_1, X_2, ..., X_n be a random sample from $G(\alpha, \beta)$ distribution. Obtain sufficient and minimal sufficient statistics for (a) α if β is known (b) β if α is known

1.7 If X_1, X_2, ..., X_n is a random sample from beta distribution, Be(α, λ) then obtain minimal sufficient statistic for

(a) α if λ is known
(b) λ if α is known
(c) (α, λ) if both are unknown.

1.8 Let X_1, X_2, ..., X_n be a random sample from Laplace distribution with p.d.f. as

$$f_X(x;\theta)=\frac{1}{2}e^{-|x-\theta|}, \quad -\infty<x<\infty$$

(a) Show that ordered sample $(Y_1, Y_2, \ldots, Y_n)$ is minimal sufficient statistic for θ.

(b) Show that maximum likelihood estimator $\hat{\theta}$ = med(X_1, X_2, ..., X_n) is a function of sufficient statistic.

1.9 Given a random sample of size n from lognormal distribution with parameters (μ, σ^2) obtain minimal sufficient statistic for (μ, σ^2).

1.10 Suppose that X_1, X_2, ..., X_n are i.i.d random variables with probability function

$$P_\theta(X=x)=(1-p)p^{x-\theta}, x=\theta,\theta+1,\ldots, 0<p<1$$

Then show that

(a) $(\min X_i, \Sigma X_i)$ is minimal sufficient for (θ, p)
(b) $\min X_i$ is minimal sufficient for θ when p is known
(c) ΣX_i is minimal sufficient for p when θ is known.

1.11 If $X_1, X_2, \ldots, X_n$ is a random sample from the p.d.f.

$$f_X(x;\theta) = \frac{\theta}{(1+x)^{1+\theta}}, \; 0 < x < \infty, \quad \theta > 0$$

then show that $\prod_{i=1}^{n}(1+X_i)$ is minimal sufficient for θ.

1.12 Let X be a single observation from the beta distribution, Be(θ, θ). Show that minimal sufficient statistic for θ is trivial sufficient statistic.

1.13 Given a random sample of size n from Laplace distribution with p.d.f.

$$f_X(x;\theta) = \frac{1}{2\theta} e^{-\frac{|x|}{\theta}}, \quad \theta < x < \infty, \quad \theta > 0$$

obtain sufficient statistic for θ.

1.14 If $X_1, X_2, \ldots, X_n$ is a random sample from displaced exponential distribution with p.d.f.

$$f_X(x;\alpha,\beta) = \frac{1}{\beta} e^{-\frac{1}{\beta}(x-\alpha)}, \; \alpha < x < \infty, \; \beta > 0$$

and if $Y_1 \le Y_2 \le \ldots \le Y_n$ is the corresponding ordered sample and if $S = \sum_{i=1}^{n}(X_i - Y_1)$, then show that (Y_1, s) is minimal sufficient for (α, β).

1.15 If $(X_1, Y_1), (X_2, Y_2), (X_3, Y_3), \ldots, (X_n, Y_n)$ denote a random sample from bivariate normal distribution $BN(\mu_1, \mu_2, \sigma_1^2, \sigma_2^2, \rho)$ then obtain sufficient statistic for

(a) $\mu_1, \mu_2, \sigma_1^2, \sigma_2^2, \rho$
(b) μ_1, σ_1^2, ρ when $\mu_2 = \mu_1, \sigma_2^2 = \sigma_1^2$
(c) ρ, when $\mu_1 = \mu_2 = 0$ and $\sigma_1^2 = \sigma_2^2 = 1$.

1.16 If $\underline{X}_1, \underline{X}_2, \ldots, \underline{X}_n$ is a random sample from p-variate normal distribution, $N_p(\underline{\mu}, \Sigma)$ then show that

$$\left(\overline{\underline{X}} = \frac{1}{n}\sum_{i=1}^{n}\underline{X}_i, S^2 = \frac{1}{n}\sum_{i=1}^{n}(\underline{X}_i - \overline{\underline{X}})(\underline{X}_i - \overline{\underline{X}})'\right)$$

is sufficient statistic for $(\underline{\mu}, \Sigma)$.

1.17 Obtain complete sufficient statistics for the parameters of the following distributions based on a random sample of size n.

(a) Poisson distribution with mean λ.
(b) Binomial distribution with parameter p when n is known.
(c) Negative Binomial distribution with parameter p.
(d) Normal distribution with parameters (μ, σ^2), $N(\mu, \sigma^2)$.
(e) Gamma distribution with parameters (α, β).
(f) Beta distribution with parameters (α, β).

1.18 Show that the family of distributions $N(0, \theta)$ is not complete.

1.19 If $X_1, X_2, \ldots, X_n$ is a random sample from $N(\xi, \sigma^2)$ and $Y_1, Y_2, \ldots, Y_n$ is an independent random sample from $N(\eta, \gamma^2)$ then obtain minimal sufficient statistics for the following three cases and determine whether it is complete or not.

(i) $\xi, \eta, \gamma, \sigma$ are arbitrary, $-\infty < \xi, \eta < \infty$, $0 < \sigma, \gamma$
(ii) $\gamma = \sigma$ and ξ, η, σ are arbitrary
(iii) $\eta = \xi$ and ξ, σ, γ are arbitrary.

1.20 Let $X_1, X_2, \ldots, X_n$ be a random sample from the distribution with p.d.f.

$$f_X(x;\theta) = \frac{(x+1)}{\theta(\theta+1)} e^{-\frac{x}{\theta}}, \; x, \theta > 0.$$

Then show that this distribution belongs to exponential family of distributions and obtain complete sufficient statistic for θ.

1.21 If $X_1, X_2, \ldots, X_n$ is a random sample from $N(\mu, \sigma^2)$, σ^2 known then by the use of Basu's theorem show that $\overline{X}$ and $S^2 = \frac{1}{n-1}\sum_{i=1}^{n}(X_i - \overline{X})^2$ are independently distributed. Does the result hold if σ^2 is unknown?

1.22 If X_1, X_2 is a random sample of size 2 from the exponential distribution with p.d.f.

$$f_X(x;\theta)=\frac{1}{\theta}e^{-\frac{x}{\theta}},\ x,\theta>0$$

then by the use of Basu's theorem show that $T_1 = X_1 + X_2$ and $T_2=\dfrac{X_1}{X_1+X_2}$ are independently distributed. Generalize this result for a sample of size n.

1.23 If $X_1, X_2, \ldots, X_n$ is a random sample from uniform distribution, $u(\theta_1, \theta_2)$ then show that $T=\left(\min_i X_i, \max_i X_i\right)$ is complete sufficient for (θ_1, θ_2).

1.24 Let $X_1, X_2, \ldots, X_n$ be a random sample from uniform distribution, $u(\theta, 3\theta)$. Show that $T=\left(\min_i X_i, \max_i X_i\right)$ is minimal sufficient for θ but it is not complete.

1.25 Show that the variance-covariance matrix Σ obtained in example (1.12) is not positive definite.

2

CHAPTER

Unbiased Estimation

2.1 INTRODUCTION

The observations under study are postulated to be values taken by the random sample, $X = (X_1, X_2, \ldots, X_n)$, of the random variable X which is assumed to follow joint probability distribution, $\{P_\theta(x), \theta \in \Omega)$. The aim of the analysis is to specify the possible values of θ with the knowledge of all possible observations $\mathscr{X} = \{(x_1, x_2, \ldots, x_n)\}$ where, $(x_1, x_2, \ldots, x_n)$ is the realized value of $(X_1, X_2, \ldots, X_n)$. One problem is to know the value of a function of the parameter θ, $g(\theta)$. Another problem is to check the truthfulness of a statement regarding the parameter θ or regarding distribution of X. This is the problem of testing of hypothesis.

The first problem is the problem of estimation which can be described as under. Consider a random sample $X = (X_1, X_2, \ldots, X_n)$ having family of distributions

$\mathbb{P} = \{P_\theta(x), \theta \in \Omega\}$. Let $\mathscr{X} = \{(x_1, x_2, \ldots, x_n)\}$ where, $x = (x_1, x_2, \ldots, x_n) \in (X_1, X_2, \ldots, X_n)$. If n is very large the sample space $\mathscr{X}$ cannot be described. In such a case a function

$T = U(x)$ is defined on $(X_1, X_2, \ldots, X_n)$ to reduce the dimensionality of $\mathscr{X}$ such that it is close to the parametric function, $g(\theta)$ to be estimated. That is, one of the closeness is

$$P_\theta\{\,|T - g(\theta)| < \epsilon\,\} = 1 \quad \text{or} \quad E_\theta(T(x) - g(\theta)) = 0, \ \forall\ \theta \in \Omega$$

$T(x)$ is called unbiased estimator of $g(\theta)$ which may be interpreted as in a long run it will estimate the true value of $g(\theta)$. This is an attractive feature of unbiasedness. Also sum of unbiased estimators is unbiased for the sum of corresponding parametric function which is widely used in sampling theory. Although these two properties are attractive but there are many situations where unbiasedness property is not advisable to use.

Some parametric functions are not unbiased estimable for example,

1. $g(\theta) = \sqrt{|\theta|}$ is not estimable.
2. Reciprocal of binomial probability $\frac{1}{p}$ of $b(n, p)$ is not estimable because if $T(x)$ is unbiased estimator of $\frac{1}{p}$ then $\sum T(x)\binom{n}{x}p^x(1-p)^{n-x} = \frac{1}{p}$. When $p \to 0$, $T(0) = \infty$.
3. Reciprocal of Poisson parameter is not estimable because if $T(x_1, x_2, \ldots, x_n)$ is unbiased estimator of θ then by definition

$$\sum_{x_1=0}\sum_{x_2=0}\cdots\sum_{x_n=0}\frac{T(x_1, x_2, \ldots, x_n)e^{-n\theta}\theta^{\sum x_i}}{x_1!\,x_2!\ldots x_n!} = \frac{1}{\theta}$$

$$\Rightarrow e^{n\theta} = \sum_{x_1=0}\sum_{x_2=0}\cdots\sum_{x_n=0}\frac{T(x_1, x_2, \ldots, x_n)\theta^{\sum_{i=0}^{n} x_i + 1}}{x_1!\,x_2!\ldots x_n!}$$

$$\Rightarrow 1 + \sum_{r=1}^{\infty}\frac{(n\theta)^r}{r!} = \sum_{r=1}^{\infty} b_r\,\theta^r$$

$\Rightarrow 1 = 0$ which is absurd.

4. Unbiased estimator is sometimes meaningless, for example, in Poisson distribution with parameter θ the unbiased estimator of $e^{2\theta}$ is $T(X) = (-1)^X$, $T(x)$ takes value -1 when x is odd even though the parameter is always positive.
5. Sometimes biased estimator is better than unbiased one in the sense of mean square error [mean square error is defined as mean of the squares of the deviations from the parameter] (MSE). The following example proves this statement.

Example 2.1

Let $X_1, X_2, \ldots, X_n$ be i.i.d random variables having $N(\mu, \sigma^2)$ distribution and let $s^2 = \frac{\sum(X_i - \bar{X})^2}{n-1}$. Then $\frac{n-1}{n+1}s^2$ has minimum MSE. As $Es^2 = \sigma^2$ and $\text{Var}(s^2) = \frac{2\sigma^4}{n-1}$, considering an estimator $ks^2, k \neq 1$, the MSE of ks^2 is

$$\text{MSE of } (ks^2) = E(ks^2 - \sigma^2)^2$$

$$= E[k(s^2 - \sigma^2) - \sigma^2(1-k)]^2$$

$$= \left[\frac{2k^2}{n-1} + (1-k)^2\right]\sigma^4$$

which is minimum when $k = \frac{n-1}{n+1}$. Thus, $s_0^2 = \frac{1}{n+1}\sum(X_i - \bar{X})^2$ has minimum MSE but s_0^2 is not unbiased estimator of σ^2. The MSE of s_0^2 is $\frac{2\sigma^4}{n+1}$ which is less than $\frac{2\sigma^4}{n-1}$, the variance of the unbiased estimator, s^2 of σ^2.

Section 2.2 deals with uniformly minimum variance unbiased (UMVU) estimation. Linear unbiased estimation is discussed in section 2.3. Defining power series distribution (PSD) and modified PSD (MPSD), their estimation is given in section 2.4. whereas after defining non-regular (truncated) family of distributions UMVU estimators of parametric function of interest are obtained in section 2.5.

2.2 UNIFORMLY MINIMUM VARIANCE UNBIASED ESTIMATION

The motive to introduce unbiasedness is in the hope that within the class of unbiased estimators there may exist an estimator whose variance is minimum.

Definition 2.1

If U_g is the class of all unbiased estimators of $g(\theta)$ and if for $T_0 \in U_g$ $\text{var}_\theta(T_0) \leq \text{var}_\theta(T)$, $\forall\ T \in U_g$ and $\forall\ \theta \in \Omega$ then T_0 is called uniformly minimum variance unbiased (UMVU) estimator of $g(\theta)$.

The following theorem characterizes the situation in which UMVU estimator exists.

Theorem 2.1: Let X have family of distributions $\mathbb{P} = \{P_\theta, \theta \in \Omega\}$ and $T(X)$ be unbiased for $g(\theta)$ such that $E_\theta(T^2(X)) < \infty$ and let $U_0(X)$ be unbiased estimator of zero such that $E_\theta(U_0^2(X)) < \infty$. Then $T(X)$ is UMVU estimator of $g(\theta)$ if and only if

$$E_\theta T U_0 = 0, \forall \theta \in \Omega$$

Proof:

Necessary Condition

As $\qquad E_\theta(T(X)) = g(\theta) \quad \text{and} \quad E_\theta(U_0(X)) = 0 \; \forall \theta \in \Omega,$

$$E_\theta(T + \lambda U_0) = g(\theta) \text{ for any } \lambda \text{ and}$$

$$\begin{aligned}\text{var}_\theta(T + \lambda U_0) &= \text{var}_\theta(T) + \lambda^2 \text{var}_\theta(U_0) + 2\lambda \text{cov}_\theta(T, U_0)\\ &\geq \text{var}_\theta(T), \text{ if } \text{cov}_\theta(T, U_0) = E(TU_0) = 0\end{aligned}$$

which establishes necessary condition.

Sufficient Condition

Let T_1 be any other unbiased estimator of $g(\theta)$. Then $E_\theta(T_1 - T) = E_\theta(U_0) = 0$ where, $U_0 = T_1 - T$. As $T_1 = T + (T_1 - T) = T + U_0$, $\text{var}_\theta(T_1) = \text{var}_\theta(T) + \text{var}_\theta(U_0) + 2\,\text{cov}_\theta(T, U_0)$ and as T is UMVU estimator $\text{var}_\theta(T) \leq \text{var}_\theta(T_1) \Rightarrow \text{cov}(T, U_0) = 0$, i.e., $E(TU_0) = 0$. Some results are given as applications of Theorem 2.1.

Remarks

(i) UMVU estimator is unique if it exists because considering two UMVU estimators, T and T_1 for $g(\theta)$, $U = T - T_1$ is unbiased for zero and hence by using Theorem 2.1, $E_\theta T(T - T_1) = ET_1(T_1 - T) = 0$

$$\begin{aligned}\Rightarrow \qquad \text{var}_\theta(T) &= \text{var}_\theta(T_1) = \text{cov}_\theta(T, T_1)\\ \Rightarrow \qquad \rho(T, T_1) &= 1\\ \Rightarrow \qquad P_\theta(aT + bT_1 = 0) &= 1\\ \Rightarrow \qquad T_1 &= T \text{ a.e by taking } a = 1, b = -1.\end{aligned}$$

(ii) If T_1 and T_2 are UMVU estimators of $g_1(\theta)$ and $g_2(\theta)$, respectively then $a_1T_1 + a_2T_2$ is UMVU estimator of $a_1g_1(\theta) + a_2g_2(\theta)$.

Proof: If $E_\theta U = 0$ then by Theorem 2.1, $E_\theta(T_1U) = E_\theta(T_2U) = 0$, $\forall \theta$ and hence $E_\theta(a_1T_1 + a_2T_2)U = a_1E_\theta(T_1U) + a_2E_\theta(T_2U) = 0$, $\forall \theta$ which implies that $a_1T_1 + a_2T_2$ is UMVU estimator of $E_\theta(a_1T_1 + a_2T_2) = a_1g_1(\theta) + a_2g_2(\theta)$.

(iii) T is UMVU estimator of $g(\theta)$ then $aT + b$ is UMVU estimator of $ag(\theta) + b$. This can be proved similarly as done in Remark (ii).

(iv) If $\{T_n\}$ is a sequence of UMVU estimators of $g(\theta)$ and if T is a statistic with $E_\theta T^2 < \infty$ such that $E_\theta(T - T_n)^2 \to 0$ as $n \to \infty$, then T is UMVU estimator of $g(\theta)$.

Proof:

As $|E_\theta(T - g(\theta))| = |E_\theta(T - T_n)| \le E_\theta|T - T_n| \le [E_\theta(T - T_n)^2]^{\frac{1}{2}}$,

by virtue of Cauchy-Swartz (C-S) inequality

$$\to 0, \text{ as } E_\theta(T - T_n)^2 \to 0 \text{ as } n \to \infty$$

$$\Rightarrow \qquad E_\theta(T - g(\theta)) = 0 \ \forall \ \theta \in \Omega.$$

Due to Theorem (2.1), if $E_\theta U = 0$ then $E_\theta(T_n U) = 0 \ \forall \ \theta \in \Omega$.

Now $$|E_\theta(TU)| = |E_\theta(TU) - E_\theta(T_n U)|$$

$$= |E_\theta(T - T_n)U| \le [E_\theta(T - T_n)^2]^{\frac{1}{2}}[E_\theta U^2]^{\frac{1}{2}},$$

by C-S inequality

$$\to 0, \text{ as } E_\theta(T - T_n)^2 \to 0 \text{ as } n \to \infty$$

$$\Rightarrow \qquad E_\theta(TU) = 0 \ \forall \ \theta \in \Omega.$$

This shows that T is UMVU estimator of $g(\theta)$.

(v) Sometimes estimator with larger variance is preferable to UMVU estimator. Let us show this by an example.

Example 2.2

Let $X_1, X_2, \ldots, X_n$ be a random sample from the exponential distribution with p.d.f. as

$$f_X(x; \theta) = \theta e^{-\theta x}, \ x > 0, \theta > 0$$

In which x denotes failure time of component of the equipment in life testing problem.

As $E_\theta X = 1/\theta$, $\text{var}_\theta(X) = 1/\theta^2$ and $\overline{X}$ being complete sufficient statistic, $\overline{X}$ is UMVU estimator of θ with variance $1/n\theta^2$. Consider $T = \min(X_i)$. Then nT having exponential distribution with $E_\theta nT = 1/\theta$ and $\text{var}_\theta(nT) = 1/\theta^2$. So $\text{var}_\theta(\overline{X}) \le \text{var}_\theta(nT)$. Even though $\overline{X}$ is better than (nT) but one prefers nT instead of $\overline{X}$ because $\overline{X}$ requires to observe n observations while nT needs only one component to fail.

In the efforts of getting UMVU estimator the first step in the direction is given by Rao (1945) and Blackwell (1946) by searching that function of sufficient statistic which has variance less than that of estimator based on observations. Later on Lehmann and Scheffe used completeness along with sufficiency and obtained UMVU estimator.

Theorem 2.2: Rao - Blackwell theorem

Let $X = (X_1, X_2, \ldots, X_n)$ be a random sample from the family of distributions, $\mathbb{P} = \{P_\theta(x), \theta \in \Omega\}$, let $E_\theta(U(X)) = g(\theta)$, $\forall\ \theta \in \Omega$, $E_\theta(U(X))^2 < \infty$ and let $T(X)$ be sufficient statistic for P_θ or θ. Then $h(T) = E_\theta[U(X)/T(X)]$ is unbiased for $g(\theta)$, and $\text{var}(h(T)) \leq \text{var}_\theta(U(X))$. Equality sign holds if and only if $U = h(T)$.

Proof: Let X and T be discrete. As $h(T) = E_\theta(U/T)$

$$E(h(T)) = E_t E_x(U/T)$$

$$= \sum_t \left\{ \sum_x U(x) P(X = x / T = t) \right\} P(T = t)$$

$$= \sum_x U(x) \sum_t P(X = x, T = t)$$

$$= E_\theta U(X) = g(\theta)$$

Since $\quad \text{var}_\theta(U/T = t) = E_\theta(U^2/t) - [E_\theta(U/T)]^2 \geq 0,$

$$\Rightarrow \qquad [E_\theta(U/T)]^2 \leq E_\theta(U^2/t)$$

$$\Rightarrow \qquad E_t[E_\theta(U/T)]^2 \leq E_t[E_\theta(U^2/t)] = E_\theta(U^2)$$

$$\Rightarrow \qquad E_\theta[h(T)]^2 \leq E_\theta(U^2)$$

$$\Rightarrow \qquad E_\theta[h(T)]^2 - [E_\theta h(T)]^2 \leq E_\theta U^2 - (E_\theta U)^2$$

$$\Rightarrow \qquad \text{var}_\theta(h(T)) \leq \text{var}_\theta(U)$$

For continuous distribution of X and T, Σ is replaced by $\int$ in the above arguments. Equality sign holds good in the above inequality if and only if $\text{var}_\theta(U/T) = 0$,

i.e., $\quad E_\theta[\{U - E(U/T)\}/T]^2 = 0$

$$\Rightarrow \qquad P_\theta[\{(U - h(T))/T\} = 0] = 1$$

$$\Rightarrow \qquad P_\theta[\{U - h(T)\} = 0] = 1$$

$$\Rightarrow \qquad U = h(t) \text{ for a.e. } t.$$

Note that $h(T) = E(U/T)$ is called Blackwell statistic which searches the function of sufficient statistic to reduce the variance. $h(T)$, the Blackwell statistic, is better than the unbiased statistic $U(X)$ based on observations. When will it be the best? The answer is given by the following theorem.

Theorem 2.3: Lehmann-Scheffe theorem

If $T(X)$ is complete sufficient statistic for the family of distributions of X, $\mathbb{P} = \{P_\theta(x),\ \theta \in \Omega\}$ then every U-estimable function estimating $g(\theta)$ has one and only one UMVU estimator of $g(\theta)$ and it is a function of T.

Proof: Let $u_1(X)$ be unbiased estimator of $g(\theta)$. Then T being sufficient $h_1(T) = E_\theta(u_1(X)/T)$ is unbiased for $g(\theta)$ and $\text{var}_\theta(h_1(T)) \le \text{var}_\theta(u_1(X))$ by virtue of the Rao-Blackwell theorem. Let $u(X)$ be any other unbiased estimator of $g(\theta)$. Then $h(T) = E_\theta(u(X)/T)$ is unbiased for $g(\theta)$ and $\text{var}_\theta h(T) \le \text{var}_\theta(X)$ due to **Rao-Blackwell theorem**

Thus, $$E_\theta(h_1(T) - h(T)) = 0 \quad \forall\ \theta \in \Omega$$

$\Rightarrow h_1(T) = h(T)$, due to completeness of T

$\Rightarrow \text{var}_\theta\, h_1(T) = \text{var}_\theta\, h(T) \le \text{var}_\theta\, u(X),\ u(X) \in u_g,$

the class of all unbiased estimators of $g(\theta)$. This shows that $h_1(T)$ is unique UMVU estimator of $g(\theta)$.

Remark

(vi) One can obtain UMVU estimator of $g(\theta)$ in any one of the following three methods.

1. If T is complete sufficient statistic for θ and if $E_\theta u(X) = g(\theta)$ is available then $E_\theta(u(X)/T)$ is UMVU estimator of $g(\theta)$.
2. If T is complete sufficient for θ and if $E_\theta h(T) = g(\theta)$ then $h(T)$ is UMVU estimator of $g(\theta)$.

3. If $\int h(t) f_T(t)dt = g(\theta)$ can be written as $\int h(t)\, \Psi(t) f_T^*(t)dt = 1$, where, $f_T^*(t)$ is of the same form as $f_T(t)$, then $E\left[h(T)\,\Psi(T) - 1\right] = 0 \Rightarrow h(T) = \dfrac{1}{\Psi(T)}$.

In the first method Rao-Blackwell theorem is utilized under completeness of T. Second method seems to be simple but in this case one has to guess the function of T, $h(T)$ which is unbiased for $g(\theta)$ but how to get $h(T)$ is given in the third method.

Example 2.3

Given a random sample $X_1, X_2, \ldots, X_n$ from Poisson distribution with parameter λ, obtain UMVU estimators of

(i) λ
(ii) λ^2
(iii) $P_\lambda(X = k)$.

Since X has exponential family of distributions $T = \sum_{i=1}^{n} X_i$ is complete sufficient statistic for λ, and T has Poisson distribution with parameter $n\lambda$. As $E(T) = n\lambda$

(i) $\dfrac{T}{n}$ is UMVU estimator of λ

(ii) As $E(T^2) = n\lambda + n^2\lambda^2$, $E\left(\dfrac{T^2}{n^2} - \dfrac{T}{n^2}\right) = \lambda^2 \Rightarrow \dfrac{T^2 - T}{n^2}$ is UMVU estimator of λ^2.

(iii) Let $u(X)$ be unbiased estimator of $P(X = k) = \dfrac{e^{-\lambda}\lambda^k}{k!}$

$$\Rightarrow \qquad u(x) = \begin{cases} 1 & \text{if } x = k \\ 0 & \text{otherwise} \end{cases}$$

Since T is complete sufficient statistic for λ, $E(u(X_1)/T = t)$ is UMVU estimator of $P(X = k)$.
Now

$$E(u(X_1)/T = t) = P\left(X_1 = k / \sum_{i=1}^{n} X_i = t\right)$$

$$= \frac{P(X_1 = k,\ X_2 + X_3 + \ldots + X_n = t - k)}{P(X_1 + X_2 + \ldots + X_n = t)}$$

$$= \frac{\dfrac{e^{-\lambda}\lambda^k}{k!} \dfrac{e^{-(n-1)\lambda}[(n-1)\lambda]^{t-k}}{(t-k)!}}{\dfrac{e^{-n\lambda} n\lambda^t}{t!}}$$

$$= \binom{t}{k}\left(\frac{1}{n}\right)^k \left(1-\frac{1}{n}\right)^{t-k}.$$

Thus, $\binom{T}{k}\left(\frac{1}{n}\right)^k \left(1-\frac{1}{n}\right)^{T-k}$ is the UMVU estimator of $P(X = k)$ particularly the UMVU estimator of $e^{-\lambda} = P(X=0)$ will be $\left(\frac{n-1}{n}\right)^T$.

Example 2.4

Let $X_1, X_2, \ldots, X_n$ be a random sample from $N(\mu, \sigma^2)$ distribution. Obtain UMVU estimators of (i) μ, (ii) μ^2 when σ^2 is known, (iii) μ^2 when σ^2 is unknown, (iv) σ^r when μ is known, and (v) σ^r when μ is unknown.

(i) As $\bar{X}$ is complete sufficient statistic for μ when σ^2 is known and as $\bar{X} \sim N\left(\mu, \frac{\sigma^2}{n}\right)$.

Taking $Y = \bar{X} - \mu$, $Y \sim N\left(0, \frac{\sigma^2}{n}\right)$ distribution. Since $\bar{X} = Y + \mu$, $E\bar{X} = \mu \Rightarrow \bar{X}$ is UMVU estimator of μ.

(ii) As $E\bar{X}^2 = E(Y+\mu)^2 = EY^2 + 2\mu EY + \mu^2 = \frac{\sigma^2}{n} + \mu^2$ which implies $T = \bar{X}^2 - \frac{\sigma^2}{n}$ is UMVU estimator of μ^2 provided $\bar{X}^2 - \frac{\sigma^2}{n} > 0$. If $\bar{X}^2 - \frac{\sigma^2}{n} \le 0$ we get an absurd estimator of μ^2. In this case one can define the estimator of μ^2 as

$$T^* = \begin{cases} \bar{X}^2 - \dfrac{\sigma^2}{n} & \text{if } \bar{X}^2 > \dfrac{\sigma^2}{n} \\ 0 & \text{otherwise} \end{cases}$$

Even though T^* is not unbiased estimator of μ^2, it can be shown that MSE of T^* is less than var(T).

(iii) To estimate μ^2 when σ^2 is unknown, we use $\frac{S^2}{\sigma^2} \sim \chi^2_{n-1}$ where,

$$S^2 = \sum_{i=1}^{n}(X_i - \bar{X})^2,\ E\left(\frac{S^2}{\sigma^2}\right) = n-1,\ \text{var}(S^2) = 2\sigma^4(n-1)$$

As $E\bar{X}^2 = \frac{\sigma^2}{n} + \mu^2,\ E\bar{X}^2 = E\frac{S^2}{n(n-1)} + \mu^2 \Rightarrow \bar{X}^2 - \frac{S^2}{n(n-1)}$

is UMVU estimator of μ^2.

(iv) Estimation of σ^r when μ is known, let $S_0^2 = \sum_{i=1}^{n}(X_i - \mu)^2$. Then $\frac{S_0^2}{\sigma^2} \sim \chi^2_n$ and hence

$$E\left(\frac{S_0^2}{\sigma^2}\right)^{r/2} = E\left(\chi^2_n\right)^{r/2} = \Gamma\left(\frac{n+r}{2}\right)2^{r/2}/\Gamma(n/2) = \frac{1}{k_{n,r}},$$

say so that $k_{n,r}S_0^r$ is UMVU estimator of σ^r.

(v) Estimation of σ^r when μ is unknown, define $S^2 = \sum_{i=1}^{n}(X_i - \bar{X})^2$. Then $N(\mu, \sigma^2)$ being exponential family of distributions, $(\bar{X}, S^2)$ is complete sufficient statistic for (μ, σ^2) and $\bar{X}$ and s^2 are independently distributed, $\bar{X}$ being UMVU estimator of μ and $\frac{S^2}{\sigma^2} \sim \chi^2_{n-1}$, $k_{n-1,r}S^r$ is UMVU estimator of σ^r. For $r = -1$, $k_{n-1,-1}S^{-1}$ is UMVU estimator of σ^{-1} and the UMVU estimator of (coefficient of variation)$^{-1} = \frac{\mu}{\sigma}$ is $\bar{X} \cdot k_{n-1,-1}S^{-1}$.

Example 2.5

Let $X_1, X_2, \ldots, X_n$ be a random sample from $N(\mu, \sigma^2)$ distribution. Obtain UMVU estimator of

(i) $P(X \le c)$ when $\sigma^2 = 1$.

(ii) $P(X \le c)$ when σ^2 is unknown, c is known.

For

(i) $u(X)$ is unbiased estimator of $P(X \le c)$ where,

$$u(X_1) = \begin{cases} 1 & \text{if } X_1 \le c \\ 0 & \text{otherwise} \end{cases}$$

Since $\bar{X}$ is complete sufficient for μ, the UMVU estimator of $P(X \le c)$ is

$$E[u(X_1)/\bar{X}] = P(X_1 \le c/\bar{X})$$

Since $X_1 - \bar{X} \sim N\left(0, \frac{n+1}{n}\right)$ being independent of μ, it is ancillary statistic and by Basu's theorem

$$P(X_1 \le c/\bar{X}) = P(X_1 - \bar{X} \le (c - \bar{X})/\bar{X})$$

$$= P(X_1 - \bar{X} \le c - \bar{X})$$

$$= \phi\left[\sqrt{\frac{n}{n+1}}(c - \bar{X})\right]$$

where $\phi(x)$ is distribution function of $N(0, 1)$.

Thus, $\phi\left[\sqrt{\frac{n}{n+1}}(c - \bar{X})\right]$ is UMVU estimator of $P(X \le c)$.

(ii) Let $S^2 = \sum_{i=1}^{n}(X_i - \bar{X})^2$. Then $(\bar{X}, S^2)$ is complete sufficient statistic for (μ, σ^2). As the distribution of $\frac{X_1 - \bar{X}}{S}$ does not depend on (μ, σ^2), it is ancillary statistic. By Basu's theorem $\frac{X_1 - \bar{X}}{S}$ is independent of $(\bar{X}, S)$. Then the UMVU estimator of $P(X \le c)$ is

$$P(X_1 \le c/\bar{X}, S) = P\left(\frac{X_1 - \bar{X}}{S} \le \frac{c - \bar{X}}{S} / \bar{X}, S\right)$$

$$= P\left(z \le \frac{c - \bar{X}}{S}\right), \text{ where } z = \frac{X_1 - \bar{X}}{S}$$

For deriving distribution of z, making incomplete orthogonal transformation from $(X_1, X_2, \ldots, X_n) \to (Y_1, Y_2, \ldots, Y_n)$ such as

$$\begin{pmatrix} Y_1 \\ Y_2 \\ \vdots \\ Y_n \end{pmatrix} = \begin{pmatrix} \frac{1}{\sqrt{n}}, & \frac{1}{\sqrt{n}}, & \ldots, & \frac{1}{\sqrt{n}} \\ \sqrt{\frac{n-1}{n}}, & -\frac{1}{\sqrt{n(n-1)}}, & \ldots, & -\frac{1}{\sqrt{n(n-1)}} \end{pmatrix} \begin{pmatrix} X_1 \\ X_2 \\ \vdots \\ X_n \end{pmatrix}$$

so that $Y_1 = \sqrt{n}\bar{X}$ and $Y_2 = \sqrt{\frac{n}{n-1}}(X_1 - \bar{X})$ and

$$s^2 = \sum_{i=1}^{n}(X_i - \bar{X})^2 = \sum_{i=1}^{n} X_i^2 - n\bar{X}^2 = \sum_{i=1}^{n} Y_i^2 - Y_1^2 = \sum_{i=2}^{n} Y_i^2$$

$$z = \frac{X_1 - \bar{X}}{s} = \frac{\sqrt{\frac{n-1}{n}}Y_2}{\sqrt{\sum_{i=2}^{n} Y_i^2}} \Rightarrow \frac{nz^2}{n-1}$$

$$= \frac{Y_2^2}{\sum_{i=2}^{n} Y_i^2} \text{ and } 1 - \frac{nz^2}{n-1} = \frac{\sum_{i=3}^{n} Y_i^2}{\sum_{i=2}^{n} Y_i^2}$$

so that

$$U = \frac{z}{\sqrt{1 - \frac{nz^2}{n-1}}} \frac{\sqrt{n-2}\sqrt{n}}{\sqrt{n-1}} = \frac{Y_2}{\sqrt{\frac{\sum_{i=3}^{n} Y_i^2}{(n-2)}}}$$

has student's t distribution on $(n-2)$ d.f. with density

$$f_U(u) = \frac{1}{\sqrt{n-2}Be\left(\frac{1}{2}, \frac{n-2}{2}\right)\left(1 + \frac{u^2}{n-2}\right)^{-\frac{n-4}{2}}}$$

which transforms to the distribution of z with density

$$f_Z(z) = \frac{1}{B\left(\frac{1}{2}, \frac{n-2}{2}\right)} \sqrt{\frac{n}{n-1}} \left(1 - \frac{nz^2}{n-1}\right)^{\frac{n-4}{2}}, 0 < |z| < \sqrt{\frac{n-1}{n}}.$$

so UMVU estimator of $P(X \le c)$ is $P\left(Z \le \frac{c - \overline{X}}{s}\right)$ with above p.d.f. of Z.

Example 2.6

Two-sample problem of normal distribution

Let $X_1, X_2, \ldots, X_m$ be a random sample from $N(\mu_1, \sigma_1^2)$ distribution and let $Y_1, Y_2, \ldots, Y_n$ be an independent random sample from $N(\mu_2, \sigma_2^2)$. Obtain UMVU estimators of

(i) $\mu_1, \mu_2, \sigma_1^r, \sigma_2^r$

(ii) $\mu_2 - \mu_1, \left(\frac{\sigma_1}{\sigma_2}\right)^r$

(iii) $\frac{\mu_2 - \mu_1}{\sigma}$, when $\sigma_1^2 = \sigma_2^2 = \sigma^2$

Let

$$\overline{X} = \frac{1}{m}\sum_{i=1}^{m} X_i,\ \overline{Y} = \frac{1}{n}\sum_{j=1}^{n} Y_j,\ s_X^2 = \sum_{i=1}^{m}(X_i - \overline{X})^2,\ s_Y^2 = \sum_{j=1}^{n}(Y_J - \overline{Y})^2,\ s^2 = s_X^2 + s_Y^2$$

Then

(i) It is easy to prove that $(\overline{X}, \overline{Y}, s_X^2, s_Y^2)$ is complete sufficient for $(\mu_1, \mu_2, \sigma_1^2, \sigma_2^2)$. $\overline{X}$ and s_X^2 being independent and $\overline{Y}$ and s_Y^2 being independent, the UMVU estimators of μ_1 and σ_1^r are $\overline{X}$ and $k_{m-1, r} s_X^r$ and those of μ_2 and σ_2^r are $\overline{Y}$ and $k_{m-1, r} s_y^r$.

(ii) The UMVU of $\mu_2 - \mu_1$ is $\overline{Y} - \overline{X}$ and that of $(\sigma_1 / \sigma_2)^r$ is the product of UMVU estimators of σ_1^r and $\left(\frac{1}{\sigma_2}\right)^r$ can be obtained by using the results of **example 2.4(v).**

(iii) It is easy to show that $(\overline{Y}, \overline{X}, s^2)$ is complete sufficient for (μ_1, μ_2, σ^2) and as s^2 / σ^2 is χ^2_{m+n-2} and as $(\overline{Y}, \overline{X})$ is independent of s^2, the UMVU estimator of $\frac{\mu_2-\mu_1}{\sigma}$ is the product of the UMVU estimator, $\bar{Y}-\bar{X}$ of $\mu_2-\mu_1$ and the UMVU estimator,

$$\sqrt{2}\frac{\Gamma\left(\frac{m+n-2}{2}\right)}{\Gamma\left(\frac{m+n-3}{2}\right)} \text{ of } \frac{1}{\sigma}.$$

Example 2.7 (Multivariate normal : One-sample problem)

Let $\underline{X}_r=(X_{1r},X_{2r},\ldots,X_{pr}), r=1,2,\ldots,n$ be a random sample from p-variate normal distribution, $N_p(\underline{\mu},\Sigma)$, Σ being positive definite. Obtain UMVU estimators of $\underline{\mu}$ and Σ.

Let $\underline{\bar{X}}=\frac{1}{n}\sum_{r=1}^{n}X_r$, $S=(s_{ij})$, $s_{ij}=\sum_{r=1}^{n}(X_{ir}-\bar{X}_i)(X_{jr}-\bar{X}_j), i,j=1,2,\ldots,p$.

The distribution of X_r being member of exponential family of distributions, $(\underline{\bar{X}},S)$ is complete sufficient for (μ,Σ) and $E(\underline{\bar{X}})=\mu$, $E\left(\frac{S}{n-1}\right)=\Sigma$, the UMVU estimators of μ and Σ are $\underline{\bar{X}}$ and $\frac{S}{n-1}$, respectively.

Example 2.8 (Multivariate normal : Two-sample problem)

Let $\underline{X}_r=(X_{1r},X_{2r},\ldots,X_{pr}), r=1,2,\ldots,n_1$ be a random sample from $N_p(\underline{\mu}_1,\Sigma)$ distribution and let $\underline{Y}_t=(X_{1t},X_{2t},\ldots,X_{pt}), t=1,2,\ldots,n_2$ be an independent random sample from $N_p(\underline{\mu}_2,\Sigma)$ distribution, Σ being unknown positive definite matrix. Obtain UMVU estimators of $\underline{\mu}_1,\underline{\mu}_2,\Sigma$.

Let $\underline{\bar{X}}=\frac{1}{n_1}\sum_{r=1}^{n_1}X_r$, $\underline{\bar{Y}}=\frac{1}{n_2}\sum_{t=1}^{n_2}Y_t$, $s_1=\sum_{r=1}^{n_1}(X_{ir}-\bar{X}_i)(X_{jr}-\bar{X}_j)$,

$s_2=\sum_{t=1}^{n_2}(Y_{it}-\bar{Y}_i)(Y_{jt}-\bar{Y}_j)$, $s=s_1+s_2$. Then it is easy to show that $(\underline{\bar{X}},\underline{\bar{Y}},s)$ is complete sufficient for $(\underline{\mu}_1,\underline{\mu}_2,\Sigma)$. Since $E(\underline{\bar{X}})=\mu_1$, $E(\underline{\bar{Y}})=\mu_2$, $E\left(\frac{s}{n_1+n_2-2}\right)=\Sigma$, the UMVU estimators of μ_1,μ_2,Σ are

$\underline{\bar{X}},\underline{\bar{Y}},\frac{s}{n_1+n_2-2}$, respectively.

Example 2.9 (Exponential Distribution)

Let $X_1, X_2, \ldots, X_n$ be a random sample from negative exponential distribution denoted by $E(\mu, \sigma)$ with p.d.f. as

$$f_X(x; \mu, \sigma) = \frac{1}{\sigma} e^{-\frac{x-\mu}{\sigma}}, \; x > \mu, \sigma > 0$$

and let $X_{(1)}, X_{(2)}, \ldots, X_{(n)}$ be the corresponding order sample. Obtain UMVU estimators of

(i) μ when σ is known
(ii) σ when μ is known
(iii) μ and σ when both are unknown
(iv) c for which $P(X \le c) = p$, where p is known
(v) $P(X \ge t)$, reliability for known t, when $\mu = 0$.

For

(i) as $X_{(1)}$ is complete sufficient statistic for μ when σ is known and as $\dfrac{n(X_{(1)} - \mu)}{\sigma} \sim E(0,1)$ distribution so that $E_{\mu,\sigma}\left(\dfrac{n(X_{(1)} - \mu)}{\sigma}\right) = 1$,

$X_{(1)} - \dfrac{\sigma}{n}$ is the UMVU estimator of μ.

(ii) Here $\Sigma(X_i - \mu)$ is complete sufficient statistic for σ, being one-parameter exponential family of distributions and as the distribution of

$$2\,\Sigma(X_i - \mu)/\sigma \sim \chi^2_{2n}$$

$$E_{\mu,\sigma} 2\Sigma(X_i - \mu)/\sigma = 2n$$

$\dfrac{\Sigma(X_i - \mu)}{n}$ is the UMVU estimator of σ.

(iii) Let $T_1 = X_{(1)}$ and $T_2 = \Sigma(X_i - X_{(1)})$. Then T_1 has exponential distribution with parameters $\left(\mu, \dfrac{\sigma}{n}\right)$ and $T_2 \sim \dfrac{1}{2}\sigma^2 \chi^2_{2n-2}$ distribution and both are independently distributed, which can be shown in the following way

[If $Y_i = (n-i-1)\{X_{(i)} - X_{(i-1)}\},\ i = 1, 2, \ldots, n,\ X_{(0)} = 0$ then

$$\sum_{i=1}^{n} Y_i = \sum_{i=1}^{n} X_{(i)} = \sum_{i=1}^{n} X_i \quad \text{and} \quad Y_1, Y_2, \ldots, Y_n$$

are independently distributed as $E(0, 1)$. Further

$$\sum_{i=1}^{n}(X_i - X_{(1)}) = \sum_{i=1}^{n}(X_i) - nX_{(1)} = \sum_{i=1}^{n} X_{(i)} - nX_{(1)} = \sum_{i=1}^{n} Y_i - Y_1 = \sum_{i=2}^{n} Y_i$$

$$\Rightarrow \qquad Y_1 = nX_{(1)} \quad \text{and} \quad \sum_{i=2}^{n} Y_i = \sum (X_i - X_{(1)})$$

are independently distributed as

$$X_{(1)} \sim E\left(\mu, \frac{\sigma}{n}\right) \quad \text{and} \quad 2\sum(X_i - X_{(1)})/\sigma \sim \chi^2_{2n-2}].$$

By the use of factorization theorem one can show that (T_1, T_2) is sufficient for (μ, σ). For proving completeness of (T_1, T_2) suppose $E_{\mu,\sigma}\, \Psi(T_1, T_2) = 0,\ \forall\ \mu, \sigma$ and for fixed σ if

$$g(t_1, \sigma) = E_\sigma\, \Psi(T_1, T_2) \quad \text{then} \quad \int_{\mu}^{\infty} g(t_1, \sigma) e^{-\frac{n+1}{\sigma}} dt_1 = 0,\ \forall\ \mu$$

$\Rightarrow g(t_1, \sigma) = 0$ for all σ,

as $g(t_1, \sigma)$ is continuous function of t_1, that is, T_1 is complete for μ when σ^2 is fixed. Similarly, by taking $h(t_2, \mu) = E_\mu\, \Psi(T_1, T_2)$ it can be shown that T_2 is complete for σ when μ is fixed. Thus,

$$E_{\mu,\sigma}\, \Psi(T_1, T_2) = 0,\ \forall\ \mu, \sigma$$

$\Rightarrow \Psi(t_1, t_2) = 0$ for all t_1 and for a.e. t_2.

So (T_1, T_2) is complete sufficient for (μ, σ) and hence UMVU estimators of μ and σ are functions of $X_{(1)}$ and $\sum_{i=1}^{n}(X_i - X_{(1)})$.

As

$$\frac{n(X_{(1)} - \mu)}{\sigma} \sim E(0, 1),\ \frac{n}{\sigma}[EX_{(1)} - \mu] = 1$$

$$\Rightarrow \qquad EX_{(1)} = \frac{\sigma}{n} + \mu,$$

$$\Rightarrow \qquad EX_{(1)} - \frac{1}{n}E\frac{\sum_{i=1}^{n}(X_i - X_{(1)})}{(n-1)} = \mu$$

$$\Rightarrow \qquad X_{(1)} - \frac{1}{n(n-1)}\sum_{i=1}^{n}(X_i - X_{(1)})$$

is UMVU estimator of μ and $\frac{1}{n-1}\sum_{i=1}^{n}(X_i - X_{(1)})$ is UMVU estimator of σ.

(iv) **Estimation of cut-off point, c**

As

$$\frac{2(X_1 - \mu)}{\sigma} \sim \chi_1^2,\ P(X_1 \le c) = p \text{ that is,}$$

$$P\left\{\frac{2(X_1 - \mu)}{\sigma} \le \frac{2(c-\mu)}{\sigma}\right\} = p$$

$$\Rightarrow \text{if} \qquad P\{\chi_1^2 \le \chi_{1,1-p}^2\} = p$$

where $\chi_{n,1-\propto}^2 = P(\chi^2 > \alpha)$ then $\frac{2(c-\mu)}{\sigma} = \chi_{1,1-p}^2$.

$$\Rightarrow \qquad c = \frac{\sigma}{2}\chi_{1,1-p}^2 + \mu$$

so that UMVU estimator of c is

$$\frac{\chi_{1,1-p}^2}{2}\frac{1}{n-1}\sum_{i=1}^{n}(X_i - X_{(1)}) + X_1 - \frac{1}{n(n-1)}\sum_{i=1}^{n}(X_i - X_{(1)})$$

(v) **Estimation of reliability, $R(t) = P(R \ge t)$.**

Here as $\mu = 0$, X has p.d.f.

$$f_X(x;\sigma) = \frac{1}{\sigma}e^{-\frac{x}{\sigma}},\ x > 0, \sigma > 0,$$

X having exponential family of distributions $S_n = \sum_{i=1}^{n} X_i$ is complete sufficient statistic for σ. The unbiased estimator of $P(X \geq t)$ is $u(t)$,

$$u(t) = \begin{cases} 1 & \text{if } X_1 \geq t \\ 0 & \text{otherwise} \end{cases}.$$

Then UMVU estimator of $P(X \geq t)$ is

$$E_\theta[u(T)/S_n] = P\{X_1 \geq t / S_n = s\}.$$

Since X_1 and S_n are not independently distributed, taking $S_n = X_1 + Y$,

where, $Y = \sum_{i=2}^{n} X_i$, X_1 and Y are independently distributed and Y has p.d.f. as

$$f_Y(y) = \frac{e^{-\frac{y}{\sigma}}(y)^{n-2}}{\sigma^{n-1}\Gamma(n-1)}, \; y > 0, \sigma > 0.$$

so that the joint p.d.f. of X_1 and Y is

$$f_{X_1,Y}(x_1, y) = \frac{1}{\sigma} e^{-\frac{x_1}{\sigma}} \frac{e^{-\frac{y}{\sigma}}(y)^{n-2}}{\sigma^{n-1}\Gamma(n-1)}$$

$$= \frac{e^{-\frac{x_1+y}{\sigma}}(y)^{n-2}}{\sigma^n \Gamma(n-1)}$$

Making the transformation $S_n = X_1 + Y, X_1 = X_1$, the joint p.d.f. of S_n and X_1 is

$$f_{S_n,X_1}(s, x_1) = \frac{e^{-\frac{s}{\sigma}}(s - x_1)^{n-2}}{\sigma^n \Gamma(n-1)}, 0 < x_1 \leq s < \infty, \sigma > 0$$

and the conditional p.d.f. of X_1 given $S_n = s$ is

$$f_{X_{1/S_n}}(x_1 / S_n = s) = \frac{(n-1)}{s}\left(1 - \frac{x_1}{s}\right)^{n-2}, x_1 \le s.$$

Using this conditional density the UMVU estimator of R is $\hat{R}(t)$,

$$\hat{R}(t) = \int_t^s \frac{(n-1)}{s}\left(1 - \frac{x_1}{s}\right)^{n-2} dx_1$$

$$= \left(1 - \frac{t}{S_n}\right)^{n-1}.$$

2.3 LINEAR UNBIASED ESTIMATION

The class of linear unbiased estimators (LUEs) of some parametric function is linear functions of the observed random variable. The problem is to obtain best linear unbiased estimator (BLUE), that is, we have to seek linear unbiased estimator whose variance is minimum amongst the class of LUE. We illustrate how to obtain BLUE.

Example 2.10

Let $\underline{X}$ be n-dimensional random variable such that $E\underline{X} = \theta\underline{1}$, where, θ is parameter and $\underline{1}' = (1, 1, \ldots, 1)$ and let $\text{var}(\underline{X}) = \Sigma$ be known and non-singular. Obtain BLUE of θ.

Let $c'X$ be a linear function such that

$$\Rightarrow \qquad E\underline{c}'\underline{X} = \theta$$

$$\Rightarrow \qquad c'\theta\underline{1} = \theta$$

$$\Rightarrow \qquad \underline{c}'\underline{1} = 1.$$

We have to minimize $\text{var}(\underline{c}'\underline{X}) = \underline{c}'\Sigma\underline{c}$ subject to $\underline{c}'\underline{1} = 1$.

Let $$\Psi(c) = c'\Sigma c - 2\lambda(\underline{c}'\underline{1} - 1)$$

where λ is Lagrange undetermined multiplier. $\dfrac{d\Psi(c)}{d\underline{c}} = 0$ yields

$$\Sigma c = \lambda\underline{1}$$

$$\Rightarrow \qquad \underline{c} = \lambda\Sigma^{-1}\underline{1}$$

$$\Rightarrow \qquad \underline{c}'\underline{1} = \lambda\underline{1}'\Sigma^{-1}\underline{1}$$

$$\Rightarrow \qquad 1 = \lambda \underline{1}'\Sigma^{-1}\underline{1}$$

$$\Rightarrow \qquad \lambda = \frac{1}{\underline{1}'\Sigma^{-1}\underline{1}}$$

Therefore, $$\underline{c} = \frac{\Sigma^{-1}\underline{1}}{\underline{1}'\Sigma^{-1}\underline{1}} \Rightarrow \underline{c}'\underline{X} = \frac{\underline{1}'\Sigma^{-1}\underline{X}}{\underline{1}'\Sigma^{-1}\underline{1}}$$

(i) If $\Sigma = \sigma^2 I$ then the BLUE of θ is $\bar{X}$

(ii) If $\Sigma = \text{diag}(\sigma_1^2, \sigma_2^2, \ldots, \sigma_n^2)$, $\Sigma^{-1} = \text{diag}(\sigma_1^{-1}, \sigma_2^{-1}, \ldots, \sigma_n^{-1})$ and BLUE of θ is

$$\hat{\theta} = \frac{\sum_{i=1}^{n} X_i^2 / \sigma_i^2}{\sum_{i=1}^{n} \frac{1}{\sigma_i^2}}.$$

Let us consider general form in which uncorrelated observations $Y_1, Y_2, \ldots, Y_n$ are such that

$$Y_i = X_{i1}\beta_1 + X_{i2}\beta_2 + \ldots + X_{ip}\beta_p + \epsilon_i, \qquad i = 1, \ldots, n$$

where $E(\epsilon_i) = 0$, $V(\epsilon_i) = \sigma^2$, $\text{cov}(\epsilon_i, \epsilon_j) = 0$, $\forall i \neq j$, $\beta_1, \beta_2, \ldots, \beta_p$ and σ^2 are parameters and $X = (X_{ij})$ is a $n \times p$ matrix of known constants.

We can rewrite the above equations in matrix notations such as

$$E(\underline{Y}) = X\underline{\beta}\,, \ \text{var}(\underline{Y}) = \sigma^2 I \qquad \ldots(2.3.1)$$

If B is a class of vectors $\underline{\beta}$ then the problem is to find a vector $\underline{b}' = (b_1, b_2, \ldots, b_p)$ from $\underline{\beta}$ that minimizes the sum of squared residuals

$$S(\underline{\beta}) = (\underline{Y} - X\underline{\beta})'(\underline{Y} - X\underline{\beta})$$

$$= \underline{Y}'\underline{Y} + \underline{\beta}'X'X\underline{\beta} - 2\underline{\beta}'X'\underline{Y}$$

$$\frac{dS(\underline{\beta})}{d\underline{\beta}} = 2X'X\underline{\beta} - 2X'\underline{Y}$$

$$\frac{d^2S(\underline{\beta})}{d\underline{\beta}^2} = 2X'X \quad \text{(non-negative definite)}$$

$$\frac{dS(\underline{\beta})}{d\underline{\beta}} = 0$$

yields equations which are called normal equations

$$X'X\underline{b} = X'\underline{Y} \qquad \ldots(2.3.2)$$

If X is of full rank p, then $X'X$ is non-singular and the unique solution of (2.3.2) is

$$\underline{b} = (X'X)^{-1}X'\underline{Y}$$

If X is not of full rank p, (2.3.2) has a set of solutions

$$\underline{b} = (X'X)^{-}X'\underline{Y} + (I - ((X'X)^{-})X'X)\underline{Z} \qquad \ldots(2.3.3)$$

where $(X'X)^{-}$ is a g-inverse [If $AA^{-}A' = A'A^{-}A = A$ then A^{-} is g-inverse of A] of $X'X$ so that

$$X'X(X'X)^{-}X'X = X'X$$

$$\Rightarrow \qquad X(X'X)^{-}X'X = X \text{ and } X'X(X'X)^{-}X' = X' \qquad \ldots(2.3.4)$$

and $\underline{Z}$ is an arbitrary vector.

Theorem 2.4: The predictor of y, that is, $\hat{y} = X\underline{b}$ has the same value for solution $\underline{b}$ of

(i) $X'X\underline{\beta} = X'\underline{Y}$

(ii) $S(\underline{\beta})$, the sum of squares of residuals $(\underline{Y} - X\underline{\beta})'(\underline{Y} - X\underline{\beta})$ attains the minimum value for any solution of $X'X\underline{b} = X'\underline{Y}$.

Proof: For proving (i) let us choose any $\underline{b}$ in (2.3.3). Then

$$X\underline{b} = X(X'X)^{-}X'\underline{Y} + X(I - (X'X)^{-}X'X)\underline{Z}$$

$$= X(X'X)^{-}X'\underline{Y}$$

in virtue of (2.3.4).

This is independent of $\underline{Z}$.

To prove (ii) observe for any $\underline{\beta}$

$$S(\underline{\beta}) = (\underline{Y} - X\underline{\beta} + X(\underline{b} - \underline{\beta}))'(\underline{Y} - X\underline{\beta} + X(\underline{b} - \underline{\beta}))$$

$$= (\underline{Y} - X\underline{\beta})'(\underline{Y} - X\underline{\beta}) + (\underline{b} - \underline{\beta})'X'X(\underline{b} - \underline{\beta}) + 2(\underline{b} - \underline{\beta})'X'(\underline{Y} - X\underline{\beta})$$

$$= (\underline{Y} - X\underline{\beta})'(\underline{Y} - X\underline{\beta}) + (\underline{b} - \underline{\beta})'X'X(\underline{b} - \underline{\beta}) \text{ using } X'X\underline{b} = X'\underline{Y}$$

$$\geq (\underline{Y} - X\hat{\underline{\beta}})'(\underline{Y} - X\hat{\underline{\beta}})$$

$$= \underline{Y}'\underline{Y} - 2Y'X\underline{\beta} + \underline{b}'X'X\underline{b}$$

$$= \underline{Y}'\underline{Y} - \underline{b}'X'X\underline{b}$$

$$= \underline{Y}'\underline{Y} - \widehat{\underline{Y}}'\widehat{\underline{Y}}$$

Also $E(\underline{b}) = (X'X)^{-1}X'E(\underline{Y})$

$$= (X'X)^{-1}X'X\underline{\beta} = \underline{\beta}$$

so that $\underline{b}$ is unbiased for $\underline{\beta}$ and

$$\text{var}(\underline{b}) = \text{var}((X'X)^{-1}X'\underline{Y})$$

$$= (X'X)^{-1}X'\,\text{var}(\underline{Y})X(X'X)^{-1}$$

$$= (X'X)^{-1}X'\sigma^2 IX(X'X)^{-1}$$

$$= \sigma^2(X'X)^{-1}.$$

If $\underline{b}^*$ is any other unbiased estimator of $\underline{\beta}$ then $\underline{b}$ is UMVU estimator if $\text{var}(\underline{b}^*) - \text{var}(\underline{b})$ is positive semi-definite. Another way to say the same principle of comparison is that the variance of any linear function of the components of $\underline{b}$ is no larger than that of the same linear function of the components of $\underline{b}^*$, i.e., if $\text{var}(\underline{c}'\underline{b}) \leq \text{var}(\underline{c}'\underline{b}^*)\ \forall \underline{c}$ then $\underline{b}$ is UMVU estimator of $\underline{\beta}$.

Theorem 2.5: Gauss-Markov theorem

If $\underline{b}$ is least square estimator of $\underline{\beta}$ in the set-up defined in (2.3.1) then $\underline{l}'\underline{b}$ has minimum variance in the class of linear unbiased estimators of $\underline{l}'\underline{\beta}$, where $\underline{l}' = (l_1, \ldots, l_p)$.

Proof: Since

$$\underline{b} = (X'X)^{-1}X'\underline{Y}$$

$$\underline{l}'\underline{b} = l'(X'X)^{-1}X'\underline{Y}$$

$$= c'X'\underline{Y} \text{ where, } c' = l'(X'X)^{-1}$$

If $\underline{l}'\underline{y}$ is any unbiased estimator of $\underline{l}'\underline{\beta}$ then

$$E(\underline{l}'\underline{y}) = \underline{l}'\underline{\beta} \Rightarrow \underline{l}'X\underline{\beta} = \underline{l}'\underline{\beta}$$

$$\Rightarrow \qquad \underline{l}'X = \underline{l}'$$

Consider

$$\text{var}(\underline{l}'\underline{y}) = \text{var}(\underline{l}'\underline{y} - c'X'\underline{Y} + c'X'\underline{Y})$$

$$= \text{var}(\underline{l}'\underline{y} - c'X'\underline{Y}) + \text{var}(c'X'\underline{Y})$$

$$= \text{var}(\underline{l}'\underline{y} - c'X'\underline{Y}) + \text{var}(\underline{l}'\underline{b})$$

because covariance of $(\underline{l}'\underline{y} - c'X'\underline{Y})$ and $c'X'\underline{Y}$ is

$$(\underline{l}' - \underline{c}'X')\text{cov}(\underline{Y}, \underline{Y}')X\underline{c} = (\underline{l}' - \underline{c}'X')X\underline{c}\sigma^2$$

$$= (\underline{l}'X - c'X'X)\underline{c}\sigma^2$$

$$= (\underline{l}' - \underline{l}'(X'X)^{-1}X'X)\underline{c}\sigma^2$$

$$= (\underline{l}' - \underline{l}')\underline{c}\sigma^2$$

$$= 0$$

Therefore, $\text{var}(\underline{l}'\underline{y}) \geq \text{var}(\underline{l}'\underline{b})$ which proves the result.

We note that (i) $\text{var}(\underline{P}'\underline{b}) = \sigma^2\underline{P}'(X'X)^{-1}\underline{P}$

(ii) $\text{var}(b_i) = \sigma^2 c_{ii}$

where c_{ii} is the (ii)th element of $(X'X)^{-1}$.

Estimation of σ^2

Let

$$R_0^2 = (\underline{Y} - X\widehat{\beta})'(\underline{Y} - X\widehat{\beta})$$

$$= \underline{Y}'\underline{Y} - 2\widehat{\underline{\beta}}'X'\underline{Y} + \widehat{\underline{\beta}}'X'X\widehat{\underline{\beta}}$$

$$= \underline{Y}'\underline{Y} - \widehat{\beta}'X'\underline{Y}$$

$$= \underline{Y}'\underline{Y} - Y'X(X'X)^{-1}X'\underline{Y} \text{ using } \hat{\beta} = (X'X)^{-1}X'\underline{Y}$$

$$= \underline{Y}'[I - X(X'X)^{-1}X']\underline{Y}$$

$$= \underline{Y}'A\underline{Y}$$

where $A = I - X(X'X)^{-1}X'$.

Note that A is symmetric and idempotent matrix. Further

$$(\underline{Y} - X\underline{\hat{\beta}})'(I - X(X'X)^{-1}X')(\underline{Y} - X\underline{\hat{\beta}}) = (\underline{Y} - X\underline{\hat{\beta}})'A(\underline{Y} - X\underline{\hat{\beta}})$$

$$= \underline{Y}'A\underline{Y} - \hat{\beta}'X'A\underline{Y} - \underline{Y}'AX\underline{\hat{\beta}} + \hat{\beta}'X'AX\underline{\hat{\beta}}$$

$$= \underline{Y}'A\underline{Y}, \text{ as } X'A = X'(I - X(X'X)^{-1}X') = 0$$

$$= R_0^2$$

Hence, $R_0^2 = \underline{Z}'A\underline{Z}$; where, $\underline{Z} = \underline{Y} - X\underline{\hat{\beta}}$ such that $EZ = 0$, $E\underline{Z}'\underline{Z} = D(\underline{Y}) = \sigma^2 I$.

As

$$R_0^2 = \sum_i \sum_j a_{ij} Z_i Z_j = \sum a_{ii} Z_i^2 + \sum\sum_{i \neq j} a_{ij} Z_i Z_j$$

so that

$$E(R_0^2) = \sum a_{ij} E(Z_i^2) + \sum\sum_{i \neq j} a_{ij} E(Z_i) E(Z_j)$$

$$= \sigma^2 \sum a_{ii} + 0$$

$$= \sigma^2 tr(A)$$

$$= \sigma^2 tr(I - X(X'X)^{-1}X')$$

$$= \sigma^2 (tr(I) - tr((X'X)^{-1}XX')), \text{ using } tr(AB) = tr(BA)$$

$$= \sigma^2 (n - r) \text{ if } R(XX') = R(X) = r$$

Thus, unbiased estimator $\hat{\sigma}^2$ of σ^2 is

$$\hat{\sigma}^2 = \frac{(\underline{Y} - X\underline{\hat{\beta}})'(\underline{Y} - X\underline{\hat{\beta}})}{(n - r)}$$

$$= \frac{\underline{Y}'\underline{Y} - YX(X'X)^{-1}X'\underline{Y}}{(n - r)}$$

Linear Model Under Normal Errors

So far we have obtained the results without assuming the distribution of $\underset{\sim}{\in}$. Now let us assume that $\underset{\sim}{\in} \sim N_n(0, \sigma^2 I)$, so that $\underset{\sim}{Y} \sim N_n(X\hat{\underset{\sim}{\beta}}, \sigma^2 I)$.

$$f_{\underset{\sim}{Y}}(\underset{\sim}{y}) = \frac{1}{(2\pi\sigma^2)^{-\frac{n}{2}}} e^{-\frac{1}{2\sigma^2}(\underset{\sim}{Y}-X\hat{\underset{\sim}{\beta}})'(\underset{\sim}{Y}-X\hat{\underset{\sim}{\beta}})}$$

$$= c(\beta, \sigma^2) e^{-\frac{\underset{\sim}{Y}'\underset{\sim}{Y}}{2\sigma^2} + \sum_{i=1}^{p} \frac{\beta_i}{\sigma^2} Z_i} \quad \ldots(2.3.5)$$

where $\underset{\sim}{Z} = X'\underset{\sim}{Y}$

writing

$$\left.\begin{aligned} t_i(y) &= Z_i, i = 1, 2, \ldots, p \\ t_{p+1}(y) &= y'\underset{\sim}{y} \end{aligned}\right\} t(\underset{\sim}{y}) = \{t_1(y), t_2(y), \ldots, t_{p+1}(y)\}$$

Reparametrizing parameters $\beta_1, \beta_2, \ldots, \beta_p, \sigma^2$ by $\theta_1, \theta_2, \ldots, \theta_{p+1}$ such as $\theta_i = \frac{\beta_i}{\sigma^2}$, $i = 1, 2, \ldots, p$ and $\theta_{p+1} = -\frac{1}{2\sigma^2}$ (2.3.5) reduces to

$$c^*(\theta) e^{\sum_{i=1}^{p+1} \theta_i t_i(y)}$$

This being exponential family of distributions $t(y)$ is complete sufficient. Since

$\beta_i = -\frac{\theta_i}{2\theta_{p+1}}$, $i = 1, 2, \ldots, p$ being function of $\theta_1, \theta_2, \ldots, \theta_{p+1}$, and since

$$\hat{\underset{\sim}{\beta}} = (X'X)^{-1} X'\underset{\sim}{Y} = (X'X)^{-1} \underset{\sim}{Z}$$

$\hat{\beta}_i$ is unbiased estimator of β_i. It follows that $\hat{\beta}_i$ is the UMVU estimator of β_i and since $R_0^2/(n-r)$ is unbiased for σ^2 it is UMVU estimator of σ^2.

We have shown that any linear function of least square estimator is the UMVU estimator of the same linear function of the parameter. Here we have proved a stronger result due to normality assumption.

Least Square Estimation with Restriction in Parameters

For the set-up $(\underline{Y}, X\underline{\beta}, \sigma^2 I)$ if $H\underline{\beta} = \underline{r}$ where H is a $q \times p$ matrix and $\underline{r}$ is q-vector the least square estimation of β is obtained by minimizing the quantity $(\underline{Y} - X\hat{\underline{\beta}})'(\underline{Y} - X\hat{\underline{\beta}})$ with restriction $H\underline{\beta} = \underline{r}$. Using Lagrange q-vector undetermined multiplier $\underline{\lambda}$ and minimizing unrestricted function Ψ

$$\Psi = (\underline{Y} - X\hat{\underline{\beta}})'(\underline{Y} - X\hat{\underline{\beta}}) - 2(\beta' H' - \underline{r}')\underline{\lambda}$$

$$\frac{d\Psi}{d\beta} = 0, \text{ gives } X'X\underline{\beta} - H'\underline{\lambda} = X'\underline{Y} \qquad \text{...(2.3.6)}$$

and

$$H\underline{\beta} + 0\underline{\lambda} = \underline{r} \qquad \text{...(2.3.7)}$$

From (2.3.6) we get $\hat{\underline{\beta}}$ as

$$\hat{\underline{\beta}} = (X'X)^{-1} X'\underline{Y} + (X'X)^{-1} H'\underline{\lambda}$$

$$= \underline{b} + (X'X)^{-1} H'\underline{\lambda} \qquad \text{...(2.3.8)}$$

where $\underline{b} = (X'X)^{-1} X'\underline{Y}$ is estimator of β without restriction.

Again from (2.3.7), (2.3.8), we have

$$H\hat{\underline{\beta}} = H\,\underline{b} + H(X'X)^{-1} H'\underline{\lambda} = \underline{r}$$

$$\Rightarrow \qquad \hat{\lambda} = (H(X'X)^{-1} H')^{-1}(r - H\underline{b}),$$

if $H(X'X)^{-1}H'$ is positive definite.

Inserting $\hat{\lambda}$ in (2.3.8) and using $S = X'X$, we get

$$\hat{\underline{\beta}} = \underline{b} + S^{-1}H'(HS^{-1}H')^{-1}(r - H\underline{b}).$$

Now, if $H\underline{\beta} = \underline{r}$ then $\hat{\underline{\beta}}$ is unbiased as

$$E(\hat{\underline{\beta}}) = \beta + S^{-1}H'(HS^{-1}H')^{-1}(r - H\underline{b})$$

$$= \beta, \text{ as } r = H\underline{b}$$

and

$$\text{var}(\hat{\underline{\beta}}) = \text{var}(\underline{b}) - S^{-1}H'(HS^{-1}H')^{-1} H \,\text{var}(\underline{b})$$

$$= \sigma^2 S^{-1} - \sigma^2 S^{-1} H'(HS^{-1}H')^{-1} HS^{-1}$$

which shows that dispersion matrix of $\hat{\underline{\beta}}$ depends on H. Also

$$\text{var}(\underline{b}) - \text{var}(\hat{\underline{\beta}}) = \sigma^2 S^{-1} H'(HS^{-1}H')^{-1} HS^{-1}$$

being positive semi-definite, $\hat{\underline{\beta}}$ has smaller variance than $\underline{b}$. Thus, restriction leads to a gain in efficiency.

Ridge Estimation of Linear Model

$E(\hat{\underline{\beta}} - \underline{\beta})'(\hat{\underline{\beta}} - \underline{\beta})$ is interpreted as the mean Euclidean distance between the vectors $\hat{\underline{\beta}}$ and $\underline{\beta}$. For controlling the parameter a number $k \geq 0$ is introduced in the solution and a ridge estimator is

$$\hat{\beta}(k) = (X'X + kI)^{-1} X'Y$$

If $k = 0$, it reduces to ordinary least square estimator.

Let
$$H_k = (X'X + kI)^{-1}$$
$$= (X'X)^{-1} - k(X'X + kI)^{-1}(X'X)^{-1}.$$
$$[\text{using } (A + BCD)^{-1} = A^{-1} - A^{-1}B(C^{-1} + DC^{-1}B)^{-1} DA^{-1}].$$

Then bias and $\text{var}(\hat{\beta}(k))$ are as follows:

$$E(\hat{\beta}(k)) = H_k X'X\beta$$
$$= \beta - kH_k\beta$$
$$\text{Bias} = -kH_k\beta$$
$$\text{var}(\hat{\beta}(k)) = \sigma^2 H_k (X'X) H_k'$$

Generalized Least Square Estimators

Let $\underline{Y}$ be $n \times 1$ vector of random variable X be $n \times p$ matrix of non-stochastic constant, $\underline{\beta}$ be $p \times 1$ vector of parameters, $\underline{\in}$ be $n \times 1$ vector of errors such that

$E(\underline{\in}) = 0$, $E(\underline{\in}\,\underline{\in}') = \sigma^2 V$. Then the generalized linear model is defined as

$$\underline{Y} = X\underline{\beta} + \underline{\in} \qquad \ldots(2.3.9)$$

we have to estimate $\underline{\beta}$, σ^2. Let us assume that V is known, positive definite matrix. As V is symmetric positive definite and V^{-1} is symmetric and positive definite and hence there exists regular symmetric matrices

$$A = V^{-\frac{1}{2}}, B = V^{\frac{1}{2}} \text{ such that } V^{-1} = AA,\ BB = V^{\frac{1}{2}}V^{\frac{1}{2}} = V.$$

By premultiplying the model given in (2.3.9) by A, we have

$$A\underline{Y} = AX\underline{\beta} + A\in$$

Let $A\underline{Y} = \underline{Z}$, $AX = W$, $A\in = \underline{\eta}$. Then $E(\eta) = E(A\in) = 0$,

$$E(\eta\eta') = E(A\in\in' A') = \sigma^2 I$$

So the model (2.3.9) is transformed to

$$\underline{Z} = W\underline{\beta} + \underline{\eta}$$

in which $E(\eta) = 0$, $E(\eta\eta') = \sigma^2 I$.

Using the results of linear model $(Y, X\beta, \sigma^2 I)$. We have

$$\hat{\beta} = (W'W)^{-1}W'\underline{Z} \quad \text{and} \quad \text{var}(\beta) = (W'W)^{-1}\sigma^2$$

that is,

$$\hat{\beta} = (X'A'AX)^{-1}X'A'A\underline{Y}$$

$$= (X'V^{-1}X)^{-1}X'V^{-1}\underline{Y} \qquad \ldots(2.3.10)$$

and $$\text{var}(\widehat{\beta}) = (X'V^{-1}X)^{-1}\sigma^2 \qquad \ldots(2.3.11)$$

As $E(\widehat{\beta}) = (X'V^{-1}X)^{-1}X'V^{-1}X\beta = \beta$, $\hat{\beta}$ is unbiased for β.

Let us show that $\hat{\beta}$ has minimum variance. Let $\tilde{\beta} = \tilde{C}\underline{Y}$ be another linear unbiased estimator of β and let

$$\tilde{C} = \hat{C} + D$$

with optimum matrix $\tilde{C} = (X'V^{-1}X)^{-1}X'V^{-1}$.

Then

$$\begin{aligned}\beta &= E(\widehat{\beta})\\ &= \widetilde{C}E(\underline{Y})\\ &= \widetilde{C}(X\underline{\beta} + E(\underline{\in})) + D(X\underline{\beta} + E(\underline{\in}))\\ &= (X'V^{-1}X)^{-1}X'V^{-1}X\underline{\beta} + DX\underline{\beta}\\ &= \beta + DX\underline{\beta}\end{aligned}$$

$$\Rightarrow \qquad DX = 0 \quad \text{and} \quad \widetilde{C}VD = 0$$

Now

$$\begin{aligned}\text{var}(\widetilde{\beta}) &= E(\widetilde{C}\,\underline{\in}\,\underline{\in}'\,\widetilde{C}')\\ &= \widetilde{C}E(\underline{\in}\,\underline{\in}')\widetilde{C}'\\ &= \sigma^2\widetilde{C}V\widetilde{C}'\\ &= \sigma^2(\widehat{C}+D)V(\widehat{C}+D)'\\ &= \sigma^2[\widehat{C}V\widehat{C}' + DVD']\\ &= \text{var}(\widehat{\beta}) + \sigma^2 DVD'\end{aligned}$$

$$\Rightarrow \qquad \text{var}(\widetilde{\beta}) - \text{var}(\widehat{\beta}) = \sigma^2 DVD'$$

is non-negative definite.

So $\widehat{\beta}$ has minimum variance. Alternatively we may get results (2.3.10), (2.3.11) by minimizing the weighted residual

$$\begin{aligned}R &= (\underline{Y} - X\underline{\beta})'V^{-1}(\underline{Y} - X\underline{\beta})\\ &= (\underline{Y}'V^{-1} - \beta'X'V^{-1})(\underline{Y} - X\underline{\beta})\\ &= \underline{Y}'V^{-1}\underline{Y} - 2\beta'X'V^{-1}\underline{Y} + \beta'X'V^{-1}X\underline{\beta}\end{aligned}$$

$$\frac{dR}{d\underline{\beta}} = 2X'V^{-1}\underline{Y} + 2X'V^{-1}X\underline{\beta} = 0 \text{ gives}$$

$$\widehat{\beta} = (X'V^{-1}X)^{-1}X'V^{-1}\underline{Y}$$

and

$$\begin{aligned}\text{var}(\hat{\beta}) &= (X'V^{-1}X)^{-1}X'V^{-1}\,\text{var}(\underline{Y})V^{-1}X(X'V^{-1}X)^{-1} \\ &= \sigma^2(X'V^{-1}X)^{-1}X'V^{-1}VV^{-1}X(X'V^{-1}X)^{-1} \\ &= \sigma^2(X'V^{-1}X)^{-1}.\end{aligned}$$

For the set-up $(\underline{Y}, X\underline{\beta}, \sigma^2 V)$ the Gauss - Markov theorem may be extended as the generalized least square estimator given in (2.3.10) is the best linear unbiased estimator of $\underline{\beta}$.

Mixed Effect Linear Model

Let

$$\underline{Y} = X\underline{\beta} + U\underline{\xi} + \underline{\in}$$

where $\underline{Y}$ is an n vector of observations, X is a given $n \times p$ matrix, $\underline{\beta}$ is a p-vector of unknown fixed parameters, U is a given $n \times m$ matrix, ξ is an m-vector of hypothetical random variable and $\underline{\in}$ is an n-vector of unknown random error with

$$E(\underline{\in}) = 0,\ E(\underline{\in}\,\underline{\in}') = V \text{ and } E(\xi) = 0,\ E(\underline{\xi}\underline{\xi}') = \Gamma,\ \text{cov}(\underline{\in}, \underline{\xi}) = 0.$$

The components of ξ are unobserved variates on individuals and our problem is to estimate linear functions of β and ξ which arise in animal breeding where, ξ represents intrinsic values of individual on the basis of which some individuals are chosen for further breeding. We assume that V, Γ are known.

Estimation of $X\underline{\beta}$ (The Fixed Effect)

Let $A'\underline{Y}$ be an unbiased estimator of $X\underline{\beta}$. Then

$$E(A'\underline{Y}) = X\underline{\beta} \text{ for every } \underline{\beta}$$

$$\Rightarrow \quad A'X = X,$$

because

$$\begin{aligned}A'\underline{Y} &= A'X\underline{\beta} + A'U\xi + A' \in \\ &= X\underline{\beta} + A'U\xi + A' \in\end{aligned}$$

and $$E[(A'\underline{Y} - X\underline{\beta})(A'\underline{Y} - X\underline{\beta})'] = E[A'(U\xi + \in)(U\xi + \in)'A]$$
$$= A'(U\Gamma U' + V)A$$
$$= A'WA$$

where $W = (U\Gamma U' + V)$. We have to minimize $A'WA$ subject to $X'A = X'$, A is the solution of the equation

$$WA + X\underline{\beta} = 0$$
$$X'A = X'$$

i.e., $$\begin{pmatrix} W & X \\ X' & 0 \end{pmatrix}\begin{pmatrix} A \\ \underline{\beta} \end{pmatrix} = \begin{pmatrix} 0 \\ X' \end{pmatrix}.$$

If $\begin{pmatrix} W & X \\ X' & 0 \end{pmatrix}^{-} = \begin{pmatrix} C_1 & C_2 \\ C_2' & -C_4 \end{pmatrix}$ is any g-inverse then

$A^* = C_2X', \underline{\beta}^* = -C_4X'$ giving the minimum dispersion estimator of $X\underline{\beta}$ as $A^{*'}\underline{Y} = XC_2'\underline{Y}$ with dispersion matrix XC_4X'.

Prediction of $U\xi$ (The Random Effect)

Let $B'\underline{Y}$ be a predictor of $U\xi$ such as

$$E(B'\underline{Y} - U\xi) = B'X\underline{\beta} = 0$$

$\Rightarrow$ $$B'X = 0$$

Now $$B'\underline{Y} = B'X\underline{\beta} + B'U\underline{\xi} + B' \underline{\in}$$
$$= B'U\underline{\xi} + B' \in, \text{ using } B'X = 0$$

$$E(B'\underline{Y} - U\underline{\xi})(B'\underline{Y} - U\underline{\xi})' = E[(B'U\underline{\xi} + B' \underline{\in} - U\underline{\xi})(B'U\underline{\xi} + B' \underline{\in} - U\underline{\xi})']$$
$$= E[\{(B' - I)U\underline{\xi} + B' \underline{\in}\}\{(B' - I)U\underline{\xi} + B' \underline{\in}\}']$$
$$= E(B' - I)U\underline{\xi}\underline{\xi}'U'(B - I) + E(B' \underline{\in}\underline{\in}' B)$$
$$= (B' - I)U\Gamma U'(B - I) + B'VB$$
$$= B'WB - B'U\Gamma U' - U\Gamma U'B + U\Gamma U'$$

where $W = U\Gamma U' + V$. We have to obtain minimum of

$$B'WB - B'Z - ZA + Z$$

where $Z = U\Gamma U'$, subject to $B'X = 0$.

The minimum attains at B^*, D^* which is the solution of equations

$$WB + XD = Z$$
$$B'X = 0$$

In terms of elements of the g-inverse the solution is

$$B^* = C_1 Z,\ D^* = C_2' Z$$

So the predictor of $U\underline{\xi}$ is

$$B^{*\prime}\underline{Y} = ZC_1'\underline{Y}$$

These results may be referred in Rao (1973).

We have used the distribution of disturbance term $\underline{\in}$ as normal but in some cases like life testing problems the distribution of $\underline{\in}$ may be taken as exponential and in this case for the estimation of $A\beta$ and prediction of $U\underline{\xi}$ refer Patel and Patel (1992c).

Simultaneous Estimation

If $\underline{T}' = (T_1, T_2, \ldots, T_k)$ is unbiased estimator of the vector parameter $\underline{\theta}' = (\theta_1, \theta_2, \ldots, \theta_k)$. Then $k \times k$ matrix

$$D(T) = E(\underline{T} - \underline{\theta})(\underline{T} - \underline{\theta})'$$

$$= \begin{pmatrix} \text{var}(T_1) & \text{cov}(T_1, T_2) & \cdots & \text{cov}(T_1, T_k) \\ \cdots\cdots & \cdots\cdots & \cdots\cdots & \cdots\cdots \\ \text{cov}(T_k, T_1) & \text{cov}(T_k, T_2) & \cdots & \text{var}(T_k) \end{pmatrix}$$

is called dispersion matrix of T. An unbiased estimator $\underline{T}_0$ is said to be uniformly minimum dispersion unbiased estimator (UMDUE) of $\underline{\theta}$ if $D(\underline{T}) - D(\underline{T}_0)$ is non-negative definite such that $E(\underline{T}) = \underline{\theta}$, for any $\underline{T}$ and $\forall\ \underline{\theta}$.

Theorem 2.6: If T_{i0} is UMVU estimator of θ_i, $i = 1, 2, \ldots, k$ then $\underline{T}_0' = (T_{10}, T_{20}, \ldots, T_{k0})$ is UMVU estimator of $\underline{\theta}$ and vice versa.

Proof: Consider a linear function $\underline{a}'\underline{T}_0$, which is unbiased for $\underline{a}'\underline{\theta}$. If U_0 is unbiased for 0, then by existence theorem proved earlier $\text{cov}(\underline{a}'\underline{T}_0, U_0) = 0$, which shows that

$$\text{var}(\underline{a}'\underline{T}_0) = \underline{a}'D(\underline{T}_0)\underline{a} \le \underline{a}'D(\underline{T})\underline{a},$$

as $\underline{T}_0$ is UMVU estimator of $\underline{\theta}$ and $\underline{T}$ is any other unbiased estimator of $\underline{\theta}$. From the above result it follows that

$$D(\underline{T}_0) \le D(\underline{T}).$$

Alternatively, if $\underline{T}_0$ is UMVU estimator of $\underline{\theta}$ then

$$\text{var}(\underline{a}'\underline{T}_0) \le \text{var}(\underline{a}'\underline{T}) \;\; \forall \; \underline{T} \text{ such that } E\underline{T} = \underline{\theta}$$

$$\Rightarrow \qquad \underline{a}'D(\underline{T}_0)\underline{a} \le \underline{a}'D(\underline{T})\underline{a}$$

$$\Rightarrow \qquad D(\underline{T}_0) \le D(\underline{T})$$

$\Rightarrow$ $\text{var}(T_{i0}) \le \text{var}(T_i)$ being i^{th} diagonal of $\underline{T}_0$ and $\underline{T}$

$\Rightarrow$ T_{i0} is UMVU estimator of $\theta_i, i = 1, 2, \ldots, k$.

This theorem, in turn, implies

$\text{cov}(U, Z/\theta_0) = 0 \;\; \forall \; Z$ such that $E(Z/\theta) = 0 \;\; \forall \; \theta$.

Now let us consider ordinary set up of linear model

$$\underline{Y} = X\underline{\beta} + \in$$

with $E(\in) = 0$, $D(\underline{\in}) = E(\underline{\in}\,\underline{\in}') = \sigma^2 I$ and let us obtain a linear function $a + \underline{b}'\underline{y}$ with zero expectation. Then

$$E(a + \underline{b}'y) = a + \underline{b}'X\underline{\beta} = 0 \quad \forall \; \beta$$

$$\Rightarrow \qquad a = 0 \text{ and } \underline{b}'X = 0 \text{ or } b \in R(Z),$$

where Z is the matrix whose columns span the space orthogonal to

$$R(X) = \{\theta : \theta = X\underline{\beta},\; \beta \in \mathbb{R}^k\}$$

with rank of $Z = n - \text{rank}(X)$. Thus, the class of all linear functions of Y with zero expectation is

$$(Z\underline{C})'Y = \underline{C}'\underline{Z}'\underline{Y}$$

where $\underline{C}$ is an arbitrary vector.

Theorem 2.7: The UMDLUE of $\underline{\beta}$ is

$$\hat{\underline{\beta}} = (X'X)^{-1}X'Y$$

which is same as the least squares estimator of $\underline{\beta}$, and minimum dispersion matrix of $\hat{\underline{\beta}}$ is $\sigma^2(X'X)^{-1}$, when $R(X) = p$.

Proof: Let $a + By$ be an unbiased estimator of $\underline{\beta}$. Then

$$E(a + B\underline{Y}) = a + BX\underline{\beta}$$

$$= \underline{\beta}$$

$$\Rightarrow \qquad a = 0,\ BX = I.$$

If $B\underline{Y}$ is UMDLUE then sufficient condition demands

$$0 = \text{cov}(B\underline{Y}, C'Z'\underline{Y}) \text{ as } E(C'Z'\underline{Y}) = 0$$

$$= \sigma^2 BZC \quad \forall\ C$$

$$\Rightarrow \qquad B\underline{Z} = 0$$

$$\Rightarrow \qquad B = AX' \text{ for some } A.$$

Thus, we have

$$BX = I, B = AX'$$

$$\Rightarrow \qquad A(X'X) = I \Rightarrow A = (X'X)^{-1},\ B = (X'X)^{-1}X'$$

and hence, UMDLUE of $\underline{\beta}$ is

$$\hat{\underline{\beta}} = B\underline{Y}$$

$$= (X'X)^{-1}X'\underline{Y}$$

and

$$D(\hat{\underline{\beta}}) = D((X'X)^{-1}X'\underline{Y})$$

$$= (X'X)^{-1}X'D(\underline{Y})X(X'X)^{-1}$$

$$= \sigma^2(X'X)^{-1}X'X(X'X)^{-1}$$

$$= \sigma^2(X'X)^{-1}.$$

Case 2 If $R(X) = r < p$ and $R(Z) = n - r$ in which case $X'X$ is singular. Let $(X'X)^-$ be g-inverse of $X'X$. Then there is no unique solution.

The deficiency in rank occurs either in some equations of $Y = X\beta + \in$ or there is deficiency in rank of X and we are unable to find B such that $BX = I$ and β cannot be estimated.

Theorem 2.8: Let $l'\underline{\beta}$ be linear functions of β such that $R(L) \subset R(X') \Rightarrow L = X'A$ for some A. Then the UMDLUE of $\underline{L}'\underline{\beta}$ is $\underline{L}'\underline{\hat{\beta}}$ where, $\underline{\hat{\beta}} = (X'X)^- X'\underline{Y}$ and the dispersion matrix of $\underline{L}'\underline{\hat{\beta}}$ is $\sigma^2\underline{L}'(X'X)^- L$, where, $(X'X)^-$ is g-inverse of $(X'X)$.

Proof: Let $C\underline{Y}$ be an unbiased estimator of $\underline{L}'\underline{\beta}$. Then

$$E(C\underline{Y}) = CX\underline{\beta}$$

$$= \underline{L}'\underline{\beta}$$

$$\Rightarrow \qquad CX = \underline{L}'.$$

Now $\qquad \text{cov}(C\underline{Y}, \underline{Z}'\underline{Y}) = \sigma^2 CZ = 0$

$$\Rightarrow \qquad C = BX' \text{ for some } B.$$

$$CX = \underline{L}'$$

$$\Rightarrow \qquad BX'X = \underline{L}'$$

$\Rightarrow B = \underline{L}'(X'X)^-$ as one solution and

$C = BX' = \underline{L}'(X'X)^- X'$ so that UMDLUE of $\underline{L}'\underline{\beta}$ is

$$C\underline{Y} = \underline{L}'(X'X)^- X'\underline{Y} = \underline{L}'\underline{\hat{\beta}}, \; D(\underline{L}'\underline{\hat{\beta}}) = \sigma^2\underline{L}'(X'X)^- \underline{L}.$$

2.4 ESTIMATION OF POWER SERIES DISTRIBUTION (PSD)

The problem of UMVU estimation of parametric function in the class of discrete distributions has been studied by many authors amongst which **Roy and Mitra (1957)** who have studied the problem associated with power series distribution (PSD).

X is said to have PSD if it is defined as

$$P(X = x) = \frac{a(x)\theta^x}{f(\theta)}, \; x \in T \qquad \ldots(2.4.1)$$

where T is subset of I, the set of non-negative integers, $a(x) > 0$ and it is independent of θ, $f(\theta)$ is positive finite and differentiable function of θ such that

$$f(\theta) = \sum_{x \in T} a(x)\theta^x .$$

The binomial distribution, $b(n, p)$, can be obtained from (2.4.1) by taking $a(x) = \binom{n}{x}$ for $x = 0, 1, 2, \ldots, n$ and $a(x) = 0$ otherwise; $\theta = \left(\frac{p}{q}\right)$, $f(\theta) = (1+\theta)^n$ and for negative binomial distribution

$$a(x) = \binom{m+x-1}{m-1}, \theta = q \quad \text{and} \quad f(\theta) = (1-\theta)^{-m}$$

If $a(x) > 0 \ \forall \ x$ then θ^r is U-estimable for any integer r and the estimator $T(x)$ is such that

$$\sum_{x=0}^{\infty} T(x)a(x)\theta^x = \theta^r f(\theta) \ \forall \ \theta \in \Omega \text{ is satisfied}$$

$$\Rightarrow \qquad \sum_{x=0}^{\infty} T(x)a(x)\theta^x = \theta^r f(\theta) = \theta^r \sum_{x=0}^{\infty} a(x)\theta^x = \sum_{x=0}^{\infty} a(x)\theta^{x+r}$$

Comparison of coefficient of θ^x gives

$$T(x) = \begin{cases} 0 & \text{if } x = 0, 1, 2, \ldots, r-1 \\ \dfrac{a(x-r)}{a(x)} & \text{if } x \geq r \end{cases} \qquad \ldots(2.4.2)$$

If $X_1, X_2, \ldots, X_n$ is a random sample from PSD defined in (2.4.1) and if

$T = X_1 + X_2 + \ldots + X_n$ then distribution of T is again PSD with p.m.f.

$$P(T = t) = \frac{A(t, n)\theta^t}{[f(\theta)]^n} \qquad \ldots(2.4.3)$$

where $A(t, n)$ is the coefficient of θ^t in the expansion of $[f(\theta)]^n$.(2.4.3) being exponential family, T is complete sufficient. So by using (2.4.2) and (2.4.3), the UMVU estimator of θ^r is

$$U(t)=\begin{cases} 0 & \text{if } t=0,1,2,\ldots,r-1 \\ \dfrac{A(t-r,n)}{A(t,n)} & \text{if } \quad t\geq r \end{cases} \qquad \ldots(2.4.4)$$

and that of $g(\theta)=P_\theta(X_1=x)$ is

$$\begin{aligned} U(t) &= P(X_1=x/X_1+X_2+\ldots+X_n=t) \\ &= P(X_1=x)P(X_2+\ldots+X_n=t-x)/P(T=t) \\ &= a(X)A\;(t-x,n-1)/A(t,n)\;, n>1,\; 0\leq x\leq t \quad \ldots(2.4.5) \end{aligned}$$

Application of these results to Poisson distribution with parameter θ with

$$a(x)=\frac{1}{x!}, f(\theta)=e^{\theta}, [f(\theta)]^n=e^{\theta n}=\sum_{t=0}^{\infty}\frac{(\theta n)^t}{t!}$$

so that $A(t,n)=\dfrac{n^t}{t!}$. Using (2.4.4) UMVU estimator of θ^r is

$$u(t)=[t(t-1)\ldots(t-r+1)]/n^r$$

and using (2.4.5) UMVU estimator of $P_\theta(X=x)$ is

$$u(t)=\binom{t}{x}\left(\frac{1}{n}\right)^x\left(1-\frac{1}{n}\right)^{t-x}, x=0,1,\ldots,t$$

which is $b\left(t,\dfrac{1}{n}\right)$.

Modified Power Series Distribution (MPSD)

Gupta (1974) has defined a new class of univariate discrete distributions called modified power series distribution (MPSD) of random variable X with p.m.f.

$$P(X=x)=a(x)[h(\theta)]^x\;/\;f(\theta), x\in T \qquad \ldots(2.4.6)$$

where T is a subset of non-negative integers, I, $a(x)>0$ and independent of parameter θ, $h(\theta)$ and $f(\theta)$ are positive finite and differentiable functions of θ,

$0 < \theta < \rho$, where, ρ is the radius of convergence, $h(\theta)$ is non-invertible and

$$f(\theta) = \sum_{x \in T} a(x)(h(\theta))^x.$$

If $h(\theta)$ is invertible then (2.4.6) reduces to generalized power series distribution (GPSD) defined by Patil and in addition if $T = I$ then it reduces to PSD.

Generalized Poisson Distribution (GPD)

If the p.m.f. of a random variable X is defined as

$$P(X = x) = \alpha(\alpha + \beta x)^{x-1}(h(\theta))^x / (x!\, e^{\alpha\theta}) \qquad \ldots(2.4.7)$$

where $h(\theta) = \dfrac{\theta}{e^{\beta\theta}}$, $x = 0, 1, 2, \ldots, \alpha, \theta > 0, |\beta\theta| < 1$

then X is said to have GPD.

This is obtained as a special case of MPSD with

$$a(x) = \frac{\alpha(\alpha + \beta x)^{x-1}}{x!}$$

$$f(\theta) = e^{\alpha\theta} = \sum_{x=0}^{\infty} \left(\frac{\theta}{e^{\beta\theta}}\right)^x \alpha(\alpha + \beta x)^{x-1}.$$

When $\alpha = 1$, $\beta = 0$, (2.4.7) reduces to classical Poisson distribution with parameter θ.

Generalized Logarithmic Series Distribution (GLSD)

If the p.m.f. of a random variable X is defined as

$$P(X = x) = \frac{\Gamma(x\beta)}{x!\,\Gamma(x\beta - x + 1)} \frac{[[\theta(1-\theta)]^{\beta-1}]^x}{-\log(1-\theta)}, 0 < \theta < 1,\ |\beta\theta| = 1, 2, \ldots \qquad \ldots(2.4.8)$$

$$= a(x)(h(\theta))^x / f(\theta)$$

then X is said to have GLSD. This is a special case of MPSD with

$$f(\theta)=[-\log(1-\theta)],\ \ h(\theta)=[\theta(1-\theta)]^{\beta-1}$$

and

$$a(x)=\frac{\Gamma(x\beta)}{x!\,\Gamma(x\beta-x+1)}.$$

For $\beta = 1$, (2.4.8) reduces to LSD with p.m.f. as

$$P(X=x)=\frac{\theta^x}{x[-\log(1-x)]}.$$

Generalized Negative Binomial Distribution (GNBD)

If the p.m.f. of a random variable X is defined as

$$P(X=x)=\frac{k\Gamma(k+x\beta)}{x!\,\Gamma(k+x\beta-x+1)}\theta^x(1-\theta)^{\beta x-x+k} \qquad \ldots(2.4.9)$$

where $k>0,\ 0<\theta<1, \beta\geq 1, \beta\theta<1,\ x\in I$, then X is said to have GNBD. This is obtained as special case of MPSD with

$$a(x)=\frac{k\Gamma(k+x\beta)}{x!\,\Gamma(k+x\beta-x+1)},\ f(\theta)=(1-\theta)^{-k}$$

and $h(\theta)=[\theta(1-\theta)]^{\beta-1}$.

When $\beta = 1$, (2.3.9) reduces to NBD with p.m.f. as

$$P(X=x)=\binom{k+x-1}{k-1}\theta^x(1-\theta)^k,\ \ x=0,1,2,\ldots.$$

UMVU Estimation of MPSD

If $X_1, X_2, \ldots, X_n$ is a random sample from MPSD defined in (2.4.1) and if $Z=\sum_{i=1}^{n}X_i$. Then the distribution of Z is given by

$$p_Z(\theta) = P(Z = z) = L(z, n)[h(\theta)]^z / [f(\theta)]^n, z \in n(T) \quad \ldots(2.4.10)$$

where $n(T) = \Sigma T$, $L(z, n) = \Sigma a(x_1), a(x_2), \ldots, a(x_n)$ and summation extends over all ordered n-tupples $(x_1, x_2, \ldots, x_n)$ such that $\Sigma X_i = Z$.

$$[f(\theta)]^n = \sum_{z \in Z} L(z, n)[h(\theta)]^Z.$$

As (2.4.10) constitutes exponential family of distributions, Z is complete sufficient. By Lagrange formula if

$$h(\theta) = \frac{\theta}{g(\theta)}$$

$$L(z, n) = \frac{1}{Z!} \frac{d^{Z-1}}{d\theta^{Z-1}} [g(\theta)]^n \frac{d}{d\theta} [f(\theta)]^n$$

$L(z, n)$ is called Lagrange number.

Theorem 2.9: If $X_1, X_2, \ldots, X_n$ is a random sample from the MPSD defined in (2.4.1) and if $Z = \sum_{i=1}^{n} X_i$ then the UMVU estimator $\hat{h}(Z, n)$ of $[h(\theta)]^k$, k is a positive integer, whenever it exists, is given by

$$\hat{h}_k(Z, n) = \frac{L(z-k, n)}{L(z, n)}, \text{ if } z > k, z \in n(T) \quad \ldots(2.4.11)$$

$$= 0, \text{ otherwise}$$

The proof being analogous to result (2.4.4) for PSD it is left. Also for MPSD, the UMVU estimator $\hat{p}_j(z, n)$ of $p_j(\theta)$ is given by

$$\hat{p}_j(z, n) = \frac{a(j)L(z-j, n-1)}{L(z, n)}, \; z > j, z \in (n-1)[T] \quad \ldots(2.4.12)$$

$$= 0, \quad \text{otherwise}$$

This can be shown just by the use of unbiasedness such as

$$E\hat{p}_j(z, n) = p_j(\theta)$$

$$\Rightarrow \qquad \sum_{Z} \hat{p}_j(Z, n) L(Z, n)[h(\theta)]^Z = \sum_{Z} a(j) L(z, n-1)[h(\theta)]^{Z+j}.$$

On comparing the coefficient of $[h(\theta)]^Z$, we get (2.4.12).

UMVU Estimation of GPD

If $X_1, X_2, \ldots, X_n$ is a random sample from GPD with p.m.f. given by (2.4.7) then $Z = \sum_{i=1}^{n} X_i$ is complete sufficient for

$$h(\theta) = \theta e^{-\beta\theta} = \frac{\theta}{g(\theta)}$$

where $g(\theta) = e^{\beta\theta}$. By using Lagrange formula for the expansion of $e^{n\alpha\theta}$ in terms of $h(\theta) = \dfrac{\theta}{g(\theta)}$,

$$e^{n\alpha\theta} = \sum_{Z=0}^{\infty} [D_\theta^{Z-1} e^{\beta\theta Z} n\alpha e^{n\alpha\theta}]_{\theta=0} \left[\frac{\theta}{e^{\beta\theta}}\right]^Z / Z!$$

$$= \sum_{Z=0}^{\infty} n\alpha(n\alpha + Z\beta)^{Z-1} \left[\frac{\theta}{e^{\beta\theta}}\right]^Z / Z!$$

so that

$$P(Z = z) = n\alpha(n\alpha + Z\beta)^{Z-1}[h(\theta)]^Z / [e^{n\alpha\theta} Z!]$$

$$= b(Z, n, \alpha, \beta)[h(\theta)]^Z / [e^{n\alpha\theta} Z!]$$

where $b(Z, n, \alpha, \beta) = n\alpha(n\alpha + Z\beta)^{Z-1}$, $Z = 0, 1, 2, \ldots$.

Since the distribution of Z form exponential family of distributions, Z is complete sufficient statistic for $h(\theta)$ when α, β are known. Using (2.4.11), we get UMVU estimator $\hat{h}(Z, n, \alpha, \beta)$ of $h(\theta)$ as

$$\hat{h}(Z, n, \alpha, \beta) = \begin{cases} 0 & \text{if } Z = 0 \\ \dfrac{b(Z-1, n, \alpha, \beta)}{b(Z, n, \alpha, \beta)} & \text{if } Z = 1, 2, \ldots. \end{cases}$$

By using (2.4.12) one can get UMVU estimator of $P(X = r)$. Taking $\propto = 1$ and $\beta = 0$, we get UMVU estimator of θ in Poisson distribution as $\frac{Z}{n}$.

UMVU Estimation of GLSD

If $X_1, X_2, \ldots, X_n$ is a random sample from GLSD defined in (2.4.8). $Z = \sum_{i=1}^{n} X_i$ is complete sufficient and its distribution is

$$P(Z = z) = b(z, n, \beta)[h(\theta)]^Z / [f(\theta)]^n, \; z = n, n+1, \ldots.$$

where $\quad [f(\theta)]^n = [-\log(1-\theta)]^n = \Sigma b(z, n, \beta)[h(\theta)]^Z$

and $b(Z, n, \beta) = \frac{1}{Z!}[D_\theta^{Z-1}\{D_\theta[f(\theta)]^n (1-\theta)^{-\beta Z + Z}\}]$, where $D_\theta = \frac{d}{d\theta}$

$$= \sum_{r=0}^{Z-1} \binom{Z-1}{r} \frac{n!(\beta Z - Z)!|S(r+1, n)|}{Z!(\beta Z - r - 1)!}$$

where $|S(m, n)|$ is the signless stirling number of the first kind.

Following **Roy and Mitra (1957),** we obtained the UMVU estimator $\hat{h}(Z, n)$ of $h(\theta)$ as

$$\hat{h}(Z, n) = \begin{cases} 0 & \text{if } Z = n \\ \dfrac{b(Z-1, n, \beta)}{b(Z, n, \beta)} & \text{if } Z = n+1, n+2, \ldots. \end{cases}$$

Taking $\beta = 1$, the UMVU estimator of θ for LSD can be obtained. This result has been given by **Patel (1980)**

UMVU Estimation of GNBD

If $X_1, X_2, \ldots, X_n$ is a random sample from GNBD defined in (2.4.9) and if $Z = \sum_{i=1}^{n} X_i$, then Z being complete sufficient statistic for

$h(\theta) = \theta(1-\theta)^{\beta-1}$ using result (2.4.10) for MPSD as GNBD is MPSD, we get the distribution of Z as

$$P(Z = z) = \begin{cases} \dfrac{b(z,n)[\theta(1-\theta)^{\beta-1}]^Z}{(1-\theta)^{-nk}}, & z \in I \\ 0 & \text{, otherwise} \end{cases}$$

where

$$b(z, n) = \frac{1}{Z!} D_\theta^{Z-1}[(1-\theta)^{-\beta Z+Z} D_\theta (1-\theta)^{-nk}]_{\theta=0}$$

$$= \frac{nk}{nk+\beta Z}\binom{nk+\beta z}{z}, \; z \in I$$

and the UMVU estimator $\hat{h}(Z, n, \beta)$ of $h(\theta) = \theta(1-\theta)^{\beta-1}$ is given by

$$\hat{h}(Z, n, \beta) = \frac{nk+\beta Z}{nk+\beta Z-\beta} \frac{\binom{nk+\beta Z-\beta}{Z-1}}{\binom{nk+\beta Z}{Z}}.$$

For $\beta = 1$, the UMVU estimator of θ is given by

$$\hat{h}(Z, n) = \frac{Z}{nk+Z-1}.$$

For decapitated distributions and for multivariate MPSD see Patel (1978a) and for left truncation multivariate MPSD see Patel (1979a) and for special cases of multivariate MPSD as multivariate generalized LSD. This result has been given by Patel (1981), multivariate GNBD is also given by Patel (1979b).

2.5 ESTIMATION BASED ON NON-REGULAR FAMILY OF DISTRIBUTIONS

A regular family of distributions is one in which the support does not depend on parameters. An exponential family of distributions is a part of regular family. A non-regular family is one in which support depends on parameters. A non-regular family of distributions may be classified into two main classes.

(1) *One-truncation parameter family*

This is defined as

(i) **Left truncated:** if p.d.f. is

$$f_1(x;\theta) = q_1(\theta)h_1(x), a < \theta < x < b \qquad \text{...(2.5.1)}$$

and

(ii) **Right truncated:** if p.d.f. is

$$f_2(x;\theta) = q_2(\theta)h_2(x),\ a < x < \theta < b \qquad \text{...(2.5.2)}$$

where $q_i(\theta), i = 1, 2$ are everywhere differentiable function of θ and $h_i(x), i = 1, 2$ are absolutely continuous functions of x, $-\infty < a < b < \infty$ are specified real numbers.

Some distributions belonging to this family are:

(i) Uniform distribution, $u(0, \theta)$ with p.d.f.

$$f(x;\theta) = \frac{1}{\theta}, 0 < x < \theta$$

(ii) Negative exponential distribution with p.d.f.

$$f(x;\theta) = e^{-(x-\theta)}, \theta < x < \infty$$

(iii) Pareto distribution with p.d.f.

$$f(x;\theta) = \frac{\alpha\, \theta^{\alpha}}{x^{\alpha+1}}, \theta > 0, \theta < x.$$

(2) *Two-truncation parameter family*

This is defined as

$$f(x;\theta_1,\theta_2) = q(\theta_1,\theta_2)h(x),\ a < \theta_1 < x < \theta_2 < b \qquad \text{...(2.5.3)}$$

where $q(\theta_1, \theta_2)$ is everywhere differentiable function of θ_1, θ_2 and $h(x)$ is absolutely continuous function of x, $-\infty < a < b < \infty$ are fixed constants.

Amongst many distributions belonging to this family some of them are given below.

(i) Uniform distribution, $u(\theta_1, \theta_2)$ defined on interval (θ_1, θ_2) with p.d.f.

$$f_X(x;\theta_1,\theta_2) = (\theta_2 - \theta_1)^{-1}, \ \theta_1 < x < \theta_2$$

(ii) Two-parameter truncated exponential distribution with p.d.f.

$$f_X(x;\theta_1,\theta_2) = \frac{\alpha e^{-\alpha x}}{e^{-\theta_1\alpha} - e^{-\theta_2\alpha}}, -\infty < \theta_1 < x < \theta_2 < \infty, \alpha > 0 \text{ (known)}$$

(iii) Truncated Pareto distribution with p.d.f.

$$f_X(x;\theta_1,\theta_2) = \frac{\theta_2\theta_1}{(\theta_2 - \theta_1)x^2}, \ 0 < \theta_1 < x < \theta_2.$$

Estimation for One-Truncated Parameter (Right) Distribution

Theorem 2.10: Let $X_1, X_2, \ldots, X_n$ be a random sample from (2.5.2) distribution and $X_{(1)}, X_{(2)}, \ldots, X_{(n)}$ be the corresponding set of ordered statistics and let $X_{(1)}, X_{(2)}, \ldots, X_{(r)}; \ 1 \le r \le n$ be type-II censored sample. Then UMVU estimator of $g(\theta)$ is

$$\hat{g}(\theta) = \frac{(r-1)!}{n!} q_2^{r-1}(X_{(r)}) D^{n-r+1} S(X_{(r)})$$

where $S(\theta) = g(\theta) q_2^{-n}(\theta)$ and D is a linear operator defined by

$$DS(\theta) = \frac{d}{d\theta}(S(\theta)) / h_2(\theta)$$

and $$D^k S(\theta) = D(D^{k-1}S(\theta)), \text{ for } k \ge 2.$$

Proof: As $X_{(1)} < X_{(2)} < \ldots < X_{(r)}, \ 1 \le r \le n$ is a type-II censored sample from (2.5.2) distribution and as

$$F_X(x) = q_2(\theta) \int_a^x h(u) du,$$

the p.d.f. of $X_{(r)}$ is

$$f_{X_{(r)}}(x;\theta)=\frac{n!}{(r-1)!(n-r)!}q_2^n(\theta)\left[\int_a^x h_2(u)du\right]^{r-1}\left[\int_x^\theta h(u)du\right]^{n-r}h_2(x)$$

which can be expressed as $g(t;\theta)h(x)$, where $T=X_{(r)}$ and hence by factorization theorem T is sufficient for θ.

Let $E\Psi(X_{(r)})=0,\ \forall\ \theta$

i.e.,

$$\frac{n!}{(r-1)!(n-r)!}q_2^n(\theta)\int_a^\theta \Psi(x)\left[\int_a^x h_2(u)du\right]^{r-1}\left[\int_x^\theta h_2(u)du\right]^{n-r}h_2(x)dx=0,\ \forall\ \theta$$

i.e., $$\int_a^\theta \Psi(x)\left[\int_a^x h_2(u)du\right]^{r-1}\left[\int_x^\theta h_2(u)du\right]^{n-r}h_2(x)dx=0,\ \forall\ \theta \qquad \text{...(2.5.4)}$$

Differentiation of (2.5.4) with respect to θ and usage of Libtnitz rule of differentiation under integral sign and usage of easy result

$$h_2(\theta)=-\frac{d}{d\theta}q_2(\theta)/q_2^2(\theta)$$

(2.5.4) results into

$$\int_a^\theta \Psi(x)\left[\int_a^x h_2(u)du\right]^{r-1}\left[\int_x^\theta h_2(u)du\right]^{n-r-1}h_2(x)dx=0,\ \forall\ \theta.$$

Repeatedly differentiating (2.5.4) with respect to θ, $(n-r+1)$ times, it implies

$$\Psi(\theta)=0 \quad \text{for a.e. } \theta$$

$$\Rightarrow\ \Psi(x_r)=0 \quad \text{for a.e. } x_{(r)}.$$

which shows that $X_{(r)}$ is complete for θ.

Let $S(\theta)=g(\theta)q_2^{-n}(\theta)$ and

$$g(\theta)=E\phi(X_{(r)})$$

$$= \frac{n!}{(r-1)!(n-r)!} q_2^n(\theta) \int_a^\theta \phi(x) \left[\int_a^x h_2(u) du \right]^{r-1} \left[\int_x^\theta h_2(u) du \right]^{n-r} h_2(x) dx$$

$$\Rightarrow \quad S(\theta) = \frac{n!}{(r-1)!(n-r)!} \int_a^\theta \phi(x) \left[\int_a^x h_2(u) du \right]^{r-1} \left[\int_x^\theta h_2(u) du \right]^{n-r} h_2(x) dx \quad \ldots(2.5.5)$$

so that differentiation with respect to θ gives

$$DS(\theta) = \frac{n!(n-r)}{(r-1)!(n-r)!} \int_a^\theta \phi(x) \left[\int_a^x h_2(u) du \right]^{r-1} \left[\int_x^\theta h_2(u) du \right]^{n-r-1} h_2(x) dx$$

Repeated differentiation of (2.5.5) with respect to θ, $(n - r)$ times, results into

$$D^{n-r} S(\theta) = \frac{n!}{(r-1)!} \int_a^\theta \phi(x) \left[\int_a^x h_2(u) du \right]^{r-1} h_2(x) dx \quad \ldots(2.5.6)$$

Again differentiating (2.5.6) with respect to θ, one gets

$$D^{n-r+1} S(\theta) = \frac{n!}{(r-1)!} \phi(\theta) q_2^{1-r}(\theta) h_2(\theta)$$

$$\Rightarrow \quad \phi(\theta) = \frac{(r-1)!}{n!} q_2^{r-1}(\theta) [D^{n-r+1} S(\theta)] \quad \ldots(2.5.7)$$

i.e.,

$$\phi(X_{(r)}) = \frac{(r-1)!}{n!} q_2^{r-1}(X_r) [D^{n-r+1} S(X_r)] \quad \ldots(2.5.8)$$

which is UMVU estimator of $g(\theta)$.

Remark

(vii) The UMVU estimator of $g(\theta)$ for uncensored sample of size n, that is, $r = n$, from one-truncation parameter family is obtained by substituting $r = n$ in (2.5.7)

$$\phi(\theta) = \frac{1}{n} q_2^{n-1}(\theta) DS(\theta)$$

where $$S(\theta) = g(\theta) q_2^{-n}(\theta).$$

Therefore,

$$\frac{d}{d\theta} S(\theta) = n g(\theta) q_2^{-n-1}(\theta) h(\theta) + q_2^{-n}(\theta) g'(\theta)$$

$$DS(\theta) = n g(\theta) q_2^{-n+1}(\theta) + \frac{g'(\theta)}{q_2^n(\theta) h_2(\theta)}$$

so that

$$\phi(\theta) = g(\theta) + \frac{g'(\theta)}{n q_2(\theta) h_2(\theta)}$$

Thus, UMVU estimator of $g(\theta)$ for complete sample is

$$\phi(X_{(n)}) = g(X_{(n)}) + \frac{g'(X_{(n)})}{n q_2(X_{(n)}) h_2(X_{(n)})} \quad \ldots(2.5.9)$$

which is obtained by Tate by different way.

Some Applications of Tate's Result

(i) For obtaining UMVU estimator of mean, $g(\theta) = \int_a^\theta x f(x;\theta) dx$, differentiating $g(\theta)$ (using Libnitz rule),

$$g'(\theta) = -q_2(\theta) h_2(\theta)[g(\theta) - \theta]$$

and keeping this value in (2.5.9) the UMVU estimator of the mean is

$$\phi(X_{(n)}) = g(X_{(n)}) - \frac{|g(X_{(n)}) - X_{(n)}|}{n}$$

$$= \left(\frac{n-1}{n}\right) g(X_{(n)}) + \frac{X_{(n)}}{n}$$

(ii) For getting UMVU estimator of reliability

$$g(\theta) = P(X \geq t)$$

$$= \int_t^\theta f(x;\theta) dx$$

differentiating $g(\theta)$,

$$g'(\theta) = -q_2(\theta)h_2(\theta)[g(\theta)-1],$$

the UMVU estimator of reliability is

$$\phi(X_{(n)}) = \left(\frac{n-1}{n}\right)g(X_{(n)}) + \frac{1}{n}.$$

Example 2.11

Let $X_{(1)}, X_{(2)}, \ldots, X_{(n)}$ be type-II censored ordered sample from $u(0,\theta)$ distribution with p.d.f.

$$f_X(x;\theta) = \frac{1}{\theta},\ 0 < x < \theta$$

and let $g(\theta) = e^{-\theta}$ to be estimated.

Here, $h_2(x) = 1,\ q_2(\theta) = \frac{1}{\theta},\ S(\theta) = \theta^n e^{-\theta}$

using

$$\frac{d^{n-r+1-k}}{d\theta^{n-r+1-k}}(e^{-\theta}) = (-1)^{n-r+1-k}.e^{-\theta}$$

$$D^{n-r+1}S(\theta) = \sum_{k=0}^{n-r+1}(-1)^{n-r+1-k}\binom{n-r+1}{k}\frac{n!}{(n-k)!}\theta^{n-k}e^{-\theta},$$

it follows that UMVU estimator of $e^{-\theta}$ is

$$\phi(X_{(r)}) = (r-1)!\sum_{k=0}^{n-r+1}(-X_{(r)})^{n-r+1-k}\binom{n-r+1}{k}\frac{e^{-X_{(r)}}}{(n-k)!}.$$

Example 2.12

Given a random sample $X_1, X_2, \ldots, X_n$ from p.d.f.

$$f_X(x;\theta) = \frac{2x}{\theta^2},\ 0 < x < \theta$$

In this case $h_2(x) = 2x$, $q_2(\theta) = \frac{1}{\theta^2}$. Consider $g(\theta) = \theta$. Then UMVU estimator of θ is

$$\phi(X_{(n)}) = X_{(n)} + \frac{X_{(n)}}{2n}$$

$$= \frac{(1+2n)}{2n} X_{(n)}$$

Analogus results can be obtained for one-truncation parameter (left) family of distributions.

Problem of UMVU Estimation for Two-, One-Truncation Parameter Family

If $X_1, X_2, \ldots, X_n$ and $Y_1, Y_2, \ldots, Y_n$ are independent random samples from one-truncation parameter families of distributions with p.d.f. as

$$f_X(x;\theta_1) = q_1(\theta_1)h_1(x), \quad a < x < \theta_1 < b$$

$$f_X(y;\theta_2) = q_2(\theta_2)h_2(y), \quad a < y < \theta_2 < b$$

where $-\infty < a < b < \infty$ are known constants, $h_1(x)$ and $h_2(y)$ are positive absolutely conditions functions and q_1, q_2 are everywhere differentiable. Suppose $X_{(1)} < X_{(2)} < \ldots < X_{(r_1)}$, $1 \le r_1 \le m$ and $Y_{(1)} < Y_{(2)} < \ldots < T_{(r_2)}$, $1 \le r_2 \le n$ are type-II censored ordered samples then by making same type of arguments as done for proving Theorem 2.10, one can prove the following theorem. So proof of the theorem is left.

Theorem 2.11: For the above set-up the UMVU estimator of $g(\theta_1, \theta_2)$ is given by

$$\hat{g}(\theta_1, \theta_2) = \frac{(r_1-1)!\,(r_2-1)!}{m!n!} q_1^{r_1-1}(X_{(r_1)}) q_2^{r_2-1}(Y_{(r_2)}) D_{11}^{m-r_1+1}$$

$$\left\{ \frac{1}{q_1^m(X_{(r_1)})} D_{01}^{n-m-r_2+r_1} \left[\frac{g((X_{(r_1)}),(Y_{(r_2)}))}{q_2^n(Y_{(r_2)})} \right] \right\} \qquad \ldots(2.5.10)$$

where $D_{ij}(W(\theta_1, \theta_2)) = \frac{d^{i+j}W(\theta_1, \theta_2)}{d^j d^i(\theta_1, \theta_2)} / (h_1(\theta)h_2(\theta))$, if $n - m - r_2 + r_1 \ge 0$.

For $n-m-r_2+r_1<0$, exchanging $n \leftrightarrow m$ and $r_1 \leftrightarrow r_2$ under operator D, in the above expression it gives the estimator of $g(\theta_1, \theta_2)$.

Using (2.5.10) with $r_1 = m$ and $r_2 = n$ that is using

$$\hat{g}(\theta_1,\theta_2) = \frac{1}{mn} q_1^{m-1}(X_{(r_1)}) q_2^{n-1}(Y_{(r_2)}) D_{11}\left(\frac{g((X_{(r_1)}),(Y_{(r_2)}))}{q_1^m(X_{(r_1)}) q_2^n(Y_{(r_2)})}\right) \qquad \ldots(2.5.11)$$

the UMVU estimators of

(i) $\theta_2 - \theta_1$

(ii) $\theta_2\theta_1$

(iii) $\dfrac{\theta_1}{\theta_2}$

for two independent uniform distributions on $(0, \theta_1)(0, \theta_2)$ intervals are, respectively:

(i) $\left(\dfrac{n+1}{n}\right)Y_{(n)} - \left(\dfrac{m+1}{m}\right)X_{(m)}$

(ii) $(m+1)(n+1)\dfrac{X_{(m)}Y_{(n)}}{mn}$

(iii) $\dfrac{(m+1)(n+1)}{mn}\dfrac{X_{(m)}}{Y_{(n)}}$.

Estimation of Reliability Functional $P_{\theta_1,\theta_2}(Y > X) = g(\theta_1, \theta_2)$ for Complete Samples

$$P_{\theta_1,\theta_2}(Y > x) = \int_a^{\theta_1} P_{\theta_1}(Y > x) f_1(x;\theta_1)dx$$

$$= \int_a^{\min(\theta_1,\theta_2)} q_1(\theta_1)h_1(x)\left[\int_x^{\theta_2} q_2(\theta_2)h_2(u)du\right]dx$$

$$= g(\theta_1, \theta_2)$$

For $\theta_1 > \theta_2$

$$g_1(\theta_1, \theta_2) = \int_a^{\theta_2} q_1(\theta_1) h_1(x) \left[\int_x^{\theta_2} q_2(\theta_2) h_2(u) du \right] dx \qquad \text{...(2.5.12)}$$

Differentiating (2.5.12) w.r.t. θ_1

$$g(\theta_1, \theta_2) = \frac{d}{d\theta_1} \int_a^{\theta_2} q_1(\theta_1) h_1(x) \left[\int_x^{\theta_2} q_2(\theta_2) h_2(u) du \right] dx$$

$$= \int_a^{\theta_2} \frac{d}{d\theta_1} q_1(\theta_1) h_1(x) \left[\int_x^{\theta_2} q_2(\theta_2) h_2(u) du \right] dx$$

$$= -\int_a^{\theta_2} q_1^2(\theta_1) h_1(\theta_1) h_1(x) \left[\int_x^{\theta_2} q_2(\theta_2) h_2(u) du \right] dx$$

where $\quad g_1(\theta_1, \theta_2) = \frac{d}{d\theta_1} g(\theta_1, \theta_2)$

$$\frac{d}{d\theta_1} q_1(\theta_1) = -q_1^2(\theta_1) h_1(\theta_1)$$

$$g_1(\theta_1, \theta_2) = -q_1(\theta_1) h_1(\theta_1) g(\theta_1, \theta_2) \qquad \text{...(2.5.13)}$$

Now differentiating (2.5.12) w.r.t θ_2

$$g_2(\theta_1, \theta_2) = \frac{d}{d\theta_2} \int_a^{\theta_2} q_1(\theta_1) h_1(x) \left[\int_x^{\theta_2} q_2(\theta_2) h_2(u) du \right] dx$$

where

$$g_2(\theta_1, \theta_2) = \frac{d}{d\theta_2} g(\theta_1, \theta_2)$$

$$= \frac{d}{d\theta_2} \int_{a<x<u<\theta_2} \int q_1(\theta_1) h_1(x) q_2(\theta_2) h_2(u) du dx$$

$$= \frac{d}{d\theta_2} \int_a^{\theta_2} q_2(\theta_2) h_2(u) \left[\int_a^u q_1(\theta_1) h_1(x) dx \right] du$$

$$= \frac{d}{d\theta_2} \int_a^{\theta_2} q_2(\theta_2) h_2(u) F_1(u, \theta_1) du$$

$$= \int_a^{\theta_2} \frac{d}{d\theta_2} q_2(\theta_2) h_2(u) F_1(u, \theta_1) du + q_2(\theta_2) h_2(\theta_2) F_1(\theta_2, \theta_1)$$

$$= \int_a^{\theta_2} - q_2^2(\theta_2) h_2(\theta_2) h_2(u) F_1(u, \theta_1) du + q_2(\theta_2) h_2(\theta_2) F_1(\theta_2, \theta_1)$$

Therefore,

$$g_2(\theta_1, \theta_2) = -q_2(\theta_2) h_2(\theta_2) g(\theta_1, \theta_2) + q_2(\theta_2) h_2(\theta_2) F_1(\theta_2, \theta_1) \qquad \ldots(2.5.14)$$

where $F(u, \theta_1) = \int_a^u q_1(\theta_1) h_1(x) dx$

Differentiating (2.5.13) w.r.t. θ_2

$$g_{12}(\theta_1, \theta_2) = -q_1(\theta_1) h_1(\theta_1) \frac{d}{d\theta_2} g(\theta_1, \theta_2)$$

$$= q_1(\theta_1) h_1(\theta_1) g_2(\theta_1, \theta_2) \qquad \ldots(2.5.15)$$

Using (2.5.14) in (2.5.15),

$$g_{12}(\theta_1, \theta_2) = q_1(\theta_1) h_1(\theta_1) q_2(\theta_2) h_2(\theta_2) [g(\theta_1, \theta_2) - F_1(\theta_2, \theta_1)] \qquad \ldots(2.5.16)$$

For $\theta_1 < \theta_2$, we have

$$g(\theta_1, \theta_2) = \int_a^{\theta_1} q_1(\theta_1) h_1(x) \left[\int_x^{\theta_2} q_2(\theta_2) h_2(u) du \right] dx \qquad \ldots(2.5.17)$$

Now differentiating (2.5.17) w.r.t. θ_1,

$$g_1(\theta_1,\theta_2)=\frac{d}{d\theta_1}\int_a^{\theta_1} q_1(\theta_1)h_1(x)[1-F_2(X,\theta_2)]dx$$

$$\int_x^{\theta_2} q_2(\theta_2)h_2(u)du=1-F_2(X,\theta_2)$$

$$g_1(\theta_1,\theta_2)=\int_a^{\theta_1}\frac{d}{d\theta_1}q_1(\theta_1)h_1(x)[1-F_2(X,\theta_2)]$$

$$dx+q_1(\theta_1)h_1(\theta_1)[1-F_2(\theta_1,\theta_2)]$$

$$=\int_a^{\theta_1}-q_1^2(\theta_1)h_1(x)[1-F_2(X,\theta_2)]$$

$$dx+q_1(\theta_1)h_1(\theta_1)[1-F_2(\theta_1,\theta_2)]$$

$$=-q_1(\theta_1)h_1(\theta_1)[g(\theta_1,\theta_2)+F_2(\theta_1,\theta_2)-1] \quad \ldots(2.5.18)$$

Now differentiating (2.5.17) w.r.t. θ_2

$$g_2(\theta_1,\theta_2)=\frac{d}{d\theta_2}\int_a^{\theta_1} q_1(\theta_1)h_1(x)\left[\int_x^{\theta_2} q_2(\theta_2)h_2(u)du\right]dx$$

$$=\int_a^{\theta_1} q_1(\theta_1)h_1(x)\frac{d}{d\theta_2}\left[\int_x^{\theta_2} q_2(\theta_2)h_2(u)du\right]dx$$

$$=\int_a^{\theta_1} q_1(\theta_1)h_1(x)\left[\int_x^{\theta_2}-q_2^2(\theta_2)h_2(\theta_2)h_2(u)du+q_2(\theta_2)h_2(\theta_2)\right]dx$$

$$=-q_2(\theta_2)h_2(\theta_2)\left\{\int_a^{\theta_1} q_1(\theta_1)h_1(x)\left[\int_x^{\theta_2} q_2(\theta_2)h_2(\theta_2)du-1\right]dx\right\}$$

$$=-q_2(\theta_2)h_2(\theta_2)[g_1(\theta_1,\theta_2)-1] \quad \ldots(2.5.19)$$

Now differentiating (2.5.19) w.r.t. θ_1

$$g_{12}(\theta_1,\theta_2) = -q_2(\theta_2)h_2(\theta_2)g_1(\theta_1,\theta_2) \quad \ldots(2.5.20)$$

Using (2.5.18) in (2.5.20),

$$g_{12}(\theta_1,\theta_2) = q_1(\theta_1)q_2(\theta_2)h_1(\theta_1)h_2(\theta_2)[g_1(\theta_1,\theta_2)+F_2(\theta_1,\theta_2)-1] \quad \ldots(2.521)$$

Using (2.5.21) with $D_{11} = g_{12}$ in (2.5.11) and simplifying,

$$\phi(\theta_1,\theta_2) = \begin{cases} \dfrac{(m-1)}{mn}[(n-1)g(\theta_1,\theta_2)+F_1(\theta_1,\theta_2)], \theta_1 > \theta_2 \\ 1+\dfrac{(n-1)}{mn}[(m-1)[g(\theta_1,\theta_2)-1]-F_2(\theta_1,\theta_2)], \theta_1 < \theta_2 \end{cases} \quad \ldots(2.5.22)$$

Substituting $X_{(m)}$ for θ_1 and $Y_{(n)}$ for θ_2 in (2.5.22) the UMVU estimator $\hat{g}(\theta_1,\theta_2)$ of $P_{\theta_1,\theta_2}(Y > X)$ is available.

Note: These results have been derived by Rohatgi (1989).

UMVU Estimation for Two-Truncation Parameter Family

The following theorem for censored type-II sample is given by Patel and Bhatt (1990) and uncensored part is also given by Patel and Bhatt (1991).

Theorem 2.12: Let X_1, X_2, ..., X_n be a random sample from two-truncation parameter family with p.d.f. $f_X(x, \theta_1, \theta_2) = q(\theta_1, \theta_2)h(x)$, $a < \theta_1 < x < \theta_2 < b$ where, $-\infty < a < b < \infty$ are known constants, $h(x)$ is absolutely continuous function of x and $q(\theta_1,\theta_2)$ is everywhere differentiable. Let $X_{(1)}, X_{(2)}, \ldots, X_{(n)}$ be the corresponding sample of ordered statistics and let $X_{(1)}, X_{(2)}, \ldots, X_{(r)}, 1 \le r \le n$ be type-II censored sample. Then UMVU estimator of any U-estimable parametric function, $g(\theta_1,\theta_2)$ is

$$\hat{g}(\theta_1,\theta_2) = -\frac{(r-2)}{n!}q^{r-2}(X_{(1)},X_{(r)})D_1(D_2^{n-r+1}S(X_{(1)},X_{(r)})) \quad \ldots(2.5.23)$$

where $S(\theta_1, \theta_2) = g(\theta_1, \theta_2) q^{-n}(\theta_1, \theta_2)$

and $$D_1 W(\theta_1, \theta_2) = \frac{1}{h(\theta_1)} \frac{d}{d\theta_1} W(\theta_1, \theta_2)$$

$$D_2 W(\theta_1, \theta_2) = \frac{1}{h(\theta_2)} \frac{d}{d\theta_2} W(\theta_1, \theta_2).$$

For uncensored sample, $r = n$, $\hat{g}(\theta_1, \theta_2)$ will be

$$\hat{g}(\theta_1, \theta_2) = g(X_{(1)}, X_{(n)}) - \frac{\frac{d}{dX_{(1)}}(g(X_{(1)}, X_{(n)}))}{(n-1)q(X_{(1)}, X_{(n)})h(X_{(1)})}$$

$$+ \frac{\frac{d}{dX_{(n)}}(g(X_{(1)}, X_{(n)}))}{(n-1)q(X_{(1)}, X_{(n)})h(X_{(n)})} - \frac{\frac{d}{dX_{(1)}} \frac{d}{dX_{(n)}}(g(X_{(1)}, X_{(n)}))}{n(n-1)q^2(X_{(1)}, X_{(n)})h(X_{(1)}, X_{(n)})}$$

...(2.5.24)

If g is a function of θ_1 only, the UMVU estimator will be the first two terms of (2.5.24) only without $X_{(n)}$ in g and if g is a function of θ_2 only, the UMVU estimator will be the first and the third term of (2.5.24), without $X_{(1)}$ in g. Particularly UMVU estimator of $g(\theta_1, \theta_2) = \theta_1^s \theta_2^t$ will be obtained by substituting the following simplification in (2.5.23).

Here, $$S(\theta_1, \theta_2) = \theta_1^s \theta_2^t q^{-n}(\theta_1, \theta_2)$$

Differentiating the above expression w.r.t. θ_2, $(n-r+1)$ times,

$$\frac{d^{n-r+1}}{d\theta_2^{n-r+1}} S(\theta_1, \theta_2) = \theta_1^s$$

$$\left[\sum_{j=0}^{n-r+1} \binom{n-r+1}{j} \frac{t!}{(t-j)!} \theta_2^{t-j} \frac{d^{n-r+1-j}}{d\theta_2^{n-r+1-j}} q^{-n}(\theta_1, \theta_2) \right]$$

That is,

$$D_2^{n-r+1} S(\theta_1, \theta_2) = \frac{1}{h(\theta_2)} \theta_1^s$$

$$\left[\sum_{j=0}^{n-r+1}\binom{n-r+1}{j}\frac{t!}{(t-j)!}\theta_2^{t-j}\frac{d^{n-r+1-j}}{d\theta_2^{n-r+1-j}}q^{-n}(\theta_1,\theta_2)\right]$$

Now differentiating the above expression w.r.t. θ_1,

$$\frac{d}{d\theta_1}D_2^{n-r+1}S(\theta_1,\theta_2)=\frac{1}{h(\theta_2)}\theta_1^s\sum_{j=0}^{n-r+1}\binom{n-r+1}{j}\frac{t!}{(t-j)!}\theta_2^{t-j}$$

$$\left[\frac{d}{d\theta_1}\frac{d^{n-r+1-j}}{d\theta_2^{n-r+1-j}}q^{-n}(\theta_1,\theta_2)+s\cdot\theta_1^{s-1}\frac{d^{n-r+1-j}}{d\theta_2^{n-r+1-j}}q^{-n}(\theta_1,\theta_2)\right]$$

Therefore,

$$D_1[D_2^{n-r+1}S(\theta_1,\theta_2)]=\frac{1}{h(\theta_1)}\frac{1}{h(\theta_2)}\theta_1^s\sum_{j=0}^{n-r+1}\binom{n-r+1}{j}\frac{t!}{(t-j)!}\theta_2^{t-j}$$

$$\left[\frac{d}{d\theta_1}\frac{d^{n-r+1-j}}{d\theta_2^{n-r+1-j}}q^{-n}(\theta_1,\theta_2)+s\cdot\theta_1^{s-1}\frac{d^{n-r+1-j}}{d\theta_2^{n-r+1-j}}q^{-n}(\theta_1,\theta_2)\right]$$

...(2.5.25)

For uncensored data, i.e., for $r = n$ expression (2.5.25) reduces to when θ_1, θ_2 are replaced by $X_{(1)}$ and $X_{(n)}$, respectively

$$\hat{g}(\theta_1,\theta_2)=X_{(1)}^sX_{(n)}^t+\frac{tX_{(1)}^sX_{(n)}^{t-1}}{(n-1)h(X_{(n)})q(X_{(1)},X_{(n)})}$$

$$-\frac{sX_{(1)}^{s-1}X_{(n)}^t}{(n-1)h(X_{(1)})q(X_{(1)},X_{(n)})}-\frac{tsX_{(1)}^{s-1}X_{(n)}^{t-1}}{n(n-1)h(X_{(1)})h(X_{(n)})q^2(X_{(1)},X_{(n)})}$$

...(2.5.26)

For the proof of Theorem 2.12 refer Patel and Bhatt (1990)

For $t = 0$, the UMVU estimator of θ_1^s will be obtained from (2.5.25) as

$$\hat{\theta}_1^s = X_{(1)}^s - \frac{sX_{(1)}^{s-1}}{(r-1)h(X_{(1)})q(X_{(1)}, X_{(r)})} \qquad \ldots(2.5.27)$$

For uncensored data, i.e., for $r = n$ and for $s = 1$ the above expression reduces to the results of Zacks (1971). For $s = 0$ the UMVU estimator of θ_2^t will be obtained from (2.5.25) as

$$\hat{\theta}_2^t = X_{(r)}^t + \frac{tX_{(r)}^{t-1}}{(r-1)h(X_{(r)})q(X_{(1)}, X_{(r)})} \qquad \ldots(2.5.28)$$

For $r = n$, we have

$$\hat{\theta}_2^t = X_{(n)}^t + \frac{tX_{(n)}^{t-1}}{(n-1)h(X_{(n)})q(X_{(1)}, X_{(n)})} \qquad \ldots(2.5.29)$$

These results are given by Patel and Bhatt(1991). For $t = 1$ the results (2.5.29) reduces to that of Zacks (1971).

UMVU Estimation of Reliability

Let $X_1, X_2, \ldots, X_n$ be a random sample from the density,

$$f_X(x, \theta_1, \theta_2) = q(\theta_1, \theta_2)h(x), \;\; a < \theta_1 < x < \theta_2 < b.$$

Then for the type-II censored data $X_{(1)} < X_{(2)} < \ldots < X_{(r)}$ we wish to obtain the UMVU estimator of $g(\theta_1, \theta_2) = P_{\theta_1, \theta_2}(X_1 > t)$, reliability. We have

$$P_{\theta_1,\theta_2}(X_1 > t) = \begin{cases} 1 & \text{if } t < \theta_1 \\ 1 - \dfrac{q(\theta_1, \theta_2)}{q(\theta_1, t)} & \text{if } \theta_1 < T < \theta_2 \\ 0 & \text{if } \theta_2 \le t \end{cases} \qquad \ldots(2.5.30)$$

we have

$$S(\theta_1, \theta_2) = q^{-n}(\theta_1, \theta_2) - q^{-(n-1)}(\theta_1, \theta_2)q^{-1}(\theta_1, t), \quad \theta_1 < \theta_2$$

Differentiating the above expression w.r.t. $\theta_2, (n-r+1)$ times we get

$$D_2^{n-r+1}S(\theta_1,\theta_2)=\frac{n!}{(r-1)!}q^{-(r-1)}(\theta_1,\theta_2)-\frac{(n-1)!}{(r-2)!}q^{-(r-2)}(\theta_1,\theta_2)q^{-1}(\theta_1,t)$$

Again differentiating the above expression w.r.t. θ_1, we get

$$D_1[D_2^{n-r+1}S(\theta_1,\theta_2)]=\frac{-n!}{(r-1)!}q^{-(r-2)}(\theta_1,\theta_2)+\frac{(n-1)!(r-2)q^{-(r-3)}(\theta_1,\theta_2)}{(r-2)!q(\theta_1,t)}$$

$$+\frac{(n-1)!}{(r-2)!}q^{-(r-2)}(\theta_1,\theta_2) \quad \dots(2.5.31)$$

Using (2.5.31) in (2.5.22) and replacing θ_1 by $X_{(1)}$ and θ_2 by $X_{(r)}$ the UMVU estimator of $P_{\theta_1,\theta_2}(X_1>t)$

$$\hat{g}(\theta_1,\theta_2)=\begin{cases}1 & \text{if } a<t<X_{(1)}\\ \left(1-\frac{1}{n}\right)-\left(\frac{r-2}{n}\right)\frac{q(X_{(1)},X_{(r)})}{q(X_{(1)},t)} & \text{if } X_{(1)}\le t<X_{(r)}\\ 0 & \text{if } X_{(r)}<t<b\end{cases}$$

$$\dots(2.5.32)$$

Following examples are based on application of result (2.5.32)

Example 2.13 **Reliability for $u(\theta_1, \theta_2)$ distribution**

Consider the uniform distribution on the interval (θ_1, θ_2). Here $a=-\infty, b=\infty, h(x)=1, q(\theta_1,\theta_2)=(\theta_2-\theta_1)^{-1}$.

Using (2.5.32) the UMVU estimator of $P_{\theta_1,\theta_2}(X_1>t)$, say $\hat{g}(\theta_1,\theta_2)$ is given by

$$\hat{g}(\theta_1,\theta_2)=\begin{cases}1 & \text{if } t<X_{(1)}\\ \left(1-\frac{1}{n}\right)-\left(\frac{r-1}{n}\right)\frac{(t-X_{(1)})}{(X_{(r)}-X_{(1)})} & \text{if } X_{(1)}\le t<X_{(r)}\\ 0 & \text{if } X_{(r)}<t\end{cases}$$

Example 2.14

Consider the two-parameter truncated exponential distribution with density

$$f_X(x,\theta_1,\theta_2)=\alpha\exp(-\alpha X)[\exp(-\alpha\theta_1)-\exp(-\alpha\theta_2)]^{-1},\alpha>0$$

where $a = -0,\ b=\infty,\ q(\theta_1,\theta_2)=[\exp(-\alpha\theta_1)-\exp(-\alpha\theta_2)]^{-1}$

$$h(x)=\alpha\exp(-\alpha x).$$

Using (2.5.32) the UMVU estimator of $P_{\theta_1,\theta_2}(X_1>t)$ say $\hat{g}(\theta_1,\theta_2)$ is given by

$$\hat{g}(\theta_1,\theta_2)=\begin{cases}1 & \text{if } t<X_{(1)}\\ \left(1-\frac{1}{n}\right)-\left(\frac{r-2}{n}\right)\dfrac{\exp(-\alpha X_{(1)})-\exp(-\alpha t)}{\exp(-\alpha X_{(1)})-\exp(-\alpha X_{(r)})} & \text{if } X_{(1)}\le t<X_{(r)}\\ 0 & \text{if } X_{(r)}<t\end{cases}$$

Example 2.15

Consider the two-parameter truncated beta distribution with density

$$f_X(x,\theta_1,\theta_2)=\frac{X^{p-1}(1-X)^{q-1}}{[\beta_{\theta_2}(p,q)-\beta_{\theta_1}(p,q)]},\ p>0,\ q>0$$

where $\beta_Z(p,q)$ is the incomplete beta function.

Here a = 0, b = 1,

$$q(\theta_1,\theta_2)=[\beta_{\theta_2}(p,q)-\beta_{\theta_1}(p,q)]^{-1},$$

$$h(x)=X^{p-1}(1-X)^{q-1}.$$

Using (2.5.32) the UMVU estimator of $g(\theta_1,\theta_2)=P_{\theta_1,\theta_2}(X_1>t)$ is given by

$$\hat{g}(\theta_1,\theta_2)=\begin{cases}1 & \text{if } t<X_{(1)}\\ \left(1-\frac{1}{n}\right)-\left(\frac{r-2}{n}\right)\dfrac{\beta_t(p,q)-\beta X_{(1)}(p,q)}{\beta X_{(r)}(p,q)-\beta X_{(1)}(p,q)} & \text{if } X_{(1)}\le t<X_{(r)}\\ 0 & \text{if } X_{(r)}<t\end{cases}$$

Refer Patel and Bhatt (1990) for further details.

UMVU Estimation in Doubly Type-II Censored Sample

Let $X_1, X_2, \ldots, X_n$ be a random sample with common probability density function

$$f_X(x;\theta_1,\theta_2) = q(\theta_1,\theta_2)h(x), \quad a < \theta_1 < x < \theta_2 < b$$

with usual assumptions on q, h.

Let $X_{(1)} < X_{(2)} < \ldots < X_{(n)}$ be the corresponding order statistics of the random sample and suppose that the sample is doubly type-II censored so that one observes only $X_{(r)}, X_{(r+1)}, \ldots, X_{(s)}$, $1 \le r < s \le n$. Without proof we state the following theorem.

Theorem 2.13: The UMVU estimator of any U-estimable parametric function $g(\theta_1,\theta_2)$ based on a doubly type-II censored sample from the p.d.f. given in (2.5.1) is of the form

$$\phi(\theta_1,\theta_2) = \frac{(s-r-1)!}{(-1)^r n!} q^{(s-r-1)}(\theta_1,\theta_2) D_1^r [D_2^r S(\theta_1,\theta_2)] \quad \ldots(2.5.33)$$

where $S(\theta_1,\theta_2) = g(\theta_1,\theta_2) q^{-n}(\theta_1,\theta_2)$ and D_1 and D_2 are operators defined in Theorem 2.12 and θ_1, θ_2 are replaced by $X_{(1)}$ and $X_{(n)}$ respectively.

For different values of r, s in (2.5.33) leads to UMVU estimator of $g(\theta_1,\theta_2)$ for type-II censored data ($r=1$), that of complete sample ($r=1, s=n$). Refer Patel and Bhatt (1991) for further details.

PROBLEMS

2.1 Given a random sample $X_1, X_2, \ldots, X_n$ from Bernoulli distribution with probability of success p, obtain UMVU estimators of

(i) p^2 (ii) p^3

(iii) pq (iv) p^r

(v) $p^r + (1-p)^{n-r}$ (vi) $P(X=r)$, $0 < r < n$

2.2 If $X_1, X_2, \ldots$ are independently random variables with $P(X_i = 1) = p$, $P(X_i = 0) = 1 - p = q$ and if N is the number of trials to get first success, find function $g(N)$ of N which is unbiased for p.

2.3 If $X_1, X_2, \ldots, X_n$ is a random sample from $N(\mu, \mu)$ distribution, $\mu > 0$ then obtain unbiased estimator of μ^2.

2.4 Let $X_1, X_2, \ldots, X_n$ be a random sample from the p.d.f.

$$f_X(x;\theta) = (1-\theta) + \frac{\theta}{2\sqrt{x}}, \; 0 < x \le 1, \; \theta \in [0,1]$$

Show that $\bar{X}$ is a biased estimator of θ and find its bias.

2.5 If $X_1, X_2, \ldots, X_n$ is a random sample from p.d.f.

$$f_X(x;\theta) = \frac{1}{2} e^{-|x-\theta|}, \; -\infty < x, \theta < \infty$$

then obtain unbiased estimator based on order statistic.

2.6 If $X_1, X_2, \ldots, X_n$ is a random sample from triangular distribution with p.d.f. as

$$f_X(x) = \begin{cases} \dfrac{4(x-a)}{(b-a)^2}, & a < x < \dfrac{a+b}{2} \\ \dfrac{4(b-x)}{(b-a)^2}, & \dfrac{a+b}{2} \le x < b, \quad a < b. \\ 0 & , \text{otherwise} \end{cases}$$

then show that $\bar{X}$ is unbiased estimator of $E(X_1)$. Obtain $\text{var}(\bar{X})$.

2.7 If $X_1, X_2, \ldots, X_n$ is a random sample from p.d.f.

$$f_X(x;\theta) = \theta e^{-\theta x}, \; x > 0, \; \theta > 0$$

then obtain UMVU estimators of θ, θ^2.

2.8 Let $X_1, X_2, \ldots, X_n$ be a random sample from probability mass function

$$P(X=x)=\frac{e^{-\theta}\theta^{x}}{x!(1-e^{-\theta})},\ x=1,2,\ldots,\theta>0$$

Obtain UMVU estimators of $\frac{\theta}{1-e^{-\theta}},\frac{1}{1-e^{-\theta}}$.

2.9 If $X_1, X_2, \ldots, X_n$ is a random sample from p.d.f.

$$f_X(x;\theta)=\frac{\theta}{(1+x)(1+\theta)},\ x>0,\theta>0$$

then obtain UMVU estimators of θ and $\frac{1}{\theta}$.

2.10 Let $X_1, X_2, \ldots, X_n$ be a random sample from p.d.f.

$$f_X(x;\theta)=\theta e^{-\theta x},\ x>0,\theta>0.$$

Then obtain UMVU estimator of $P(X>k)=e^{-\theta k}$.

2.11 If $X_1, X_2, \ldots, X_n$ is a random sample from p.d.f.

$$f_X(x;\theta)=\frac{x+1}{\theta(\theta+1)}e^{-\frac{x}{\theta}},\ x>0,\theta>0$$

then obtain UMVU estimator of $\frac{\theta(2\theta+1)}{(1+\theta)}$.

2.12 Based on a random sample of size n from $N(\theta,\theta^2)$ distribution check whether UMVU estimator of θ exists?

2.13 Let $X_1, X_2, \ldots, X_n$ be a random sample from truncated binomial distribution with p.d.f.

$$P(X=x)=\frac{\binom{n}{x}p^{x}q^{n-x}}{(1-q^{n})},\ x=1,2,..,n\ ,\ q=1-p$$

Obtain UMVU estimator of $\frac{p}{(1-q^{n})}$.

2.14 If $X_1, X_2, \ldots, X_n$ is a random sample from $N(\theta,\theta)$ find UMVU estimator of θ (if it exists) for θ.

2.15 If $X_1, X_2, \ldots, X_n$ is a random sample from p.d.f.

$$f_X(x;\theta) = \frac{1}{2\theta} e^{-\frac{|x|}{\theta}}, -\infty < x < \infty, \theta > 0$$

then obtain UMVU estimator of θ, if it exists.

2.16 Let $X_1, X_2, \ldots, X_n$ be i.i.d. random variable with common p.d.f.

$$f_X(x;\theta) = \frac{1}{\theta} e^{-\frac{x}{\theta}}, x > 0, \theta > 0$$

Obtain complete sufficient statistic T for θ and UMVU estimator $\phi(T)$ of θ and by using Rao-Blackwell theorem find $\text{var}(\phi(T))$.

2.17 If X has binomial distribution with parameter p, find complete sufficient statistic T, UMVU estimator $\phi(T)$ of p and $\text{var}(\phi(T))$, using Rao-Blackwell theorem.

2.18 If the joint p.d.f. of X and Y is

$$f_{X,Y}(x, y;\theta) = \frac{2}{\theta^2} e^{-\frac{x+y}{\theta}}, \quad 0 < x < y < \infty$$

then find $\phi(x) = E(Y/X = x)$, using Rao-Blackwell theorem and obtain $\text{var}(\phi(x))$.

2.19 If $\bar{X}$ and S^2 are respectively mean and variance of a random sample of size n from Poisson distribution with parameter θ, find $E(S^2/\bar{X})$.

2.20 Obtain UMVU estimator of $P(X = 0)$ for the negative binomial distribution on the basis of random sample of size n.

2.21 Determine the variance of UMVU estimator of $P(X = r)$ of Poisson distribution with parameter θ.

2.22 Determine the variance of UMVU estimator of $P(X > r)$ on the basis of a random sample from p.d.f.

$$f_X(x;\theta) = \frac{1}{\theta} e^{-\frac{x}{\theta}}, x > 0, \theta > 0.$$

2.23 Let $X_i \sim N(\alpha + \beta t_i, \sigma^2)$, $i = 1, 2, \ldots, n$ be independently distributed where, α, β, σ^2 are unknown and $t_i = i = 1, 2, \ldots, n$ are known constants. Then obtain UMVU estimators of α and β.

2.24 If $Y_1, Y_2, \ldots, Y_n$ are such that

$$E(Y_i) = \beta_0 + \beta_1 X_i, \; i = 1, 2, \ldots, n$$

where, X_i are known and $\text{var}(X_i) = \sigma^2$ for $i = 1, 2, \ldots, n$ then obtain the least square estimators $\hat{\beta}_0, \hat{\beta}_1, \hat{\sigma}^2$ of $\beta_0, \beta_1, \sigma^2$. Find (β_0), $\text{var}(\beta_1)$, $\text{cov}(\beta_0, \beta_1)$.

2.25 If $(Y_i) = \beta_0 + \beta_1 X_i + \beta_2 X_i^2$, $i = 1, 2, \ldots, n$,$\text{var}(Y_i) = \sigma^2, i = 1, 2, \ldots, n$ $\text{cov}(Y_i, Y_j) = \rho\sigma^2, i \neq j$, then obtain the least square estimator of $2\beta_0 - 3\beta_1 + 5\beta_1$ and the variance of the estimator.

2.26 Aerial observations Y_1, Y_2, Y_3, Y_4 are made of the angles $\theta_1, \theta_2, \theta_3, \theta_4$ of a quadrilateral on the ground. If the observations are subject to independent estimators with zero means and common variance σ^2, determined least square estimators of $\theta_1, \theta_2, \theta_3, \theta_4, \sigma^2$. If quadrilateral is parellelogram with $\theta_1 = \theta_3$ and $\theta_2 = \theta_4$. Then what are least square estimators of angels and how would you estimate σ^2.

2.27 Instead of other error functions such as $\max_i |e_i|$ or $\sum_{i=1}^{n} |e_i|$ why function $\underline{e}'\underline{e}$ is taken for minimization?

2.28 Let $Y_1 = \alpha_1 \underline{1} + X\beta + \epsilon_1$, and $Y_2 = \alpha_2 \underline{1} + X\beta + \epsilon_2$ be two linear models. Then assuming ϵ_1, ϵ_2 to be independent with same distribution properties, find the least square estimators of α_1, α_2 and β.

2.29 Let $\hat{\underline{\beta}}_1$ and $\hat{\underline{\beta}}_2$ be least square estimators of β from $Y_1 = X\beta + \epsilon_1$ and $Y_2 = X\beta + \epsilon_2$, respectively. If β is estimated by

$$\hat{\beta} = w\hat{\underline{\beta}}_1 + (1 - w)\hat{\underline{\beta}}_2 \quad \text{with} \quad 0 < w < 1,$$

Determine the value of w that minimizes the trace of matrix of $\text{Var}(\hat{\beta})$. Does this value change if we minimize $E(\hat{\underline{\beta}} - \underline{\beta})'X'X(\hat{\underline{\beta}} - \underline{\beta})$?

2.30 If the model $Y = \alpha_1 \underline{1} + X\beta + \epsilon_1$ with $\underline{1}$ denoting column vector having each element 1, show that generalized least square estimator of α is given by

$$\frac{1}{\underline{1}'\Sigma\underline{1}} \underline{1}'[\Sigma - \Sigma X(X'\Sigma_*^{-1} X)^{-1} X'\Sigma_*^{-1}]Y$$

where, $E(\underline{\epsilon}\,\underline{\epsilon}') = \Sigma^{-1}$ and $\Sigma_*^{-1} = \Sigma - \dfrac{1}{\underline{1}'\Sigma\underline{1}} \Sigma\underline{1}\,\underline{1}'\Sigma$.

2.31 In the linear model $Y = X\underline{\beta} + \in$ with $E(\underline{\in}) = 0$, $E(\underline{\in}\,\underline{\in}') = \sigma^2 w$, show that $\alpha = \underline{\beta}'X'(\sigma^2 w + X\underline{\beta\beta}'X')^{-1}Y$ is an unbiased estimator of $\left(1 + \dfrac{\sigma^2}{\underline{\beta}'X'w^{-1}X\underline{\beta}}\right)^{-1}$. Find its variance.

2.32 In the linear model $= X\underline{\beta} + \in$, $E(\underline{\in}) = 0$, $E(\underline{\in}\,\underline{\in}') = \sigma^2 w$, show that σ^2 is estimated by

$$\hat{\sigma}^2 = \frac{1}{n-p}\underline{Y}'(I - X(X'X)^{-1}X)\underline{Y}.$$

Show that $\hat{\sigma}^2$ is not unbiased estimator of σ^2.

3

CHAPTER

Efficiency of Estimator

3.1 INTRODUCTION

In chapter 2 we have obtained UMVU estimator of a parametric function when such an estimator exists. The condition of existence for UMVU estimator is stated and derived. In the case of non-existence of UMVU estimator, one would like to know that how far the variance of the unbiased estimator falls down, that is, the lower bound of the variance of any unbiased estimator of the given parametric function is obtained through Rao-Cramer inequality and it is discussed in section 3.2 of this chapter. To avoid the *R-C* regularity conditions Chapman-Robbins inequality is established in section 3.3. Further to sharpen the *R-C* inequality Bhattacharya bounds have been derived in section 3.4. *R-C* inequality is extended for several parameters in section 3.5.

3.2 RAO-CRAMER (*R-C*) INEQUALITY

When UMVU estimator does not exit then one may find an estimator whose variance is bounded (lower).

Theorem 3.1: Let $L(\theta / x)$ be function of θ for fixed x. Then $L(\theta / x)$ is called likelihood of x. $p_\theta(x)$ being function of x given θ, $L(\theta / x) = p_\theta(x)$, is probability function for a random sample $X = (X_1, X_2, \ldots, X_n)$. Let $T(x)$ be unbiased estimator of a real valued function $g(\theta)$.

For $\theta \in (a, b) \subset R$ let us assume that

I. $\dfrac{\partial^i L}{\partial \theta^i} < \infty$ for i = 1, 2

II. $\dfrac{\partial^i}{\partial \theta^i} \int L dx = \int \dfrac{\partial^i L}{\partial \theta^i}$, $i = 1, 2$

III. $0 < E_\theta \left(\frac{\partial \log L}{\partial \theta}\right)^2 = I(\theta) < \infty$

IV. $\frac{\partial}{\partial \theta}\int TLdx = \int T\frac{\partial L}{\partial \theta}dx$.

Then $\text{Var}_\theta(T) \geq [g'(\theta)]^2 / I(\theta)$ and equality sign holds iff

$$T - g(\theta) = \frac{g'(\theta)}{I(\theta)}\frac{\partial \log L}{\partial \theta} \text{ or } f_T(t) = e^{a(\theta)t + b(\theta) + c(t)}$$

Proof: Differentiating $\int Ldx = 1$ with respect to θ we have $E_\theta \frac{\partial \log L}{\partial \theta} = 0$ and differentiating $\int TLdx = g(\theta)$ with respect to θ, we have $\text{cov}\left(T, \frac{\partial \log L}{\partial \theta}\right) = g'(\theta)$.

Since

$$\rho^2\left(T, \frac{\partial \log L}{\partial \theta}\right) \leq 1, \text{var}_\theta T \,\text{var}_\theta \frac{\partial \log L}{\partial \theta}$$

$$\geq \left[\text{cov}\left(T, \frac{\partial \log L}{\partial \theta}\right)\right]^2 \Rightarrow \text{var}_\theta(T) \geq \frac{[g'(\theta)]^2}{I(\theta)},$$

the equality sign holds if and only if T is a linear function of $\frac{\partial \log L}{\partial \theta}$

i.e., iff $T = A(\theta)\frac{\partial \log L}{\partial \theta} + B(\theta)$ taking expectation, $B(\theta) = g(\theta)$ and using $B(\theta) = g(\theta)$the condition reduces to $T - g(\theta) = A(\theta)\frac{\partial \log L}{\partial \theta}$, which when multiplied by $\frac{\partial \log L}{\partial \theta}$ and taking expectation, we shall have $A(\theta) = \frac{g'(\theta)}{I(\theta)}$, so that the necessary and sufficient condition for equality sign to hold is $T - g(\theta) = \frac{g'(\theta)}{I(\theta)}\frac{\partial \log L}{\partial \theta}$.

Remarks

(i) The statistic T which satisfies equality condition in R-C inequality is called minimum variance bound (MVB) estimator.

(ii) $[g'(\theta)]^2 / I(\theta)$ is called the lower bound of the variance of unbiased estimator of $g(\theta)$ and $I(\theta)$ is called the fisher information.

(iii) An MVB estimator is always UMVU estimator but the converse may not be true.

(iv) An MVB estimator is sufficient but the converse may not be true.

(v) An MVB estimator is unique if it exists.

(vi) Any linear function of an MVB estimator is also an MVB estimator.

(vii) Only linear function of an MVB estimator is an MVB estimable.

(viii) $E(T-\theta)^2 \geq \dfrac{[1+b'(\theta)]^2}{I(\theta)} + [b(\theta)]^2$, where, $b(\theta)$ is bias, $ET = \theta + b(\theta)$.

Proof of (vii): Let T be MVB estimator of $g(\theta)$ and U be MVB estimator of $h(\theta)$.

Then
$$\begin{cases} U - h(\theta) = \dfrac{h'(\theta)}{I(\theta)} \dfrac{\partial \log L}{\partial \theta} \\ T - g(\theta) = \dfrac{g'(\theta)}{I(\theta)} \dfrac{\partial \log L}{\partial \theta} \end{cases}$$

$$\Rightarrow \quad g'(\theta)[U - h(\theta)] = h'(\theta)[T - g(\theta)]$$

$$\Rightarrow \quad g''(\theta)[U - h(\theta)] - g'(\theta)h'(\theta) = h''(\theta)[T - g(\theta)] - h'(\theta)g'(\theta)$$

$$\Rightarrow \quad \frac{g''(\theta)}{g'(\theta)} = \frac{h''(\theta)}{h'(\theta)} \Rightarrow \frac{\partial \log h'(\theta)}{\partial \theta} = \frac{\partial \log g'(\theta)}{\partial \theta}$$

$$\Rightarrow \quad \log h'(\theta) = \log g'(\theta) + \log k$$

$$\Rightarrow \quad h'(\theta) = kg'(\theta) \Rightarrow U - h(\theta) = k[T - g(\theta)]$$

$$\Rightarrow \quad u = kT + l, \text{ where } l = h(\theta) - kg(\theta).$$

Proof of (viii): If $g(\theta) = \theta + b(\theta)$ then R-C inequality will give

$\text{var}_\theta(T) \geq [1 + b'(\theta)]^2 / I(\theta)$ and

$$E(T-\theta)^2 = E[T-g(\theta)+g(\theta)-\theta]^2$$

$$= \text{var}(T)+(b(\theta))^2 \geq \frac{[1+b'(\theta)]^2}{I(\theta)} + [b(\theta)]^2$$

If $g(\theta)=\theta$, $\text{var}(T) \geq \frac{1}{I(\theta)}$, $I(\theta)$ is called information and it is given by

$$I(\theta)=E_\theta\left(\frac{\partial \log L}{\partial\theta}\right)^2 = -E_\theta\left(\frac{\partial^2 \log L}{\partial\theta^2}\right) = nE\left(\frac{\partial \log f}{\partial\theta}\right)^2 = -nE\left(\frac{\partial^2 \log f}{\partial\theta^2}\right)$$

Differentiating $\left[\int \frac{\partial \log L}{\partial\theta} L\,dx = 0\right]$, we get second form of $I(\theta)$, and taking $L=\Pi f(x_i,\theta)$ we get third and fourth forms of $I(\theta)$.

Theorem 3.2: If T is a statistic with p.d.f. $g(t)$ satisfying R-C condition (i) to (iv) given in Theorem (3.1) and $p_\theta(x)=L(\theta/x)$ also satisfies the R-C regularity conditions then,

$$I^T(\theta) \leq I^X(\theta),$$

and equality sign holds iff T is sufficient statistic for θ.

Proof: Let A be measurable set in the T space and A' be the corresponding set in the X-space. Then

$$\int_{A'} \frac{\partial \log L}{\partial\theta} L\,dx = \frac{\partial}{\partial\theta}\int_{A'} L\,dx$$

$$= \frac{\partial}{\partial\theta}\int_{A} g(t,\theta)\,dt$$

$$= \int_{A} \frac{\partial}{\partial\theta} g(t,\theta)\,dt$$

$$= \int_{A} \frac{\partial \log g(t,\theta)}{\partial\theta} g(t,\theta)\,dt \quad \ldots(3.2.1)$$

Also by condition of expectation,

$$\int_{A'} \frac{\partial \log L}{\partial \theta} L dx = \int_{A} E\left(\frac{\partial \log L}{\partial \theta} / T = t\right) g(t;\theta) dt \qquad \ldots(3.2.2)$$

From (3.2.1) and (3.2.2), we have

$$\frac{\partial \log g(t;\theta)}{\partial \theta} = E\left(\frac{\partial \log L}{\partial \theta} / T = t\right) \qquad \ldots(3.2.3)$$

By definition,

$$I^T(\theta) = E_\theta\left[\frac{\partial}{\partial \theta} \log g(T;\theta)\right]^2$$

so,

$$0 \le E\left[\frac{\partial}{\partial \theta} \log L - \frac{\partial}{\partial \theta} \log g(T,\theta)\right]^2$$

$$= E_\theta\left[\frac{\partial}{\partial \theta} \log L\right]^2 + E\left[\frac{\partial}{\partial \theta} \log g(T,\theta)\right]^2$$

$$-2E_\theta\left[\frac{\partial}{\partial \theta} \log L \frac{\partial}{\partial \theta} \log g(T,\theta)\right]$$

$$= E\left[\frac{\partial}{\partial \theta} \log L\right]^2 - E\left[\frac{\partial}{\partial \theta} \log g(T,\theta)\right]^2,$$

Using (3.2.2) and (3.2.3)

$$\Rightarrow \qquad I^T(\theta) \le I^X(\theta).$$

If T is sufficient,

$$L(\theta / x) = h(x) g^*(T;\theta) \Rightarrow \frac{\partial \log L}{\partial \theta} = \frac{\partial g^*(T,\theta)}{\partial \theta}$$

$$\Rightarrow \qquad I^X(\theta) = E\left[\frac{\partial}{\partial \theta} g^*(T,\theta)\right]^2 \ge I^T(\theta) \qquad \ldots(3.2.4)$$

If there exists a function $Z(T)$ whose d.f. is $J(t)g^*(t;\theta)$ then by the above arguments we shall have

$$E_\theta\left[\frac{\partial \log g^*(T;\theta)}{\partial\theta}\right]^2 \le I^T(\theta) \qquad \ldots(3.2.5)$$

From (3.2.4) and (3.2.5), we have $I^X(\theta)=I^T(\theta)$

This shows that the only statistic which reduces the data in the sample without losing information is the sufficient statistic.

Example 3.1

If $X_1, X_2, \ldots, X_n$ is a random sample from $N(\theta, \sigma^2)$, θ is known then show that $\frac{\Sigma(x_i-\theta)^2}{n}$ is MVB estimator of σ^2.

Here,

$$L(\theta/x)=2\pi^{-\frac{n}{2}}(\sigma^2)^{\frac{-n}{2}}e^{-\frac{\Sigma(x_i-\theta)^2}{2\sigma^2}} \quad \text{and} \quad \frac{\partial \log L}{\partial\sigma^2}=\frac{n}{\sigma^3}\left[\frac{\Sigma(x_i-\theta)^2}{n}-\sigma^2\right]$$

$\Rightarrow \quad \frac{\Sigma(x_i-\theta)^2}{n}$ is MVB estimator of σ^2,

$$\text{var}(T)=\frac{[g'(\theta)]^2}{I(\theta)}=\frac{g'(\theta)}{\frac{I(\theta)}{g'(\theta)}}=\frac{g'(\theta)}{\text{coeff of }[T-g(\theta)]}=\frac{2\sigma}{\frac{n}{\sigma^3}}=\frac{2\sigma^4}{n}.$$

Example 3.2

Let $X_1, X_2, \ldots, X_n$ be a random sample from Poisson distribution, $P(\theta)$ with parameter θ. Then as $T=\bar{X}$ is complete sufficient and as density of $\bar{X}$ belong to exponential family, $\bar{X}$ is MVB estimator of θ and its variance is $\frac{\theta}{n}$, so that R-C lower bound is $\frac{n}{\theta}$. The UMVU estimator of $g(\theta)=e^{-\theta}$ being $\left(1-\frac{1}{n}\right)^T$, whose distribution does not belong to exponential family, its variance does not attain R-C lower bound. As the p.g.f. of $P(\lambda)$ is $e^{-\lambda(1-t)}$ and

$$\operatorname{var}_\theta\left(1-\frac{1}{n}\right)^T = E\left[\left(1-\frac{1}{n}\right)^2\right]^T - e^{-2\theta}$$

$$= e^{-n\theta\left[1-\left(1-\frac{1}{n}\right)^2\right]} - e^{-2\theta}$$

$$= e^{-2\theta+\frac{\theta}{n}} - e^{-2\theta}$$

$$= e^{-2\theta}(e^{\frac{\theta}{n}} - 1)$$

the R-C lower bound is $[g'(\theta)^2 / I(\theta)] = \frac{\theta}{n}e^{-2\theta}$. Thus, both are not same. Generally, efficiency of the estimator T w.r.t T' is defined as $E = \operatorname{var}(T')/\operatorname{var}(T) \cdot T'$ is taken as MVB estimator of $g(\theta)$. Thus the efficiency of UMVU estimator of $e^{-\theta}$ is

$$E = \frac{R\text{-}C \text{ lower bound}}{\operatorname{var}(T)} = \frac{\theta}{n}(e^{\frac{\theta}{n}} - 1)^{-1}.$$

Example 3.3

Let $X_1, X_2, \ldots, X_n$ be a random sample from the distribution with p.d.f.

$$f_X(x) = \theta^2 x e^{-\theta x},\ x > 0,$$

Then as $T = \sum_{i=1}^{n} X_i$ is complete sufficient for θ and

$E\left(\frac{1}{T}\right) = \frac{\theta}{2n-1}$, $\frac{2n-1}{T}$ is UMVUE of θ. Let us obtain $\operatorname{var}\left(\frac{2n-1}{T}\right)$. As

$T = \sum_{i=1}^{n} X_i \sim G(\theta, 2n)$,

$$E\left(\frac{1}{T^r}\right) = \frac{\theta^{2n}}{\Gamma(2n)}\int_0^\infty e^{-\theta t} t^{2n-r-1} dt = \frac{\theta^{2n}}{\Gamma(2n)}\frac{\Gamma(2n-r)}{\theta^{2n-r}} = \frac{\theta^r \Gamma(2n-r)}{\Gamma(2n)},$$

so

$$E\left(\frac{1}{T}\right)=\frac{\theta}{2n-1},\ E\left(\frac{1}{T^2}\right)=\frac{\theta^2}{(2n-1)(2n-2)},$$

$$\text{Var}\left(\frac{1}{T}\right)=\frac{\theta^2}{(2n-1)^2(2n-2)},\ \Rightarrow \text{Var}\left(\frac{2n-1}{T}\right)=\frac{\theta^2}{2(n-1)},$$

Let us obtain *R-C* lower bound for θ. As

$$\log f(x_i,\theta)=2\log\theta+\log x_i-\theta x_i,\quad \frac{\partial \log f}{\partial\theta}=\frac{2}{\theta}-x_1\ ,$$

$$E\left(\frac{\partial \log f}{\partial\theta}\right)^2=E\left(X_1-\frac{2}{\theta}\right)^2=\text{Var}(X_i)=\frac{2}{\theta^2}\Rightarrow E\left(\frac{\partial \log L}{\partial\theta}\right)^2=\frac{2n}{\theta^2}$$

so *R-C* lower bound is $\frac{\theta^2}{2n}$.

Thus, variance of the UMVU estimator of θ does not attain *R-C* lower bound. The efficiency of the UMVU estimator of θ is $\frac{n-1}{n}\to 1$ as $n\to\infty$ but as $n=2$, efficiency is 0.5.

The *R-C* inequality is derived for regular family. If the family does not satisfy *R-C* regular conditions then the inequality is obtained by Chapman-Robbins. By Cauchy-Schwartz inequality of an estimator T of $g(\theta)$ and a function $h(x,\theta)$ is

$$\text{Var}(T)\ge[\text{cov}(T,h)]^2/\text{var}(h) \qquad \ldots(3.2.6)$$

This does not provide lower bound of T, as (3.2.6) depends on T. When $\text{cov}(T, h)$ depends on T only through $E(T)=g(\theta)$ the inequality has some meaning.

Theorem 3.3: A necessary and sufficient condition for $\text{cov}(T,h)$ to depend on T only through $g(\theta)$ is that for all, $\text{cov}(U,h)=0, \forall\ U\in u$, where, u is class of unbiased estimator of zero.

Proof: If T_1, T_2 are two estimators with finite variance and common expectation then $T_2 - T_1 \in u$ and by sufficient condition cov($T_2 - T_1$, h) = 0 $\Rightarrow$ cov(T_1, h) = cov(T_2, h) which shows that cov(T, h) depend on only T through $g(\theta)$.

Necessary Condition

As cov(T, h) depends on only T through $g(\theta)$, for any $U \in u$.

$$\text{cov}(T+U, h) = \text{cov}(T, h) \Rightarrow \text{cov}(U, h) = 0.$$

R-C inequality is established under *R-C* regular conditions. We shall establish it without regularity conditions, known as Chapman-Robbins inequality.

3.3 CHAPMAN-RABBINS-KIEFFER INEQUALITY

Theorem 3.4: Let $L(\theta/x)$ be likelihood of X, $\theta \in \Omega$ and let $\phi \in \Omega$, $\phi \neq \theta$, $L(\theta/x) \neq L(\phi/x)$.

If T is an unbiased estimator of $g(\theta)$ then

$$\text{Var}_\theta(T) \geq \sup_{\phi} \frac{[g(\phi) - g(\theta)]^2}{\text{Var}_\theta[L(\phi/x)/L(\theta/x)]}.$$

Proof: As T is unbiased for

$$g(\theta), \int TL(\theta/x)dx = g(\theta), \forall\, \theta \text{ and } \int TL(\phi/x)dx = g(\phi), \ \forall\ \phi.$$

$$\Rightarrow \qquad \int T\left[\frac{L(\phi/x) - L(\theta/x)}{L(\theta/x)}\right] L(\theta/x)dx = g(\phi) - g(\theta)$$

$$\Rightarrow \qquad E\left[T\frac{L(\phi/x) - L(\theta/x)}{L(\theta/x)}\right] = g(\phi) - g(\theta)$$

$$\Rightarrow \qquad \text{cov}\left[T, \frac{L(\phi/x) - L(\theta/x)}{L(\theta/x)}\right] = g(\phi) - g(\theta) \text{ as}$$

$$E\left[\frac{L(\phi/x) - L(\theta/x)}{L(\theta/x)}\right] = 0$$

Therefore, by using C-S inequality $[\text{cov}(x, y)]^2 \leq \text{Var}(X) \cdot \text{Var}(Y)$

$$\text{var}_\theta(T)\,\text{var}\left(\frac{L(\phi/x)}{L(\theta/x)} - 1\right)$$

$$\geq \left[\text{cov}\left(T, \frac{L(\phi/x) - L(\theta/x)}{L(\theta/x)}\right)\right]^2 = \{g(\phi) - g(\theta)\}^2$$

$$\Rightarrow \qquad \text{var}_\theta(T) \geq \{g(\phi) - g(\theta)\}^2 / \text{Var}_\theta\left(\frac{L(\phi/x)}{L(\theta/x)}\right)$$

$$\geq \sup_\phi \frac{\{g(\phi) - g(\theta)\}^2}{\text{Var}_\theta\left(\frac{L(\phi/x)}{L(\theta/x)}\right)} \qquad \ldots(3.3.1)$$

Remark

(ix) Without any R-C regularity conditions we have established (3.3.1)

(x) Let $g(\theta)$ admit the first derivation $g'(\theta)$ and

$$\text{Var}\left(\frac{L(\phi/x)}{L(\theta/x)}\right) / (\phi - \theta)^2 \to J(\theta) \text{ as } \phi \to \theta.$$

Then dividing the numerator and denominator of the r.h.s. of inequality (3.3.1) by $(\phi - \theta)^2$ and taking limit as

$\phi \to \theta$, we get $\text{var}_\theta(T) \geq \dfrac{[g'(\theta)]^2}{J(\theta)}$, moreover let $L(\theta/x)$ admit the first derivative and for a fixed θ and $|\phi - \theta| < \in$ (given),

let $\dfrac{L(\phi/x) - L(\theta/x)}{(\phi - \theta)L(\theta/x)} < G(x/\theta)$, $E\left[G(x/\theta)\right]^2 < \infty$.

Then $\text{var}_\theta(T) \geq [g'(\theta)]^2 / I(\theta)$ because under the assumed conditions

$$J(\theta)=\lim_{\phi\to\theta}\frac{\text{Var}_{\theta}[L(\phi/x)/L(\theta/x)]}{(\phi-\theta)^2}$$

$$=\lim_{\phi\to\theta}\frac{\int[L(\phi/x)-L(\theta/x)]^2}{(\phi-\theta)^2[L(\theta/x)]^2}\cdot L(\theta/x)dx$$

$$=\int\left[\frac{L'(\theta/x)^2}{L(\theta/x)}\right]L(\theta/x)dx$$

taking limit under integral sign

$$=E\left[\frac{\partial\log L(\theta/x)}{\partial\theta}\right]^2$$

$$=I(\theta).$$

Example 3.4

Let X have $u(0,\ \theta)$ distribution. It is not R-C regular family. Let

$$g(\theta)=\theta\cdot E_{\theta}\left[\frac{[L(\phi/x)}{L(\theta/x)}\right]^2=\int_0^{\phi}\left(\frac{\theta}{\phi}\right)^2\frac{1}{\theta}dx=\frac{\theta}{\phi}\Rightarrow \text{var}\left[\frac{[L(\phi/x)}{L(\theta/x)}\right]=\frac{\theta}{\phi}-1.$$

Then by Chapman-Robbin's inequality,

$$\text{var}_{\theta}(T)\geq\sup_{\phi}\frac{(\phi-\theta)^2}{\left(\dfrac{\theta}{\phi}-1\right)}=\sup_{\phi}[\phi(\theta-\phi)]=\frac{\theta^2}{4}.$$

As the UMVU estimator of θ is $2X$ with $\text{var}_{\theta}(2X)=\frac{\theta^2}{3}$, which is greater then $\frac{\theta^2}{4}$, Chapman lower bound.

Example 3.5

Let X have p.m.f $P_N(X=k)=\frac{1}{N}, k=1,2,,\ldots,N$ and let $g(N)=N$. Then by Chapman-Robbin's inequality

$$\text{Var}(T) \geq \sup_{N'<N} \frac{(N-N')^2}{\text{Var}\left(\frac{P_{N'}}{P_N}\right)}.$$

Now as $\frac{P_{N'}}{P_N} = \frac{N}{N'}$,

$$E_N\left[\frac{P_{N'}(x)}{P_N(x)}\right]^2 = \frac{1}{N}\sum_{k=1}^{N'}\left(\frac{P_{N'}}{P_N}\right)^2 = \frac{N}{N'}$$

$$\text{var}_N\left(\frac{P_{N'}(x)}{P_N(x)}\right) = \frac{N}{N'} - 1$$

so,

$$\text{var}_N(T) \geq \sup_{N'<N} \frac{(N-N)^2 N'}{N-N'} = \sup_{N'<N} [(N-N')N']$$

if $\qquad k < \frac{N+1}{2}, \frac{k(N-k)}{(k-1)(N-k+1)} > 1$

so that $(N-N')\,N'$ increases if $N' > \frac{N+1}{2}$ the maximum is achieved at $\frac{N+1}{2}$.

Therefore, $\text{var}(T) \geq \left[\frac{N+1}{2}\right]\left[N - \frac{N+1}{2}\right]$, where, $[x]$ is the largest integer in x.

3.4 BHATTACHARYA LOWER BOUNDS

If the variance of an unbiased estimate does not attain *R-C* lower bound then one can attain it by introducing more variables in the linear function in addition to $\frac{1}{L}\frac{\partial L}{\partial \theta}$ whose expectations vanish such as

$$\frac{1}{L}\frac{\partial^2 L}{\partial \theta^2}, \frac{1}{L}\frac{\partial^3 L}{\partial \theta^3}, \ldots.$$

Theorem 3.5: If T is unbiased for $g(\theta)$ and if $L(\theta / x) = p_\theta(x)$ satisfy the following conditions for $\theta \in (a, b) \subset I$ and if

(i) $\frac{\partial^i L}{\partial \theta^i} = L^{(i)} < \infty,\ i = 1, 2,, \ldots, k\ , \forall \theta$

(ii) $\frac{\partial^i L}{\partial \theta^i} \int L dx = \int \frac{\partial^i L}{\partial \theta^i} dx,\ i = 1,2, \ldots, k, \forall\ \theta$

(iii) $V = (V_{ij}) = (\text{cov}(S_i, S_j))$ for $i, j = 1, 2, \ldots, k$ is positive definite where

$$S_i = \frac{1}{L} L^{(i)},\ i = 1, 2, \ldots, k.$$

(iv) $\frac{\partial^i}{\partial \theta^i} \int T L dx = \int T \frac{\partial^i L}{\partial \theta^i} dx,$

then $\text{var}_\theta T \geq \underline{N}' V^{-1} \underline{N}$ where $\underline{N} = (g^{(1)}, g^{(2)}, \ldots, g^{(k)})'$,

$g^{(i)} = \frac{\partial^i g(\theta)}{\partial \theta^i},\ i = 1, 2, \ldots, k$ and equality sign holds iff T has exponential family of distributions.

Proof: As $E(S_i) = \int \frac{\partial^i L}{\partial \theta^i} dx = 0$, $\text{cov}(T, S_i) = \frac{\partial^i g(\theta)}{\partial \theta^i} = g^{(i)}$ the covariance matrix Σ of $T, S_1, S_2, \ldots, S_k$ will be $\Sigma = \begin{pmatrix} \text{var}(T) & \underline{N}' \\ \underline{N} & V \end{pmatrix}$ which being p.s.d.,

$$|\Sigma| = |V| (\text{var}(T) - \underline{N}' V^{-1} \underline{N}) \geq 0 \Rightarrow \text{var}_\theta(T) \geq \underline{N}' V^{-1} \underline{N}.$$

$\underline{N}' V^{-1} \underline{N}$ is called k^{th} Bhattacharya lower bound, L_k.

Since the multiple correlation coefficient $\rho_T(S_1, \ldots, S_k)$ is given by

$$\rho_T(S_1, \ldots, S_k) = \frac{\underline{N}' V^{-1} \underline{N}}{\text{Var}(T)} = \frac{L_k}{\text{var}(T)},$$

so $L_i = \text{var}_\theta(T)\rho_T(S_1,\ldots,S_i)$, $i=1,2,\ldots,k$. L_i are non-decreasing. The equality sign in Bhattacharya inequality holds iff $\rho_T^2(S_1,\ldots,S_i)=1$, which implies T and S_i are linearly related. Thus, T has exponential family of distributions.

Remarks

(xi) When $k = 1$ L_1, is R-C lower bound because $N = g'(\theta)$ and $V_{11} = E(S_1^2) = I(\theta)$.

(xii) When $g(\theta)=\theta$, $N' = (1,0,0,\ldots,0)$, so that

$$L_k = (1,0,0,\ldots,0)V^{-1}\begin{pmatrix}1\\0\\\vdots\\0\end{pmatrix}, V^{11} = \begin{vmatrix}V_{22} & \cdots & V_{2k}\\ \vdots & & \vdots\\ V_{k2} & \cdots & V_{kk}\end{vmatrix} \Big/ \begin{vmatrix}V_{11} & \cdots & V_{1k}\\ \vdots & & \vdots\\ V_{k1} & \cdots & V_{kk}\end{vmatrix}$$

$$L_1 = \frac{1}{V_{11}},\ L_2 = V_{22}/(V_{11}V_{22} - V_{12}^2)$$

Example 3.6

Let $X_1, X_2, \ldots, X_n$ be a random sample from Poisson distribution, $P(\theta)$. Then obtain L_1 and L_2 of $g(\theta) = e^{-\theta}$. As $\hat{g}(\theta) = \left(1 - \frac{1}{n}\right)^T$ is UMVU estimator of $g(\theta)$ and $\text{var}(\hat{g}) = e^{-2\theta}\left(e^{\frac{\theta}{n}} - 1\right)$, $L_1 =$ R-C lower bound $= \frac{\theta}{n}e^{-2\theta}$.

Since $T = \Sigma X_i \sim P(n\theta)$ is sufficient, L is replaced by probability function of T.
As

$$\frac{\partial f}{\partial \theta} = f(t)\left(\frac{t}{\theta} - n\right),$$

$$\frac{\partial^2 f}{\partial \theta^2} = f(t)\left[\left(\left(\frac{t}{\theta} - n\right)\right)^2 - \frac{t}{\theta^2}\right],$$

$$S_1 = \frac{1}{f}\frac{\partial f}{\partial \theta} = \left(\frac{t}{\theta} - n\right),$$

$$S_2 = \left[\left(\left(\frac{t}{\theta} - n\right)\right)^2 - \frac{t}{\theta^2}\right],$$

$$V_{11} = E(S_1^2) = E\left(\frac{t}{\theta} - n\right)^2 = \frac{\text{var}(T)}{\theta^2} = \frac{n\theta}{\theta^2} = \frac{n}{\theta}, V_{12} = E(S_1 S_2)$$

$$= E\left[\left(\frac{T - n\theta}{\theta}\right)\left\{\left(\frac{T - n\theta}{\theta}\right)^2 - \frac{T - n\theta}{\theta^2} - \frac{\theta}{n}\right\}\right]$$

$$= \frac{1}{\theta^3}\left[\mu_3 - \mu_2 - \frac{n\theta}{\theta^2}\mu_1\right] = 0$$

$$V_{22} = E(S_2^2) = \frac{1}{\theta^4}(2n^2\theta^2) = \frac{2n^2}{\theta^2},$$

using $\mu_1' = \theta$, $\mu_2 = \mu_3 = \theta$, $\mu_4 = 3\theta^2 + \theta$.

Thus,

$$V = \begin{pmatrix} \frac{n}{\theta} & 0 \\ 0 & \frac{2n^2}{\theta^2} \end{pmatrix}, V^{-1} = \begin{pmatrix} \frac{\theta}{n} & 0 \\ 0 & \frac{\theta^2}{2n^2} \end{pmatrix},$$

$$g^{(1)} = e^{-\theta}, g^{(2)} = e^{-\theta}, \underset{\sim}{N} = (-e^{-\theta}, e^{-\theta}),$$

$$L_1 = \frac{[g^{(1)}\theta]^2}{V_{11}} = \frac{e^{-2\theta}}{\frac{n}{\theta}} = \frac{\theta}{n}e^{-2\theta},$$

$$L_2 = (-e^{-\theta}e^{-\theta})\begin{pmatrix} \frac{\theta}{n} & 0 \\ 0 & \frac{\theta^2}{2n^2} \end{pmatrix}\begin{pmatrix} -e^{-\theta} \\ e^{-\theta} \end{pmatrix} = \frac{n}{\theta}e^{-2\theta}\left(\frac{\theta}{2n} + 1\right)$$

which does not attain var $(\hat{g})$.

Example 3.7

Given a random sample $X_1, X_2, \ldots, X_n$ from $f(x;\theta) = \frac{1}{\theta} e^{-\frac{x}{\theta}}$, obtain L_1 and L_2 of θ.

The complete sufficient statistics $T = \Sigma X_i \sim G\left(\frac{1}{\theta}, n\right)$ with density

$$f_T(t) = \frac{t^{n-1} e^{-\frac{t}{\theta}}}{\theta^n \Gamma(n)},$$

$$S_1 = \frac{\partial \log f}{\partial \theta} = \frac{1}{\theta^2}(t - n\theta), \frac{\partial^2 \log f}{\partial \theta^2} = -\frac{2t}{\theta^3} + \frac{2n}{\theta^2},$$

$$S_2 = \frac{1}{f}\frac{\partial^2 f}{\partial \theta^2} = \frac{\partial^2 \log f}{\partial \theta^2} + \left(\frac{1}{f} \frac{\partial f}{\partial \theta}\right)^2 = -\frac{2}{\theta^3}(t - n\theta) - \frac{n}{\theta^2} + \frac{1}{\theta^4}(t - n\theta)^2,$$

using the rth moment

$$M_r' = n(n+1), \ldots, (n+r-1)\theta^r, E(T) = n\theta, \operatorname{var}(T) = n\theta^2,$$

$$\mu_3 = 2n\theta^3, \mu_4 = 3n(n+2)\theta^4, V_{11} = \frac{n}{\theta^2},$$

$$V_{12} = 0, V_{22} = \frac{n(3n+1)}{\theta^4}, L_1 = \frac{\theta^2}{n}, L_2 = \frac{\theta^2}{n}.$$

Example 3.8

Let $f(x) = q^x p$, $x = 0, 1, 2, \ldots,$ $0 \le p \le 1, q = 1 - p$. Then obtain L_k of p and show that it is decreasing.

As $u(x) = \begin{cases} 1 & \text{if} \quad x = 0 \\ 0 & \text{otherwise} \end{cases}$, is UMVU estimator of p

$$\operatorname{var}(u(x)) = pq, S_1 = -\frac{1}{q}\left(x - \frac{q}{p}\right), S_2 = \frac{x^2}{p^2} - \frac{x(1+q)}{pq^2},$$

$$E[X(X-1)\dots(X-m+1)] = m!\,q^m(1-q)^{-m},$$

$$E(X) = \frac{q}{p},\; E(X^2) = \frac{q+q^2}{p^2}, E(X^3) = \frac{q+4q^2+q^3}{p^3},$$

$$E(X^4) = \frac{q+11q^2+11q^3+q^4}{p^4}, V_{11} = \frac{1}{p^2q}, V_{12} = -\frac{2}{p^3q},$$

$$V_{22} = \frac{4(1+q)}{q^2p^4}, L_1 = p^2q, L_2 = p^2q(1+q), \dots.$$

3.5 THE RAO-CRAMER INEQUALITY FOR SEVERAL PARAMETERS

Theorem 3.6: Let $p_\theta(x) = L(\theta / x) = L$ be p.f. of X with $\theta = (\theta_1, \dots, \theta_r)$ and let $T_1, \dots, T_k$ be unbiased for $g_1(\theta), g_2(\theta), \dots, g_k(\theta)$, respectively, denoting $k \times k$ matrix $U = (u_{ij})$, where $u_{ij} = \text{cov}(T_i, T_j)$. Let us assume that

(i) $\frac{\partial L}{\partial \theta_j} < \infty,\; j = 1, 2, \dots, r$.

(ii) $\frac{\partial}{\partial \theta_j} \int L dx = \int \frac{\partial L}{\partial \theta_j},\; j = 1, 2, \dots, r$

(iii) $V = (V_{ij}) = E(S_i S_j)$ is positive definite (p.d.) matrix

where $S_j = \frac{\partial \log L}{\partial \theta_j}, i = 1, 2, \dots, k$ and $j = 1, 2, \dots, r$

(iv) $\frac{\partial}{\partial \theta_j} \int TL dx = \int T \frac{\partial L}{\partial \theta_j} dx.$

Then $U - DV^{-1}D'$ is positive semi-definite where $D = (d_{ij})$, $i = 1, 2, \dots, k$ and $j = 1, 2, \dots, r$,

$$d_{ij} = \frac{\partial g_i(\theta)}{\partial \theta_j}.$$

Proof: Since $E(S_j) = \int \frac{\partial L}{\partial \theta_j} dx = 0, \operatorname{cov}(T_i, S_i) = d_{ij}$ then $(k+r)*(k+r)$ covariance matrix Σ of $T_1, \ldots, T_k, S_1, \ldots, S_r$ will be $\Sigma = \begin{pmatrix} U & D \\ D' & V \end{pmatrix}$ and Σ being p.s.d., $|\Sigma| \geq 0$,

$$\Rightarrow \quad |V|\,|U - DV^{-1}D'| \geq 0$$

$$\Rightarrow \quad |U - DV^{-1}D'| \geq 0$$

this being true for every k and r, $U - DV^{-1}D'$ is p.s.d.

Remarks

(xiii) $DV^{-1}D'$ is called Rao-Cramer lower bound matrix. Since $U - DV^{-1}D'$ is p.s.d. $|U| \geq |DV^{-1}D'|$, which shows that the generalized variance is never less than a quantity which can be computed, independently without the knowledge of an estimator or the method of estimation.

(xiv) If one wishes to find lower bound for the variance of the unbiased estimator of $g_i(\theta)$ only. Then D being $\left(\frac{\partial g_i}{\partial \theta_1}, \ldots, \frac{\partial g_i}{\partial \theta_r}\right)$,

$$\operatorname{var}(T_i) = u_{ii},\ \Sigma = \begin{pmatrix} u_{ii} & d_{i1} \ldots d_{ir} \\ d_{i1} & \\ \vdots & V \\ d_{ir} & \end{pmatrix},\ |\Sigma| \geq 0 \Rightarrow u_{ii} \geq \sum_s \sum_t d_{is} V^{st} d_{it}.$$

moreover, if equality sign holds, that is $u_{ii} = \sum_s \sum_t d_{is} V^{st} d_{it}$, u_{ii} is the variance of UMVU estimator of $g_i(\theta)$. Since $U - DV^{-1}D'$ is p.s.d. any principle minor is nonnegative, that is,

$$\begin{vmatrix} u_{ii} - \sum_s \sum_t d_{is} V^{st} d_{it} & u_{ij} - \sum_s \sum_t d_{is} V^{st} d_{jt} \\ u_{ji} - \sum_s \sum_t d_{is} V^{st} d_{jt} & u_{jj} - \sum_s \sum_t d_{js} V^{st} d_{jt} \end{vmatrix} \geq 0$$

$$\Rightarrow \quad -\left(u_{ij} - \sum_s \sum_t d_{is} V^{st} d_{jt}\right)^2 \geq 0 \text{ as } u_{ii} = \sum_s \sum_t d_{is} V^{st} d_{it}$$

$$\Rightarrow \quad u_{ij} = \sum_s \sum_t d_{is} V^{st} d_{jt}$$

(xv) If $g_i(\theta) = \theta$ then $D = I$ and the Rao-Cramer lower bound matrix is V^{-1} and *R-C* lower bound of θ_i is V^{ii}, where V^{ii} is $(ii)^{th}$ element of V^{-1}. As $V^{ii} > \dfrac{1}{V_{ii}}$, lower bound θ_i then *R-C* lower bound for θ_i in the presence of nuisance parameters is larger than that of θ_i in the absence of other parameters.

Example 3.9

Let $X_1, X_2, \ldots, X_n$ be a random sample from $N(\theta_1, \theta_2)$. Obtain *R-C* lower bound matrix for (θ_1, θ_2)

Here $\log L = \text{Constant} - \dfrac{n}{2}\log\theta_2 - \dfrac{1}{2\theta_2}\Sigma(X_i - \theta_1)^2.$

$$S_1 = \frac{\partial \log L}{\partial \theta_1} = \frac{n(\bar{x} - \theta_1)}{\theta_2},\ S_2 = \frac{\partial \log L}{\partial \theta_2} = -\frac{n}{2\theta_2} + \frac{\Sigma(x_i - \theta_1)^2}{2\theta_2} = \frac{U - n}{2\theta_2},$$

where $U = \dfrac{\Sigma(x_i - \theta_1)^2}{\sqrt{\theta_2}} \sim \chi^2_n.$

$$V_{11} = E(S_1^2) = \frac{n^2}{\theta_2^2} E(\bar{X} - \theta_1)^2 = \frac{n}{\theta_2},\ V_{12} = E(S_1 S_2) = 0$$

as U is independent of $\bar{X}$.

$$V_{22} = E(S_2^2) = \frac{1}{4\theta_2^2} E(U - n)^2 = \frac{\text{Var}(U)}{4\theta_2^2} = \frac{2n}{4\theta_2^2} = \frac{n}{2\theta_2^2}.$$

$$V = \begin{pmatrix} \dfrac{n}{\theta_2} & 0 \\ 0 & \dfrac{n}{2\theta_2^2} \end{pmatrix} \Rightarrow V^{-1} = \begin{pmatrix} \dfrac{\theta_2}{n} & 0 \\ 0 & \dfrac{2\theta_2^2}{n} \end{pmatrix},$$

which is lower bound matrix of θ_1 and θ_2.

Example 3.10

If $p_\theta(x) = L = \dfrac{n!}{x_1!, x_2!, x_3!}\theta_1^{x_1}\theta_2^{x_2}\theta_3^{x_3}, \theta_1+\theta_2+\theta_3=1, x_1+x_2+x_3=n$ then

Obtain *R-C* lower bound matrix for (θ_1, θ_2)

From $\dfrac{\partial \log L}{\partial \theta_i}$ for $i = 1, 2$, we have

$$V_{11} = n\left(\frac{1}{\theta_1}+\frac{1}{1-\theta_1-\theta_2}\right),$$

$$V_{12} = \frac{n}{1-\theta_1-\theta_2},$$

$$V_{22} = n\left(\frac{1}{\theta_2}+\frac{1}{1-\theta_1-\theta_2}\right)$$

$$V^{-1} = \frac{1}{n}\begin{pmatrix}\theta_1(1-\theta_1) & -\theta_1\theta_2\\ -\theta_1\theta_2 & \theta_2(1-\theta_2)\end{pmatrix}.$$

PROBLEMS

3.1 Prove the results given in (3.2.3), (3.2.4), (3.2.5), (3.2.6) remark 3.1 {(i), (ii), (iii), (iv), (v), (vi)}.

3.2 Let $X_1, \ldots, X_n (n \geq 3)$ be a random sample from Poisson distribution with parameter θ. For the estimators

(i) $T_1 = \bar{X}$,

(ii) $T_2 = \dfrac{1}{2}\sum_{i=1}^{n-1}\dfrac{X_i}{n-1}+\dfrac{X_n}{2}$,

(iii) T_3 = unbiased estimator of $e^{-\theta}(1+\theta)$, obtain *R-C* lower bound for each of them and examine whether variance of any one of them attains *R-C* lower bound.

3.3 Obtain $I(\theta)$ for the following distribution

(i) $N(\mu, \sigma^2)$ for μ (ii) $N(\mu, \sigma^2)$ for σ^2

(iii) $b(n, p)$ for p (iv) poisson (θ) for θ

(v) Gamma (α, p) for α

3.4 If X has $P(\theta)$, then show that the information contained about (i) $\log \theta$ and (ii) $\sqrt{\theta}$ are independent of θ.

3.5 Obtain a function of θ for which the amount of information is independent of θ (i) for the gamma distribution $G(\alpha \cdot P)$, where P is known $\alpha = \theta$. (ii) for the binomial distribution $b(n, p)$ with $\theta = p$

3.6 For the location parameter family with density $f(x - \theta)$ show that $I(\theta)$ is independent of θ.

3.7 If $X_1, \ldots, X_n$ is a random sample from $b(1, \theta)$ obtain MVB estimators for θ, $\theta(1 - \theta)$.

3.8 If $X_1, \ldots, X_n$ is a random sample from $P(\theta)$ and if $T_n = \bar{X}_n^2 - \bar{X}_n$, obtain the R-C lower bound for $E(T_n - \theta^2)^2$. Find a function of T_n which is unbiased for θ^2 and find its R-C lower bound.

3.9 Let $X_1, \ldots, X_n$ be a random sample from $N(\mu, \mu)$, $\mu > 0$. Find the R-C lower bound for $E(\bar{X}_n S_n^2 - \mu^2)^2$, where, S_n^2 is variance of the sample.

3.10 If $X_1, \ldots, X_n$ is a random sample from $G\left(\frac{1}{\theta}, 1\right)$ distribution, what is the R-C lower bound for the variance of an unbiased estimator of $g(\theta) = e^{-\frac{1}{\theta}}$? Does the variance of the UMVU estimator of $g(\theta)$ attain the R-C lower bound?

3.11 If $X_1, \ldots, X_n$ is a random sample from $u(0, \theta)$ distribution, obtain R-C lower bound for the variance of the unbiased estimator of θ and compare it with the variance of the UMVU estimator of θ.

3.12 If $X_1, \ldots, X_n$ is a random sample from p.d.f. $f(x; \theta) = e^{-(x-\theta)}$, $x > \theta$, obtain R-C lower bound for the variance of the unbiased estimator of (i) θ^2 (ii) e^{θ} (iii) θ (iv) $\theta^2 + \theta$.

3.13 Obtain R-C lower bound for the variance of unbiased estimator of θ where, θ is the parameter of Cauchy distribution on the basis of a random sample of size n from it.

3.14 For the scale parameter family with density as $\frac{f(x;\theta)}{\theta}$, $\theta > 0$, obtain the amount of information based on single observation and show that $I(\theta)$ is independent of θ.

3.15 A random sample of size n is available from the distribution of positive real numbers with density $\frac{x+1}{\theta(\theta+1)}e^{-\frac{x}{\theta}}$, where, θ is an unknown positive parameter. Obtain unbiased estimator of $\frac{3+2\theta}{6}$.

4

CHAPTER

Criteria and Methods of Estimation in Large Sample

4.1 INTRODUCTION

In earlier chapter we have discussed the problem of estimation for small sample but however sometimes it is difficult to find efficient estimator so one has to increase the sample size and obtain the estimator. We shall discuss methods of consistency in this chapter. Also large sample estimation methods are discussed in this chapter. Section 4.2 is devoted for consistency and in section 4.3 asymptotic normality is discussed. Different methods of estimation such as moment method, maximum likelihood method, minimum chi-square method have been discussed in section 4.4.

4.2 CONSISTENCY

Let $\{X_n\}, n = 1, 2, \ldots$ be a sequence of i.i.d. random variables having probability function $f_{X_n}(x; \theta)$ which provide information on θ by defining estimator T_n based on the first n variables. Our object is to examine the properties of the sequence T_n as $n \to \infty$.

Definition 4.1

The estimator T_n is said to be consistent for θ if $T_n \to \theta$ in probability, that is, $P_\theta\{|T_n - \theta| < \in\} \to 1$ as $n \to \infty$.

This criteria of consistency is not valid with reference to a particular sample size. One of the desirable properties to be satisfied, however, is that with increase in the observations, the estimator tends to a definite quantity which is the true value to be estimated, for instance, if $X_1, X_2, \ldots, X_n$ are i.i.d. random variables from Cauchy distribution with p.d.f.

$$f_X(x;\theta) = \frac{1}{\pi[1+(x-\theta)^2]}, -\infty < x < \infty, \theta > 0,$$

it is well known that $\bar{X}_n = \frac{\Sigma X_i}{n}$ has the same distribution as a single random variable has. Hence, $P_\theta\{|\bar{X}_n - \theta| < \in\}$ remains same for all n, so $\bar{X}_n$ is not consistent estimator of θ. But, however, the sample median, $\tilde{X}_n$ is consistent estimator of θ, because $\tilde{X}_n \to N\left(\theta, \frac{\pi^2}{4n}\right)$, where θ is the population median.

By strong law of large numbers, it follows that the sample mean, $(\bar{X}_n)$ is consistent estimator of the population mean if it exists and if $\{X_n\}$ is a sequence of i.i.d. random variables. In the case of Cauchy distribution mean of the distribution does not exist.

To check consistency following theorems are useful.

Theorem 4.1: An unbiased estimator, T_n for θ is consistent if $\text{var}_\theta(T_n) \to 0$ as $n \to \infty$.

Proof: Using Chebychev's inequality, for $k > 0$,

$$P_\theta\{|T_n - \theta| \geq k\sigma\} \leq \frac{1}{k^2}$$

where σ is standard deviation of T_n and taking $k = \frac{\in}{\sigma}$, the theorem is established.

Theorem 4.2: If T_n is a statistic such that $E(T_n) = \theta_n \to \theta$ and $\text{var}(T_n) \to 0$ as $n \to \infty$ then T_n is consistent estimator of θ.

Proof: As $|T_n - \theta| < \delta$, $|\theta_n - \theta| < \delta_1$ then

$$|T_n - \theta| < |T_n - \theta_n| + |\theta_n - \theta| < \delta + \delta_1$$

$$\Rightarrow \qquad P_\theta\{|T_n - \theta| < \delta + \delta_1\} \geq P\{|T_n - \theta_n| < \delta\} \geq 1 - \frac{\text{var}(T_n)}{\sigma^2},$$

by Chebychev's inequality $\geq 1 - \in$

as $\text{var}(T_n) \to 0, T_n \to \theta, \forall\, \theta.$

Theorem 4.3: If $g(T)$ is a continuous function of T and if T is a consistent estimator of θ then $g(T)$ is a consistent estimator of $g(\theta)$.

Proof: Since $g(T)$ is a continuous function of T, there exits $\in > 0, \delta > 0$ such that $|g(T) - g(\theta)| < \in$ whenever $|T - \theta| < \delta$ and hence

$$P_\theta |g(T) - g(\theta)| < \in \geq P\{|T - \theta| < \delta\} \geq 1 - \in$$

which proves the theorem. A more general result is given in the following theorem

Theorem 4.4: If $g(T_1, T_2, \ldots, T_r)$ is a continuous function of $T_1, T_2, \ldots, T_r$ at $\theta_1, \theta_2, \ldots, \theta_r$ and if $T_1, T_2, \ldots, T_r$ are consistent estimators of $\theta_1, \theta_2, \ldots, \theta_r$ respectively then $g(T_1, T_2, \ldots, T_r)$ is a consistent estimator of $g(\theta_1, \theta_2, \ldots, \theta_r)$.

Proof: Since $g(T_1, T_2, \ldots, T_r)$ is a continuous function of $T_1, T_2, \ldots, T_r$ at $\theta_1, \theta_2, \ldots, \theta_r$ $|g(T_1, T_2, \ldots, T_r) - g(\theta_1, \theta_2, \ldots, \theta_r)| < \in$ whenever $\{T_i - \theta_i| < \delta_i$, $i = 1, 2, \ldots, r$ that is, whenever $\{T_i - \theta_i| < \delta$, $i = 1, 2, \ldots, r$, where δ is the smallest of $(\delta_1, \delta_2, \ldots, \delta_r)$, $\delta > 0$. Thus,

$$P\{|g(T_1, T_2, \ldots, T_r) - g(\theta_1, \theta_2, \ldots, \theta_r)| < \in\} \geq P\{|T_1 - \theta_1| < \delta_1 \ldots |T_r - \theta_r| < \delta_r\}$$

$$= P\{A_1 \cap A_2, \ldots, \cap A_r\},$$

where $$A_i = |T_i - \theta_i| < \delta_i$$

$$= P(B), B = A_1 \cap A_2, \ldots, \cap A_r$$

$$= 1 - P(B^c) \geq 1 - P\{\Sigma A_i^c\},$$

using DeMorgan's law and using Boole's inequality

$$= 1 - rP\{|T_i - \theta_i| \geq \delta\} \geq 1 - r\frac{\eta}{r}$$

$$= 1 - \eta$$

as T_i is consistent estimator of θ_i, $P\{|T_i - \theta_i| \geq \delta\} < \dfrac{\eta}{r}$ for

$$\eta \geq N = \max(N_1, N_2, \ldots N_r).$$

So $g(T_1, T_2, \ldots, T_r)$ is a consistent estimator of $g(\theta_1, \theta_2, \ldots, \theta_r)$. Particularly if T_1 and T_2 are consistent estimators of θ_1 and θ_2, respectively then

(i) $T_1 + T_2$ is consistent for $\theta_1 + \theta_2$

(ii) $T_1 T_2$ is consistent for $\theta_1 \theta_2$.

Example 4.1

Let $X_1, X_2, \ldots, X_n$ be a random sample from $N(\mu, \sigma^2)$. Then obtain usual estimators of μ and σ^2 and show that they are consistent. Also obtain consistent estimator of pth factile.

The usual estimator of μ is $\bar{X}_n = \frac{1}{n}\sum_{i=1}^{n} X_i$ and $\text{var}(\bar{X}_n) = \frac{\sigma^2}{n} \to 0$ as $n \to \infty$, so $\bar{X}_n$ is consistent for μ.

Further $s^2 = \frac{1}{n}\Sigma(X_i - \bar{X})^2$ is an usual estimator of σ^2. Also we know that

$$Es^2 = \frac{n-1}{n}\sigma^2 \to \sigma^2 \text{ as } n \to \infty. \text{ var}\left(\frac{ns^2}{\sigma^2}\right) = 2(n-1) \text{ as } \frac{ns^2}{\sigma^2} \sim \chi^2_{(n-1)}.$$

Then $\text{var}(s^2) = \frac{2(n-1)\sigma^4}{n^2} \to 0$ as $n \to \infty$. Hence, s^2 is consistent for σ^2.

Also for estimating θ_p, the pth quantile of the population of $N(\mu, \sigma^2)$ given by $\mu + \theta_p \sigma$ which is a continuous function of (μ, σ^2). Then $\bar{X} + Z_p \frac{s}{\sqrt{n}}$ is a consistent estimator of θ_p where, Z_p is the pth sample quantile.

4.3 CONSISTENT ASYMPTOTICALLY NORMAL (CAN) ESTIMATORS

Consistency is a weak property in the sense that for large n, the error $T_n - g(\theta)$ is made small but to which order $\left(\frac{1}{n}, \frac{1}{\sqrt{n}}, \frac{1}{n^2} \ldots\right)$ of smallness is not stated. In order to get convergence of a consistent estimator, T_n such that

$$P\left\{|T_n - g(\theta)| \le \frac{a}{b_n}\right\} = P_n(a) \to P_n(a) = \begin{cases} 1 & \text{if } b_n \text{ is bounded} \\ 0 & \text{if } b_n \to \infty \end{cases}$$

which suggests that there is a value of b_n for which $P_n(a)$ tends to a value between 0 and 1. Thus,

$$P\{b_n |T_n - g(\theta)| \le a\} \to K(a) \text{ as } n \to \infty,$$

where $K(a) = T_n P_n(a)$ and $|T_n - g(\theta)| = 1$ at the rate $\frac{1}{b_n}$.

This suggests that there is a normalizing factor b_n so that T_n is consistent and when $b_n \to \infty$ it magnifies the minute error. Hence b_n, normalizing factor plays an important role.

Definition 4.2

If estimator T_n is such that $b_n(T_n - g(\theta)) \xrightarrow{d} N(0, J^2(\theta))$ for some function $J^2(\theta)$ it is called a consistent asymptotically normal (CAN) estimator.

***Example* 4.2**

If $X_1, X_2, \ldots, X_n$ is a random sample from a distribution with mean μ and variance σ^2 and kurtosis α_4 then $\hat{\mu} = \bar{X}_n$, $\hat{\sigma}^2 = \frac{1}{n}\Sigma(X_i - \bar{X})^2$ are CAN estimators of μ and σ^2, respectively.

By law of large numbers, we have

$$\hat{\mu} \xrightarrow{p} \mu \quad \text{and} \quad \hat{\sigma}^2 \xrightarrow{p} \sigma^2$$

so by central limit theorem

$$\sqrt{n}(\hat{\mu} - \mu) \xrightarrow{d} N(0, \sigma^2) \quad \text{and} \quad \sqrt{n}(\hat{\sigma}^2 - \sigma^2) \xrightarrow{d} N(0, (\alpha_4 - 1)\sigma^4)$$

$\Rightarrow$ $\hat{\mu}$ is CAN estimator for μ and $\hat{\sigma}^2$ is CAN estimator for σ^2.

Remark (i)

If $J^2(\theta) = \dfrac{1}{I(\theta)}$, $I(\theta) = E_\theta\left(\dfrac{d\log L}{d\theta}\right)^2$, Fisher information then m.l.e. T_n of θ is CAN estimator because

$$\sqrt{n}|(T_n - \theta)| \xrightarrow{d} N\left(0, \frac{1}{I(\theta)}\right)$$

Definition 4.3

The asymptotic relative efficiency (ARE) of two CAN estimators, T_1 and T_2, with $J^2(\theta)$'s as $J_1^2(\theta)$ and $J_2^2(\theta)$, is

$$e_\theta(T_1, T_2) = \frac{J_1^2(\theta)}{J_2^2(\theta)}.$$

Theorem 4.5: If T_n is CAN estimator of θ and if $g(\theta)$ is a differentiable function of θ, that is, $\dfrac{dg(\theta)}{d\theta}$ is continuous and non-vanishing then $g(T_n)$ is CAN estimator of $g(\theta)$ and if $\dfrac{dg}{d\theta} = 0$ and $\dfrac{d^2g}{d\theta^2} \neq 0$ at $\theta = \theta_0$ and continuous then $b_n^2(g(T_n) - g(\theta)) \rightarrow \dfrac{J^2(\theta_0)}{2}\dfrac{d^2g}{d\theta_0^2}\chi_1^2$.

Proof: Expanding $g(T_n)$ around $g(\theta)$, $g(T_n) = g(\theta) + (T_n - \theta)[g'(\theta) + R_n]$ where, R_n is remainder.

Now $b_n[g(T_n) - g(\theta)] = b_n(T_n - \theta)g'(\theta) + b_n(T_n - \theta)R_n$ and theorem is proved if we show that $b_n(T_n - \theta)R_n \xrightarrow{d} 0$. In this case

$$b_n[g(T_n) - g(\theta)] = b_n(T_n - \theta)g'(\theta) \xrightarrow{d} N(0, J^2(\theta)[g'(\theta)]^2)$$

as $\quad b_n(T_n - \theta) \rightarrow N(0, J^2(\theta))$.

As $T_n \xrightarrow{p} \theta, b_n(T_n - \theta)R_n \xrightarrow{p} p$, using the result that if $A_n \xrightarrow{d} A$ and $B_n \xrightarrow{p} 0$, then $A_nB_n \xrightarrow{p} 0$.

For the other part of the theorem if $\frac{dg(\theta)}{d\theta} = 0$, expand $g(T_n)$ around $g(\theta)$ as

$$g(T_n) = g(\theta) + (T_n - \theta)g'(\theta) + \frac{1}{2}(T_n - \theta)^2[g''(\theta) + R_n]$$

$$\Rightarrow \quad b_n^2(g(T_n) - g(\theta)) = \frac{1}{2}(T_n - \theta)^2[g''(\theta) + R_n] \text{ as } g'(\theta) = 0.$$

Since $T_n - \theta \xrightarrow{p} 0, b_n^2 R_n (T_n - \theta)^2 \xrightarrow{p} 0$ and since

$$b_n(T_n - \theta_0) \xrightarrow{d} N(0, J^2(\theta_0)) \xrightarrow{d} N(0, \sigma_{T_n}^2(\theta_0))$$

$$J^2(\theta_0) = \text{var}_{\theta_0}(T_n), b_n^2(T_n - \theta_0)^2 \xrightarrow{d} \sigma_{T_n}^2(\theta_0)\chi_1^2$$

$$\Rightarrow \quad b_n^2(g(T_n) - g(\theta_0)) \xrightarrow{d} \frac{\sigma_{T_n}^2(\theta_0)}{2} g''(\theta_0)\chi_1^2$$

If $$b_n = \sqrt{n}, \; n(g(T_n) - g(\theta_0)) \to \frac{\sigma^2(\theta_0)}{2} g''(\theta_0)\chi_1^2.$$

Example 4.3

If $X_1, X_2, \ldots, X_n$ is a random sample from $N(\theta, \sigma^2)$ and if we wish to obtain CAN estimator of θ^2, we have m.l.e. of θ^2 as $T_1 = \bar{X}^2$, UMVU estimator of θ^2 as $T_2 = \bar{X}^2 - \frac{\sigma^2}{n}$ if σ^2 is known.

Case I Let $\theta \neq 0$.

As $\bar{X} \sim N\left(\theta, \frac{\sigma^2}{n}\right)$ and as $g(\theta) = \theta^2$, $g'(\theta) = 2\theta \neq 0$, for any θ except $\theta = 0$, $\bar{X}^2$ is CAN estimator of θ^2 with asymptotic variance $\frac{4\theta^2}{n}$, that is, $\sqrt{n}(\bar{X}^2 - \theta^2) \xrightarrow{d} N(0, 4\sigma^2\theta^2)$. Now $T_2 = \bar{X}^2 - \frac{\sigma^2}{n}$, as

$$\sqrt{n}\left(\bar{X}^2 - \frac{\sigma^2}{n} - \theta^2\right) = \sqrt{n}(\bar{X}^2 - \theta^2) - \frac{\sigma^2}{\sqrt{n}},$$

$$\sqrt{n}(T_2 - \theta^2) \xrightarrow{d} N(0, 4\sigma^2\theta^2), \text{ because } \frac{\sigma^2}{\sqrt{n}} \to 0 \text{ as } n \to 0.$$

Case II Let $\theta = 0$.

Then at $\theta = 0$, $g'(\theta) = 0$, $g''(\theta) = 2$ and as

$$\frac{\sqrt{n}}{\sigma}(\bar{X} - 0) \sim N(0, 1)$$

$$n(T_1 - \theta^2) = n(\bar{X}^2 - \theta^2) = n\bar{X}^2 \to \sigma^2 2\chi_1^2$$

using the Theorem (4.5). We have used $b_n = n$ instead of $b_n = \sqrt{n}$, which enhances the convergence faster.

For T_2, we have

$$n\left(\bar{X}^2 - \frac{\sigma^2}{n} - \theta^2\right) = n\bar{X}^2 - \sigma^2 \xrightarrow{d} \sigma^2(\chi_1^2 - 1).$$

Example 4.4

If $X_1, X_2, \ldots, X_n$ is a random sample from $b(1, p)$ then show that $\bar{X}(1-\bar{X})$ is CAN estimator for $p(1-p)\ \forall\ p$ except $p = \frac{1}{2}$. Obtain asymptotic distribution of $\bar{X}(1-\bar{X})$, when $p = \frac{1}{2}$. Show that UMVU estimator of $p(1-p)$ is CAN estimator.

Since $X = \sum_{i=1}^{n} X_i \sim b(n, p)$ distribution, $E\bar{X} = p$ and $\text{var}(\bar{X}) = \frac{pq}{n}$ and hence $\sqrt{n}(\bar{X} - p) \xrightarrow{d} N(0 \cdot pq)$, which shows that $\bar{X}$ is CAN estimator for p.

Consider $g(\bar{X}) = \bar{X}(1-\bar{X})$ so that $g(p) = p(1-p)$, $g'(p) = 1-2p$, $g''(p) = -2$.

If $p \neq \frac{1}{2}$, $n[g(\bar{X}) - g(p)] \xrightarrow{d} N(0, pq(1-2p)^2) = N(0, pq(q-p)^2)$

and if $p = \frac{1}{2}$, $n[g(\bar{X}) - g(p)] \xrightarrow{d} N\left(0, \frac{1}{2}\frac{1}{2}\frac{1}{2}(-2)^2\right) = N\left(0, \frac{1}{2}\right)$.

The UMVU estimator of $p(1 - p)$ being

$$g(\bar{X}) = \frac{X(n-X)}{n(n-1)} = \frac{n}{n-1}\bar{X}(1-\bar{X}),$$

$$g(p) = \frac{n}{n-1}p(1-p),\ g'(p) = \frac{n}{n-1}(1-2p)$$

$$g''(p) = \frac{-2n}{n-1},$$

for $p \neq \frac{1}{2}$,

$$n\left(\frac{X(n-X)}{n(n-1)} - \frac{n}{n-1}p(1-p)\right) \to N(0, pq(pq)^2)$$

and for $p = \frac{1}{2}$,

$$n\left(\frac{X(n-X)}{n(n-1)} - \frac{n}{n-1}p(1-p)\right) \to N\left(0, \frac{1}{2}\right)$$

as $\frac{n}{n-1} \to 1$ as $n \to \infty$.

If $X_1, X_2, \ldots, X_n$ are i.i.d. random variables with $E(X_i) = \mu(\theta)$, $\text{var}(X_i) = \sigma^2(\theta)$ then by central limit theorem $\sqrt{n}(\bar{X} - \mu(\theta)) \to N(0, \sigma^2(\theta))\bar{X}$ is CAN estimator for $\mu(\theta)$ with asymptotic variance $\frac{\sigma^2(\theta)}{n}$. Moment estimator requires $\mu^{-1}(\bar{X})$. So by using the above theorem one can obtain asymptotic distribution of $\mu^{-1}(\bar{X})$ as $N\left(\theta, \frac{\frac{\sigma^2(\theta)}{n}}{\left(\frac{d\mu}{d\theta}\right)^2}\right)$.

If X is a random variable with p.d.f. $f(x;\theta)$, $\theta \in \Omega$ then p^{th} quantile, θ_p is defined as the solution of

$$F_X(\theta_p) = P(X_1 \le \theta_p)$$

$$\Rightarrow \qquad \int_{-\infty}^{\theta_p} f(x;\theta)\,dx = p, \; 0 < p < 1$$

$$\Rightarrow \qquad \theta_p = F_X^{-1}(p).$$

If $X_1, X_2, \ldots, X_n$ is a random sample from $f(x;\theta)$ then p^{th} sample quantile, Z_p is defined as $r = [np] + 1$th order statistic $X_{(r)}$ in the corresponding order sample $X_{(1)}, X_{(2)}, \ldots, X_{(n)}$. So one method of estimating θ_p is to solve the equation

$$X_{(r)} = \theta_p, \quad \text{if} \quad \frac{d\theta_p}{d\theta} \ne 0$$

If θ_p is the pth quantile, $0 < p < 1$ and if $f(\theta_p) > 0$ it can be shown that

$$X_{([np]+1)} \xrightarrow{d} N\left(0, \frac{p(1-p)}{n[f(\theta_p)]^2}\right)$$

$\Rightarrow X_{([np]+1)}$ is CAN estimator for θ_p.

If $\left(\frac{d\theta_p}{d\theta}\right)^{-1} \ne 0$ and continuous then

$$\theta_p^{-1}(X_{([np]+1)}) \xrightarrow{d} N\left(\theta, \frac{\dfrac{p(1-p)}{n[f(\theta_p)]^2}}{\left(\dfrac{d\theta_p}{d\theta}\right)^2}\right)$$

Multivariate Central Limit Theorem

Let $\underline{X}_r = (X_{1r}, X_{2r}, \ldots, X_{pr}), r = 1, 2, \ldots, n$ be i.i.d. with mean vector

$$\underline{\mu} = (\mu_1, \mu_2, \ldots, \mu_p)$$

and covariance matrix Σ and let

$$\bar{X}_i = \frac{1}{n}\sum_{r=1}^{n} X_{ir}.$$

Then

$$[\sqrt{n}(\bar{X}_{1n} - r_1), \ldots, \sqrt{n}(\bar{X}_{pn} - \mu_p)] \xrightarrow{d} N_p(\underline{0}, \Sigma)$$

From this multivariate central limit theorem we state the following asymptotic result without proof.

Theorem 4.6: Suppose $Y_i = f_i \underline{X}_i$, $EX_i = \mu_i$, $D(\underline{X}_i) = \Sigma$, $i = 1, 2, \ldots, p$ and suppose $[\sqrt{n}(Y_{1n} - \mu_1), \ldots, \sqrt{n}(Y_{pn} - \mu_p)] \to N_p(0, \Sigma)$ and suppose that $f_1, \ldots, f_p$ are real vector functions of $\underline{\mu} = (\mu_1, \mu_2, \ldots, \mu_p)$ and continuously differentiable in the neighbourhood w of the parameter point $\underline{\mu}$ and such that the matrix $A = \left(\frac{df_i}{d\mu_i}\right)$ is non-singular in w. Then

$$[\sqrt{n}(f_1(Y_n) - f_1(\mu_1)), \ldots, \sqrt{n}(f_p(Y_n) - f_p(\mu_p))] \to N_p(\underline{0}, A\Sigma A').$$

Example 4.5

Let X_1, X_2, …, X_n be a random sample from Pareto distribution with p.d.f.

$$f(x;\theta) = \frac{\theta}{x^{\theta+1}}, \; x > 1, \theta > 0.$$

Obtain moment estimator of θ and its asymptotic variance.

(i) Moment Estimation

Since $\mu(\theta) = \frac{\theta}{(\theta-1)}$, $\sigma^2(\theta) = \frac{\theta}{(\theta-2)(\theta-1)^2}$, $\frac{d\mu(\theta)}{d\theta} = -\frac{1}{(\theta-1)^2}$ and

$$\left(\frac{d\mu}{d\theta}\right)^{-1} = -(\theta-1)^2 \neq 0,$$

except at $\theta = 1$. Moment equation $\bar{X} = \dfrac{\theta}{(\theta - 1)}$

$\Rightarrow$ moment estimator

$$\mu^{-1}(\bar{X}) = \frac{\bar{X}}{(\bar{X}-1)} \xrightarrow{d} N\left(\theta, \frac{\theta}{(\theta-2)(\theta-1)^2}(\theta-1)^4\right)$$

i.e., $$\hat{\theta} = \mu^{-1}(\bar{X}) \xrightarrow{d} N\left(\theta, \frac{\theta(\theta-1)^2}{n(\theta-2)}\right)$$

(ii) Quantile Estimation

Consider median $M(\theta) = 2^{\frac{1}{\theta}}$ of the distribution. Then

$$\frac{dM(\theta)}{d\theta} = -\frac{2^{\frac{1}{\theta}} \log(2)}{\theta^2} \quad \text{and} \quad f(M(\theta)) = \frac{\theta}{2\left(2^{\frac{1}{\theta}}\right)}, p$$

being $\dfrac{1}{2}, \dfrac{p(1-p)}{n[f(M)]^2} = \dfrac{2^{\frac{2}{\theta}}}{n\theta^2}$.

So sample median $X_{\left[\frac{n}{2}+1\right]} \xrightarrow{d} N\left(2^{\frac{2}{\theta}}, \dfrac{2^{\frac{2}{\theta}}}{n\theta^2}\right)$ and for solution of quantile equation, we have asymptotic distribution

$$M^{-1}\left(X_{\left[\frac{n}{2}+1\right]}\right) \xrightarrow{d} N\left(\theta, \frac{\theta^2}{n(\log(2))^2}\right)$$

Example 4.6

If $X_1, X_2, \ldots, X_n$ has multinomial distribution with p.m.f.

$$p(x_1, x_2, \ldots, x_k) = P\{X_1 = x_1, \ldots, X_k = x_k\} = \frac{n!}{x_1!, x_2!, \ldots, x_k!} x_1^{p_1}, \ldots, x_k^{p_k},$$

$$\Sigma x_i = n, \Sigma p_i = 1, 0 < p_i < 1$$

Obtain asymptotic distribution of $x_1, x_2, \ldots, x_k$ (cell frequencies).

Since $(X_i) = np_i$, $\text{var}(X_i) = np_i(1-p_i)$, $\text{cov}(X_i, X_j) = -np_ip_j$, it can be shown that the moment matrix

$$\Sigma = \begin{pmatrix} np_1(1-p_1) & \cdots & -np_1p_k \\ \vdots & \cdots & \vdots \\ -np_kp_1 & \cdots & np_k(1-p_k) \end{pmatrix}$$

is singular and hence we define $X_1, X_2, \ldots, X_{n-1}$ by defining $X_k = n - X_1 - \ldots - X_{k-1}$ non-singular multinomial distribution of $p_k = 1 - p_1 - p_2 -, \ldots, - p_{k-1}$. Using multivariate central limit theorem to $X_1, X_2, \ldots, X_{k-1}$, its asymptotic distribution is $(k-1)$ variate normal distribution with mean vector as $(np_1, np_2, \ldots, np_{k-1})$ and covariance matrix as $\frac{1}{n}V$, where V is

$$V = \begin{pmatrix} np_1(1-p_1) & \cdots & -np_1p_{k-1} \\ \vdots & \cdots & \vdots \\ -np_{k-1}p_1 & \cdots & np_{k-1}(1-p_{k-1}) \end{pmatrix}$$

4.4 SOME METHODS OF ESTIMATION IN LARGE SAMPLES

(A) Method of Moments

If $X_1, X_2, \ldots, X_n$ is a random sample from p.f. $f(x;\theta)$, $\bar{X} = \frac{1}{n}\sum_{i=1}^{n} X_i$ and if $EX_i = \alpha(\theta)$ then $\bar{X} \xrightarrow{p} \alpha(\theta)$, and $\bar{X}$ is consistent estimator of $\alpha(\theta)$. If α^{-1} exists and is continuous then $\alpha^{-1}(\bar{X}) = \alpha^{-1}(\alpha(\theta)) = \theta$. Thus, $\alpha^{-1}(\bar{X})$ is consistent. The method of moments demands to solve $\bar{X} = \alpha(\theta)$ and this requires that α^{-1} must exist, that is, $\frac{d\alpha}{d\theta} \neq 0$ is sufficient condition.

Example 4.7

Let $X_1, X_2, \ldots, X_n$ be a random sample from Pareto distribution with p.d.f.

$$f(x;\theta) = \frac{\theta}{x^{\theta+1}}, \; x \geq 1, \theta > 0$$

If $\theta > 1$, $E(\bar{X}) = E(X_i) = \dfrac{\theta}{\theta - 1} = \alpha(\theta)$, say. Since $\dfrac{d\alpha(\theta)}{d\theta} = -\dfrac{1}{(\theta-1)^2} \neq 0$ except at $\theta = 1$, $\alpha^{-1}(\theta)$ exists, so we solve the equation

$$\bar{X} = \frac{\theta}{\theta - 1} \Rightarrow \hat{\theta} = \frac{\bar{X}}{\bar{X} - 1},$$

if $0 < \alpha < 1$, $E(X_i)$ does exist.

For vector parameter $\underline{\theta} = (\theta_1, \theta_2, \ldots, \theta_k)$ let $(T_1, T_2, \ldots, T_k)$ be such that $E(T_r(X_i)) = \alpha_r(\theta_1, \theta_2, \ldots, \theta_k)$, $r = 1, 2, \ldots, k$. Then by the law of large numbers

$$\frac{1}{n}\sum_{i=1}^{n} T_r(X_i) = \bar{T}_r(x)$$

is consistent for $\alpha_r(\theta_1, \theta_2, \ldots, \theta_k)$ if $\alpha_1, \alpha_2, \ldots, \alpha_k$ are such that

$$\begin{vmatrix} \dfrac{d\alpha_1}{d\theta_1} & \cdots & \dfrac{d\alpha_1}{d\theta_k} \\ \vdots & \cdots & \vdots \\ \dfrac{d\alpha_k}{d\theta_1} & \cdots & \dfrac{d\alpha_k}{d\theta_k} \end{vmatrix} \neq 0$$

Then the method of moments demands to solve

$$\bar{T}_r(x) = \alpha_r(\theta_1, \theta_2, \ldots, \theta_k), r = 1, 2, \ldots, k.$$

Generally, $\bar{T}_r(x)$ is taken up as r^{th} raw moment, $\bar{T}_r(x) = \dfrac{1}{n}\Sigma X_i^r$.

Example 4.8

Let $X_1, X_2, \ldots, X_n$ be a random sample from gamma distribution with p.d.f.

$$f(x; \alpha, p) = \frac{\alpha^p e^{-\alpha x} x^{p-1}}{\Gamma(p)}, x > 0, \alpha > 0, p > 0.$$

Obtain $\hat{\alpha}, \hat{p}$ by method of moments.

Here $\mu_1' = \dfrac{p}{\alpha}$, $\mu_2 = \dfrac{p}{\alpha^2}$. Then

$$\begin{vmatrix} \dfrac{d\mu_1'}{d\alpha} & \dfrac{d\mu_1'}{dp} \\ \dfrac{d\mu_2}{d\alpha} & \dfrac{d\mu_2}{dp} \end{vmatrix} = \frac{p}{\alpha^4} \neq 0$$

So moment estimators of α, p exist and they are obtained by solving the equations,

$$m_1' = \frac{p}{\alpha}, \; m_2 = \frac{p}{\alpha^2},$$

where $$m_1' = \frac{1}{n}\Sigma X_i, \; m_2 = \frac{1}{n}\sum_{i=1}^{n}(X_i - \bar{X})^2,$$

which leads to $$\hat{\alpha} = \frac{m_1'}{m_2}, \; \hat{p} = \frac{m_1'^2}{m_2}.$$

(B) Method of Maximum Likelihood Estimation

The estimators can be obtained by some estimation procedures which do not guarantee whether the obtained estimators possess necessary properties or not. That procedure is advisable to use which produces good estimator having desirable properties of good estimators. One such procedure is maximum likelihood method of estimation which provides consistent, CAN, etc., estimator, known as maximum likelihood estimator (m.l.e.).

Definition 4.4

Let $X = (X_1, X_2, \ldots, X_n)$ be a random sample from family $\mathbb{P} = \{P_\theta(x), \theta \in \Omega\}$ of X, with joint probability function $P_\theta(x)$ denoted by $L(\theta/x) = L$, function of θ for given x. It is called by Fisher as likelihood and $\hat{\theta}(X_1, X_2, \ldots, X_n)$ is called m.l.e. of θ if

$$L(\hat{\theta}/x) \geq L(\theta/x) \; \forall \; \theta \in \Omega.$$

If $L(\theta/x)$ is differential function of θ then $\frac{dL}{d\theta}=0$ or $\frac{d\log L}{d\theta}=0$ provides m.l.e. provided $\frac{d^2L}{d\theta^2}<0.$

The following examples reflect some adverse situations arising in the use of m.l.e.

Example 4.9

By counter example show that m.l.e. is lying outside the range.

Consider a random sample $X_1, X_2, \ldots, X_n$ from Poisson distribution with parameter θ. Then the likelihood L is

$$L=\frac{e^{-n\theta}\theta^{\sum x_i}}{(\prod x_i!)}$$

If $\sum x_i=0$, $L(\theta)=e^{-n\theta}$, which is a decreasing function of θ and hence $L(\theta)$ is maximum at $\hat{\theta}=0$, which is outside the parametric space, Ω.

Example 4.10

By counter example show that the solution of maximum likelihood equation does not provide maximum.

Let $X_i \sim b(1,\theta), i=1,2,\ldots,n$ be i.i.d. random variables. Then

$L=\theta^{\sum x_i}(1-\theta)^{n-\sum x_i}$. If $\sum x_i=0$, $L=(1-\theta)^n$, $\frac{dL}{d\theta}=0$ seems to give m.l.e.

but since $\hat{\theta}=1$, $\frac{d^2L}{d\theta^2}>0$, $\hat{\theta}=1$ is not m.l.e. of θ. As L is decreasing function of θ, maximum occurs at $\theta=0$. Thus $\hat{\theta}=0$ which is not the solution of the likelihood equation. When $\sum x_i=n$, $\hat{\theta}=1$ which is not the m.l.e. of θ, as $\frac{d^2L}{d\theta^2}>0$. When $0<\sum x_i<n$, $\hat{\theta}=\bar{X}$ is the m.l.e. of θ.

Example 4.11

(i) If $X_1, X_2, \ldots, X_n$ is a random sample from

$$f(x;\theta)=\frac{1}{2}e^{-|x-\theta|},\ L=\frac{1}{2^n}e^{-\sum_{i=1}^{n}|x_i-\theta|}$$

L is not differentiable so likelihood equation does not exist. But maximum of L occurs only at $\hat{\theta}=\text{med } X_i$.

(ii) Let $X_1, X_2, \ldots, X_n$ be a random sample from $u(0, \theta)$. Then $L=\theta^{-n}$ but $\frac{dL}{d\theta}=0$ does not give m.l.e. but the value that maximizes L is minimum value of θ and it is max X_i.

Example 4.12

If $X_1, X_2, \ldots, X_n$ is a random sample from $u(\theta, \theta + 1)$ then m.l.e of θ is not unique because $\hat{\theta}$ = min X_i and $\hat{\theta}$ = max X_i – 1. Also m.l.e. is not a function of sufficient statistic (min X_i, max X_i) which shows that m.l.e. is not unique

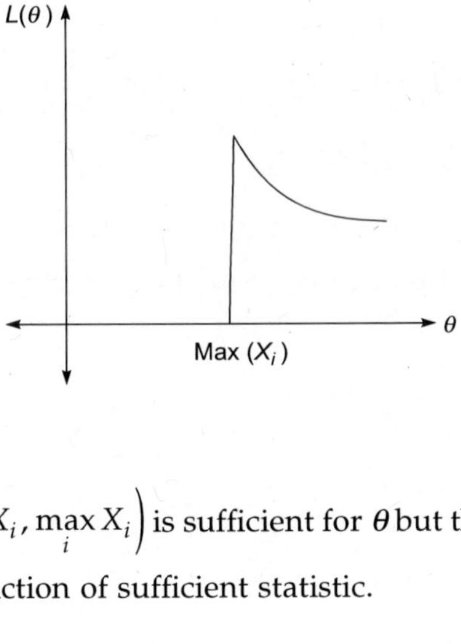

Example 4.13

For $u(\theta, 2\theta)$, $\left(\min_i X_i, \max_i X_i\right)$ is sufficient for θ but the m.l.e. is $\frac{\max X_i}{2}$ which is not a function of sufficient statistic.

Example 4.14

Show that MVB estimator may be m.l.e. but it is not always true.

If $X \sim b(1, \theta)$, $\theta \in \left[\frac{1}{4}, \frac{3}{4}\right]$ then $L=\theta^x(1-\theta)^{1-x}$, $\frac{d\log L}{d\theta}=0$ gives $\hat{\theta}=X$,

its value lies outside $\left[\frac{1}{4}, \frac{3}{4}\right]$. Let us obtain real m.l.e.

$$L = \begin{cases} \theta & \text{when } x = 1 \\ 1-\theta & \text{when } x = 0 \end{cases}$$

$\max L = \frac{3}{4}$ occurs at $\theta = \frac{3}{4}$ when $x = 1$ and $\max L = \frac{3}{4}$ at $\theta = \frac{1}{4}$ when $x = 0$. Thus $\frac{2X+1}{4}$ is m.l.e. of θ.

As

$$\frac{d\log L}{d\theta} = \frac{x}{\theta} - \frac{1-x}{1-\theta}$$

$$= \frac{x-\theta}{\theta(1-\theta)}$$

$$= \frac{1}{\theta(1-\theta)}(x-\theta)$$

$\Rightarrow$ X is MVB estimator of θ.
Thus, MVB estimator is not m.l.e.

Invariance Property of m.l.e.

Theorem 4.7: Let $f(x/\theta)$ be p.f. and $L = L(\theta/x)$ be likelihood of x and g be a continuous function of θ. If $\hat{\theta}$ is m.l.e. of θ then $g(\hat{\theta})$ is m.l.e. of $g(\theta)$.

Proof: Let $\xi = g(\theta)$. Then $L(\theta)$ is likelihood function of X for θ, $f(x/\theta)$ and if $L^*(\xi)$ is likelihood function of X for ξ. Thus, $f^*(x/\xi) = L^*(\xi)$

$$= f(x/g^{-1}(\xi))$$

$$= L(g^{-1}(\xi))$$

$$= L(\theta) \le L(\hat{\theta})$$

$$= L^*(g(\hat{\theta}))$$

because when $g^{-1}(\xi) = \hat{\theta}$ that is $\xi = g(\hat{\theta})$.

Example 4.15

Obtain m.l.e. of $\theta(1 - \theta)$ for Bernouli parameter θ.

As $\hat{\theta} = \bar{X}$ is m.l.e. of θ in $b(1, \theta)$, $\bar{X}(1-\bar{X})$ is m.l.e. of $\theta(1 - \theta)$.

Example 4.16

Obtain m.l.e. of mean and variance of log normal distribution.

If $X_1, X_2, \ldots, X_n$ is a random sample from log normal distribution

$$\log X \sim N(\mu, \sigma^2)$$

then as $\xi = EX = e^{\mu+\frac{\sigma^2}{2}}$, $D^2 = \text{var}(X) = \xi^2(e^{\sigma^2} - 1)$ and as m.l.e. of μ and σ^2 are $\hat{\mu} = \frac{1}{n}\Sigma \log X_i = \bar{Y}$, $\hat{\sigma}^2 = \frac{1}{n}\Sigma(Y_i - \bar{Y})^2$,

where $Y_i = \log X_i$, $i = 1, \ldots, n$, the m.l.e. of ξ and D^2 are $\hat{\xi} = e^{\hat{\mu}+\frac{\hat{\sigma}^2}{2}}$ and $\hat{D}^2 = \hat{\xi}^2(e^{\hat{\sigma}^2} - 1)$, respectively.

Large Sample Properties of m.l.e.

Theorem 4.8: If first two derivatives of L exist and if $\left|\frac{d^3 \log L}{d\theta^3}\right| < H(x)$ and $\int H(x) p_\theta(x) dx = M(\theta) < \infty$ then

(i) m.l.e. is consistent
(ii) m.l.e. is asymptotical normally distributed
(iii) It maximizes the likelihood
(iv) It is unique.

Proof:

(i) Let θ_0 be true value of θ and θ^* be any other value of θ. Then by Jensen's inequality

$$E_{\theta_0}\left[\log \frac{L(X/\theta^*)}{L(X/\theta_0)}\right] \le \log E_{\theta_0}\left[\frac{L(X/\theta^*)}{L(X/\theta_0)}\right]$$

i.e.,
$$E_{\theta_0}\left[\frac{1}{n}\log(L(X/\theta^*))\right] \le E_{\theta_0}\frac{1}{n}\log(L(X/\theta_0)) \qquad \ldots(4.4.1)$$

Now

$$\frac{1}{n}\log(L(X/\theta_0)) = \frac{1}{n}\left[\log \prod_{i=1}^{n}(f(X_i, \theta_0))\right]$$

$$= \frac{1}{n}\Sigma Y_i, \text{ where } Y_i = \log(f(X_i, \theta_0))$$

$$\rightarrow \frac{1}{n}\Sigma E(Y_i), \text{ by W.L.L.N.}$$

Then (4.4.1) gives

$$\log L(X / \theta^*) \leq \log L(X / \theta_0)$$

Therefore, $\lim_{n\to\infty} P\{\log L(X/\theta^*) < \log L(X/\theta_0)\} = 1$

Therefore, $\lim_{n\to\infty} P\{\log L(X/\theta) \leq \log L(X/\theta_0)\} = 1,$

but $\log L(X/\theta) \geq \log L(X/\theta_0) P(\hat{\theta} = \theta_0) = 1.$

(ii) Let $R^2(\theta) = E\left(\frac{d\log L}{d\theta}\right)^2$. Using Taylor's expansion of $\left.\frac{d\log L}{d\theta}\right|_{\theta}$ about θ_0, we have

$$\left.\frac{d\log L}{d\theta}\right|_{\theta=\hat{\theta}} = \left.\frac{d\log L}{d\theta}\right|_{\theta_0} + (\hat{\theta} - \theta_0)\frac{d^2 \log L}{d\theta'^2}, \ \hat{\theta} < \theta' < \theta_0$$

since $\hat{\theta} \xrightarrow{p} \theta_0, \theta' \xrightarrow{p} \theta_0,$

$$0 = \frac{d\log L}{d\theta}$$

$$\Rightarrow \qquad \hat{\theta} - \theta_0 = -\frac{\frac{d\log L}{d\theta_0}}{\frac{d^2 \log L}{d\theta'^2}}$$

$$\Rightarrow \ R(\theta_0)(\hat{\theta} - \theta_0)$$

$$= \frac{\frac{d\log L}{d\theta_0} \Big/ R(\theta_0)}{\frac{d^2 \log L}{d\theta_0^2} \Big/ \{-R^2(\theta_0)\}} \qquad \ldots(4.4.2)$$

By W.L.L.N. denominator of (4.4.2) is 1 as $n \to \infty$.

$$\frac{d\log L}{d\theta_0} = \Sigma Y_i,\ Y_i = \frac{d\log f(x_i;\theta)}{d\theta_0}, i = 1, 2, \ldots, n$$

are independently distributed with mean zero and variance $i(\theta_0)$.

By Lindberglevy form of CLT it follows that $\frac{d\log L}{d\theta_0} \to N(0, R^2(\theta_0))$, therefore

$$\frac{d\log L}{d\theta_0} / R(\theta_0) \to N(0, 1)$$

$$\Rightarrow \qquad R(\theta_0)(\hat{\theta} - \theta_0) \to N(0, 1)$$

$$\hat{\theta} \to N\left(\theta_0, \frac{1}{R(\theta_0)}\right),$$

$\hat{\theta}$ is efficient.

(iii) Let $\hat{\theta}$ be m.l.e. of θ then by mean value theorem

$$[f(b) - f(a) = (b - a)f'(a)]$$

$$\frac{1}{n}\frac{d^2\log L}{d\theta^2}\bigg|_{\theta=\hat{\theta}} - \frac{1}{n}\frac{d^2\log L}{d\theta^2}\bigg|_{\theta=\theta_0} = (\hat{\theta} - \theta_0)\frac{1}{n}\frac{d^3\log L}{d\theta_0^3}$$

As $\hat{\theta} \to \theta_0$, $\frac{1}{n}\frac{d^2\log L}{d\theta^2}\bigg|_{\theta=\hat{\theta}} \to \frac{1}{n}\frac{d^2\log L}{d\theta^2}\bigg|_{\theta=\theta_0} = -I(\theta_0)/n = -i(\theta_0)$

We have

$$P\left\{\frac{1}{n}\frac{d^2\log L}{d\theta^2} < i(\theta_0) + \in\right\} \to 1 \text{ as } n \to \infty$$

$$\frac{1}{n}\frac{d^2\log L}{d\theta_0^2} < 0$$

(iv) Let $\hat{\theta}_1, \hat{\theta}_2$ be two consistent solutions of $\frac{d\log L}{d\theta_0} = 0$ then

$$P\{|\hat{\theta}_1 - \theta_0| < \in\} \to 1, P\{|\hat{\theta}_2 - \theta_0| < \in\} \to 1$$

$$\frac{d\log L}{d\theta_1} = \frac{d\log L}{d\hat{\theta}_2} = 0.$$

Using Rolle's theorem $\left.\frac{d^2\log L}{d\theta^2}\right|_{\theta^*} = 0, \hat{\theta}_1 < \theta^* < \hat{\theta}_2$

As $\theta^* \to \theta_0$,

$$P\left\{\frac{d^2\log L}{d\theta_0^2} = 0\right\} = 1 \text{ which centradicts}$$

$$P\left\{\frac{d^2\log L}{d\theta_0^2} < \in\right\} = 1 \text{ and hence } \hat{\theta}_1 = \hat{\theta}_2.$$

Example 4.17

m.l.e. is not consistent (Neymann and Scott)

Let $\{X_1, X_2, \ldots\}$ be a sequence of i.i.d. random variables having $N(\mu_i, \sigma^2)$ distribution and let $\{Y_1, Y_2, \ldots\}$ be a sequence of i.i.d. random variables having $N(\mu_i, \sigma^2)$ distribution and are independent of $X_1, \ldots, X_n$. Then

$$L = \frac{1}{(2\pi)^n \sigma^{2n}} e^{-\frac{1}{2\sigma^2}\{\Sigma(x_i-\mu_i)^2 + \Sigma(y_i-\mu_i)^2\}}$$

$$\frac{d\log L}{d\mu_i} = 0 \Rightarrow \hat{\mu}_i = \frac{X_i + Y_i}{2}$$

$$\frac{d\log L}{d\sigma^2} = 0 \Rightarrow \hat{\sigma}^2 = \frac{1}{4n}\Sigma(X_i - Y_i)^2$$

$$X_i - Y_i \sim N(0, 2\sigma^2)$$

So $\quad \Sigma(X_i - Y_i)^2 \sim 2\sigma^2\chi_n^2,$

Therefore, $E\hat{\sigma}^2 = \dfrac{1}{4n}2\sigma^2 n \to \dfrac{\sigma^2}{2}$. Thus, $\hat{\sigma}^2$ does not tend to σ^2

Example 4.18

Let $X_1, X_2, \ldots, X_n$ be a random sample from Cauchy distribution with p.d.f.

$$\frac{1}{\pi[1+(x-\theta)^2]}, -\infty < x, \theta < \infty$$

The likelihood equation is

$$\frac{d\log L}{d\theta} = 2\Sigma\frac{(x_i - \theta)}{1+(x_i-\theta)^2} = 0$$

Using $x - \theta = \tan\alpha$ and

$$I(\theta) = nE\left(\frac{d\log L}{d\theta}\right)^2 = \frac{4n}{\pi}\int_{-\infty}^{\infty}\frac{(x-\theta)^2}{[1+(x-\theta)^2]}dx = \frac{n}{2}$$

which is independent of θ. So $\theta_{k+1} = \theta_k + \dfrac{2}{n}\dfrac{d\log(\theta_k, x)}{d\theta}$

(a) Computational Method for m.l.e. (Method of Scoring)

In many cases it is difficult to solve likelihood equations. The general method is to assume a trial value and derive linear equation for small additive correction. The process is repeated till the correction becomes negligible. The method of scoring is one of them in which correction is made till the score $\dfrac{\partial\log L}{\partial\theta}$ vanishes. Let us expand $\dfrac{\partial\log L(\hat{\theta}, X)}{\partial\theta}$ around θ_1 in the equation $\dfrac{\partial\log L}{\partial\theta} = 0$

$$0 = \frac{\partial \log L(\hat{\theta}, X)}{\partial \theta} = \frac{\partial \log L(\theta_1, X)}{\partial \theta} + (\hat{\theta} - \theta_1)\frac{\partial^2 \log L}{\partial \theta^2}(\theta_1 + k(\hat{\theta} - \theta_1); x)$$

for some $k(0 \le k \le 1)$,where, $\hat{\theta}$ is the root of the likelihood equation and θ_1 is an initial solution. If we take $k = 0$, we get approximation for θ as

$$\theta_2 = \theta_1 - \frac{\dfrac{\partial \log L(\theta_1, X)}{\partial \theta}}{\dfrac{\partial^2 \log L(\theta_1, X)}{\partial \theta^2}} = \theta_1 - \frac{\dfrac{\partial \log L(\theta_1, X)}{\partial \theta}}{I(\theta_1)},$$

if we approximate $\left(-\frac{\partial^2 \log L}{\partial \theta^2}\right)$ by its expected value.

The value of θ_2 can be substituted in the above equation to obtain another value θ_3 and so on. Thus, we generate a sequence $\{\theta_k, k \ge 1\}$ which is determined successively by the above correction such as

$$\theta_{k+1} = \theta_k + \frac{\dfrac{\partial \log L(\theta_k, X)}{\partial \theta}}{I(\theta_k)}.$$

(b) Method of Scoring for Grouped Data when Several Parameters are Involved

If the joint density contains q-vector parameters $\underline{\theta} = (\theta_1, \theta_2, \ldots, \theta_q)$ and if $S = (S_1, S_2, \ldots, S_q)$ is q-vector of scores, $S_i = \frac{\partial \log L}{\partial \theta_i}, i = 1, 2, \ldots, q$ and if $q \times q$ information matrix $I(\theta) = (ES_iS_j)$, the method of scoring requires to improve the first trial values $\underline{\theta}_0$, $[I(\theta)]^{-1}$, $S(\theta_0)$ and the corrections are made until the correction is negligible. Sometimes information matrix at each stage is kept constant.

For grouped data with π_i and f_i as probability and frequency of ith class respectively then calculate at t^{th} stage

$$S_t = \sum_{i=1}^{k} \frac{f_i}{\pi_i} \frac{\partial \pi_i}{\partial \theta_t}$$

$$I_{st} = f \sum_{i=1}^{k} \frac{1}{\pi_i} \frac{\partial \pi_i}{\partial \theta_s} \frac{1}{\pi_i} \frac{\partial \pi_i}{\partial \theta_t}$$

Example 4.19

In grouped data if f_i is the observed frequency of ith group with probability $\pi_i(\theta)$, $i = 1, 2, \ldots, k$ then

$$L(\theta, x) = \prod_{i=1}^{k} \pi_i^{f_i}$$

$$\frac{\partial \log L}{\partial \theta} = \sum_{i=1}^{k} \frac{f_i}{\pi_i} \frac{\partial \pi_i}{\partial \theta}$$

and

$$I(\theta) = E\left(\frac{\partial \log L}{\partial \theta}\right)^2$$

$$= \text{var}\left(\frac{\partial \log L}{\partial \theta}\right)$$

$$= \text{var}(l_1 f_1 + l_2 f_2 + \ldots + l_k f_k),$$

where $l_i = \dfrac{\partial \pi_i}{\partial \theta} \dfrac{1}{\pi_i}$

$$= \sum_{i=1}^{k} l_i^2 \,\text{var}(f_i) + 2 \sum_{i \neq j}\sum l_i l_j \,\text{cov}(f_i, f_j)$$

$$= \sum_{i=1}^{k} l_i^2 f \pi_i (1 - \pi_i) - 2 \sum_{i \neq j}\sum f l_i l_j \pi_i \pi_j, \text{ as } \text{var}(f_i) = f \pi_i (1 - \pi_i),$$

$$\text{cov}(f_i, f_j) = -f \pi_i \pi_j$$

$$= f\left[\sum_{i=1}^{k} l_i^2 \pi_i - \left(\sum_{i=1}^{k} l_i^2 \pi_i^2 + 2 \sum_{i \neq j}\sum l_i l_j \pi_i \pi_j\right)\right], \; f = \sum_{i=1}^{k} f_i$$

$$= f\left[\sum_{i=1}^{k} l_i^2 \pi_i - (\Sigma l_i \pi_i)^2\right]$$

$$= f\sum_{i=1}^{k} l_i^2 \pi_i \text{, using } \Sigma l_i \pi_i = \Sigma \frac{\partial \pi_i}{\partial \theta} = \frac{\theta}{\partial \theta}\Sigma \pi_i = \frac{\partial 1}{\partial \theta} = 0$$

$$= f\Sigma \frac{1}{\pi_i}\left(\frac{\partial \pi_i}{\partial \theta}\right)^2.$$

Hence,

$$\theta_{k+1} = \theta_k + \frac{\Sigma \frac{f_i}{\pi_i}\frac{\partial \pi_i}{\partial \theta_k}}{f\Sigma \frac{1}{\pi_i}\left(\frac{\partial \pi_i}{\partial \theta_k}\right)^2}$$

Illustration 4.1

Let $\pi_1 = \frac{2+\theta}{4}$, $\pi_2 = \pi_3 = \frac{1-\theta}{4}$, $\pi_4 = \frac{\theta}{4}$. Then $\frac{\partial \pi_1}{\partial \theta} = \frac{\partial \pi_4}{\partial \theta} = \frac{1}{4}$,

$$\frac{\partial \pi_2}{\partial \theta} = \frac{\partial \pi_3}{\partial \theta} = -\frac{1}{4}$$

so that

$$\Sigma \frac{f_i}{\pi_i}\frac{\partial \pi_i}{\partial \theta_k} = \frac{f_1}{2+\theta} - \frac{f_2+f_3}{1-\theta} + \frac{f_4}{\theta} = -\frac{f\theta^2 + \theta[2(f_2+f_3)+f_4-f_1]-2f_4}{\theta(1-\theta)(2+\theta)}$$

and

$$f\Sigma \frac{1}{\pi_i}\left(\frac{\partial \pi_i}{\partial \theta}\right)^2 = \frac{4f(2+2\theta-\theta^2)}{\theta(1-\theta)(2+\theta)}$$

Hence,

$$\theta_{k+1} = \theta_k - \frac{f\theta^2 + \theta[2f_2+2f_3+f_4-f_1]-2f_4}{4f(2+2\theta-\theta^2)}$$

Example 4.20

For truncated Poisson distribution with p.m.f.

$$P\{X=r\}=\frac{e^{-\lambda}\lambda^r}{r!(1-e^{-\lambda})}, x = 1, 2, \ldots$$

$$=\frac{\lambda^r}{r!(e^{\lambda}-1)}$$

and for grouped data (r^{th} class contains n_r observations), obtain m.l.e. by scoring method.

$$L=\prod_{r=1}^{k}\left[\frac{\lambda^r}{r!(e^{\lambda}-1)}\right]^{n_r}$$

$\Rightarrow$

$$\log L=\text{constant}-\Sigma n_r\log(e^{\lambda}-1)+\Sigma rn_r\log\lambda$$

Therefore,

$$\frac{\partial\log L}{\partial\lambda}=-\frac{ne^{\lambda}}{e^{\lambda}-1}+\frac{\Sigma rn_r}{\lambda}$$

and

$$\frac{\partial^2\log L}{\partial\lambda^2}=-\frac{\Sigma rn_r}{\lambda^2}+\frac{ne^{\lambda}}{(e^{\lambda}-1)^2}, n=\sum_{r=1}^{k}rn_r$$

Therefore,

$$I(\lambda)=-E\frac{\partial^2\log L}{\partial\lambda^2}=\frac{ne^{\lambda}}{e^{\lambda}-1}\left[\frac{1}{\lambda}-\frac{1}{e^{\lambda}-1}\right]$$

So that correction of initial value of λ is

$$\frac{\partial\log L}{\partial\lambda}/I(\lambda)=\frac{T(e^{\lambda}-1)}{n\lambda e^{\lambda}\left[\frac{1}{\lambda}-\frac{1}{e^{\lambda}-1}\right]}-\frac{1}{\left[\frac{1}{\lambda}-\frac{1}{e^{\lambda}-1}\right]}, \text{ where } T=\sum_{r=1}^{k}rn_r$$

Illustration 4.2

For four blood group classes let

Class	Frequency	Probability π_i	$\frac{\partial \pi_i}{\partial p}$	$\frac{\partial \pi_i}{\partial q}$
O	176	r^2	$-2r$	$-2r$
A	182	$p^2 + 2pr$	$2r$	$-2p$
B	80	$q^2 + 2qr$	$-2q$	$2r$
AB	17	$2pq$	$2q$	$2p$

where $p + q + r = 1$.

Taking initial values of p, q, r as

$$r = \sqrt{\frac{O}{f}}$$

$$p = 1 - \sqrt{\frac{O+B}{f}}$$

$$q = 1 - \sqrt{\frac{O+A}{f}}$$

where $f = \sum_{i=1}^{4} f_i$, scores and information matrix at different stages can be obtained by using

$$\sum_{i=1}^{4} \frac{f_i}{\pi_i} \frac{\partial f_i}{\partial p}, \sum_{i=1}^{4} \frac{f_i}{\pi_i} \frac{\partial f_i}{\partial q}$$

$$I_{pp} = f \sum_{i=1}^{4} \left(\frac{1}{\pi_i} \frac{\partial f_i}{\partial p} \right)^2$$

$$I_{pq} = f \sum_{i=1}^{4} \left(\frac{1}{\pi_i} \frac{\partial f_i}{\partial p} \right) \left(\frac{1}{\pi_i} \frac{\partial f_i}{\partial q} \right)$$

Example 4.21

If $X_1, X_2, \ldots, X_n$ are i.i.d. random variables and if $Y_i = \log X_i \sim N(\mu, \sigma^2)$, $i = 1, 2, \ldots, n$, X_i follows log normal distribution. Obtain asymptotic distribution of mean and variance of Y.

As
$$Y_i = \log X_i \sim N(\mu, \sigma^2)$$

$$\bar{Y}_n = \frac{1}{n}\Sigma Y_i \sim N\left(\mu, \frac{\sigma^2}{n}\right)$$

and

$$Q_n = \frac{1}{n}\Sigma(Y_i - \bar{Y})^2 \sim \chi^2_{n-1}$$

$(\bar{Y}_n, Q_n)$ is m.l.e. of (μ, σ^2). By the property of asymptotic normality of m.l.e. we have

$$\begin{pmatrix} \bar{Y}_n - \mu \\ Q_n - \sigma^2 \end{pmatrix} \xrightarrow{d} N\left(\begin{bmatrix} 0 \\ 0 \end{bmatrix}, \begin{bmatrix} \frac{\sigma^2}{n} & 0 \\ 0 & \frac{2\sigma^4}{n} \end{bmatrix}\right)$$

which shows that $\bar{Y}_n \xrightarrow{d} N\left(0, \frac{\sigma^2}{n}\right)$ and $Q_n \xrightarrow{d} N\left(\sigma^2, \frac{2\sigma^4}{n}\right)$ are independently distributed. Since X follows log normal,

$$EX = \xi = e^{\mu + \frac{\sigma^2}{2}}$$

and we want to obtain asymptotic distribution of m.l.e. of $\xi = g(\mu, \sigma^2)$, say.

The m.l.e. $\hat{\xi}$ of ξ is

$$\hat{\xi} = e^{\bar{Y}_n + \frac{Q}{2}}$$

and hence $\hat{\xi}$ is asymptotically normal with mean

$$\xi = e^{\mu + \frac{\sigma^2}{2}}$$

and variance-covariance matrix as

$$\left(\frac{\partial g}{\partial \mu}, \frac{\partial g}{\partial \sigma^2}\right) I(\theta_0) \begin{pmatrix} \frac{\partial g}{\partial \mu} \\ \frac{\partial g}{\partial \sigma^2} \end{pmatrix}$$

where $\frac{\partial g}{\partial \mu} = \xi$, $\frac{\partial g}{\partial \sigma^2} = \frac{1}{2}\xi$, so that asymptotic variance of $\hat{\xi}$ is

$$\text{Var}(\xi) = \left(1, \frac{1}{2}\right) \begin{pmatrix} \frac{\sigma^2}{n} & 0 \\ 0 & \frac{2\sigma^4}{n} \end{pmatrix} \begin{pmatrix} 1 \\ \frac{1}{2} \end{pmatrix}$$

$$= \frac{\xi^2 \sigma^2}{n}\left(1 + \frac{\sigma^2}{2}\right)$$

Similarly, m.l.e. of $\text{var}(Y) = \xi^2(e^{\sigma^2} - 1)$ being $\hat{\xi}^2(e^{\hat{\sigma}^2} - 1)$ one can obtain asymptotic variance matrix for it.

Example 4.22

If $(X_i, Y_i), i = 1, 2, \ldots, n$ is a random sample from bivariate normal distribution with parameters $(0, 0, \sigma^2, \sigma^2, \rho)$. Then obtain m.l.e. of σ^2, ρ and find its asymptotic distribution.

The p.d.f. of (X, Y) is

$$f(x, y) = \frac{1}{2\pi\sigma^2\sqrt{1-\rho^2}} e^{-\frac{1}{2(1-\rho^2)}\left(\frac{x^2+y^2}{\sigma^2} - \frac{2\rho xy}{\sigma^2}\right)}$$

$$= \frac{1}{2\pi} e^{\phi_1 T_1(x, y) + \phi_2 T_2(x, y) + \xi(\phi_1, \phi_2)}$$

where $\qquad \phi_1 = \frac{1}{2\sigma^2(1-\rho^2)}, \; T_1(x, y) = x^2 + y^2$

$$\phi_2 = \frac{\rho}{\sigma^2(1-\rho^2)}, \; T_2(x, y) = xy$$

$$\xi(\phi_1\ \phi_2) = e^{\left\{\log \frac{1}{[\sigma^4(1-\rho^2)]^{\frac{1}{2}}}\right\}}$$

$$= e^{-\frac{1}{2}\log[\sigma^4(1-\rho^2)]}$$

$$= e^{-\frac{1}{2}\log[4\phi_1^2 - \phi_2^2]}$$

Then $\dfrac{\partial \log f}{\partial \phi_1} = 0$ gives $T_1(x, y) + \dfrac{\partial \xi}{\partial \phi_1} = 0$

$$\Rightarrow \qquad T_1(x, y) + \frac{4\phi_1}{(4\phi_1^2 - \phi_2^2)} = 0$$

$$\Rightarrow \qquad x^2 + y^2 = -\frac{4\phi_1}{(4\phi_1^2 - \phi_2^2)}$$

and $\dfrac{\partial \log f}{\partial \phi_2} = 0$ gives $T_2(x, y) - \dfrac{\phi_2}{(4\phi_1^2 - \phi_2^2)} = 0$

$$\Rightarrow \qquad xy = \frac{\phi_2}{(4\phi_1^2 - \phi_2^2)}$$

If $\hat{\phi}_1$ and $\hat{\phi}_2$ are m.l.e. of ϕ_1 and ϕ_2, respectively,

$$\frac{1}{n}\sum_{i=1}^{n}(X_i^2 + Y_i^2) = -\frac{4\hat{\phi}_1}{4\hat{\phi}_1^2 - \hat{\phi}_2^2}$$

and

$$\frac{1}{n}\Sigma X_i Y_i = \frac{\hat{\phi}_2}{4\hat{\phi}_1^2 - \hat{\phi}_2^2}$$

As

$$\sigma^2 = -\frac{2\phi_1}{4\phi_1^2 - \phi_2^2},\ \rho = -\frac{\phi_2}{2\phi}$$

by invariance principle of m.l.e., we get m.l.e. of σ^2, ρ as

$$\hat{\sigma}^2 = \frac{1}{2n}\sum_{i=1}^{n}(X_i^2 + Y_i^2)$$

$$\hat{\rho} = \frac{2\Sigma X_i Y_i}{\sum_{i=1}^{n}(X_i^2 + Y_i^2)}$$

Now let us obtain asymptotic distribution of $\hat{\sigma}^2$ and $\hat{\rho}$. We have

$$\frac{\partial \log f}{\partial \sigma^2} = T_1(x, y)\frac{\partial \phi_1}{\partial \sigma^2} + T_2(x, y)\frac{\partial \phi_2}{\partial \sigma^2} + \frac{\partial \xi}{\partial \sigma^2}$$

$$= \frac{T_1(x, y) - 2\rho T_2(x, y)}{2\sigma^4(1-\rho^2)} - \frac{1}{\sigma^2}$$

and

$$\frac{\partial \log f}{\partial \rho} = \frac{(1+\rho^2)T_2(x, y) - \rho T_1(x, y)}{\sigma^2(1-\rho^2)^2} + \frac{\rho}{1-\rho^2}$$

Since the marginal distributions of X and Y are $N(0, \sigma^2)$ and since the conditional distribution of X given Y is $N(\rho Y, \sigma^2(1-\rho^2))$ and that of Y given X is

$$N(\rho X, \sigma^2(1-\rho^2)),$$

we obtain

$$EX^4 = EY^4 = 3\sigma^4$$

$$\begin{aligned} EX^2Y^2 &= EX^2E(Y^2/X^2) \\ &= EX^2[\text{var}(Y/X^2) + [E(Y/X^2)]^2] \\ &= EX^2[\sigma^2(1-\rho^2) + \rho^2X^2] \\ &= \sigma^2(1-\rho^2)EX^2 + \rho^2EX^4 \\ &= \sigma^4(1-\rho^2) + 3\rho^2\sigma^4 \\ &= \sigma^4(1+2\rho^2) \end{aligned}$$

$$E(X^3Y) = EX^3E(Y/X) = \rho EX^4 = 3\rho\sigma^4$$

and

$$E(T_1 - 2\rho T_2)^2 = 8\sigma^4(1-\rho^2)^2,$$

$$E(T_1 - 2\rho T_2) = 2\sigma^2(1-\rho^2).$$

Hence, $E\left(\dfrac{\partial \log f}{\partial \sigma^2}\right)^2 = \dfrac{1}{\sigma^4}.$

Moreover $E[(1+\rho^2)T_2 - \rho T_1] = -\rho\sigma^2(1-\rho^2)$

and

$$E[(1+\rho^2)T_2 - \rho T_1]^2 = \sigma^4(1-\rho^2)(1+\rho^2-2\rho^4)$$

$$E[((1+\rho^2)T_2 - \rho T_1)(T_1 - 2\rho T_2)] = \rho\sigma^4(2\rho^4 - 11\rho^2 + 3)$$

Therefore,

$$E\left(\frac{\partial \log f}{\partial \sigma^2}\frac{\partial \log f}{\partial \rho}\right) = -\frac{\rho}{\sigma^2(1-\rho^2)} \quad \text{and} \quad E\left(\frac{\partial \log f}{\partial \rho}\right)^2 = \frac{1+\rho^2}{(1-\rho^2)^2}$$

and Fisher information matrix is

$$I = \begin{pmatrix} \dfrac{1}{\sigma^4} & -\dfrac{\rho}{\sigma^2(1-\rho^2)} \\ -\dfrac{\rho}{\sigma^2(1-\rho^2)} & \dfrac{1+\rho^2}{(1-\rho^2)^2} \end{pmatrix}$$

So that asymptotic distribution of $(\hat{\sigma}^2, \hat{\rho})$ is bivariate normal with mean vector $\underline{0}$ and variance matrix as nI^{-1}.

(c) Minimum Chi-Square and Other Methods of Estimation

These methods are applicable when the observations are continuous measurements but are grouped in suitable class intervals or observations are themselves frequencies of finite number of mutually exclusive events. In either case we have a set of observed frequencies

$n_1, n_2, \ldots, n_k$ in k cells with hypothetical probability $p_1, p_2, \ldots, p_k$ as functions of r unknown parameters. We define different measures of discrepancies between observed frequencies $n_1, n_2, \ldots, n_k$ and the hypothetical expectations $n\pi_1(\theta), n\pi_2(\theta), \ldots, n\pi_k(\theta)$ where, $n = \Sigma n_i$. The estimators are obtained by minimizing such a measure w.r.t. $\theta_j, j = 1, 2, \ldots, r$. Some of the measures are as follows:

(i) Minimum chi-square method:

$$\chi^2 = \sum_{i=1}^{k} \frac{(n_i - n\pi_i)^2}{n\pi_i}$$

Here,

$$\frac{\partial \chi^2}{\partial \theta_j} = \sum_{i=1}^{k} \frac{-2n(n_i - n\pi_i)n\pi_i \dfrac{\partial \pi_i}{\partial \theta_j} - (n_i - n\pi_i)^2 n \dfrac{\partial \pi_i}{\partial \theta_j}}{(n\pi_i)^2} = 0, j = 1, 2, \ldots, n$$

gives minimum χ^2 estimators of θ_j by solving the above equations.

(ii) Modified minimum χ^2 method:

Modified χ^2 is defined as

$$\text{mod } \chi^2 = \sum_{i=1}^{k} \frac{(n_i - n\pi_i)^2}{n_i}$$

replacing n_i by 1 if it is zero. For minimizing we differentiate w.r.t. θ_j and setting the derivative to zero we have

$$\frac{\partial \text{mod } \chi^2}{\partial \theta_j} = \sum_{i=1}^{k} \frac{-2(n_i - n\pi_i)^n \dfrac{\partial \pi_i}{\partial \theta_j}}{n_i} = 0$$

gives

$$\sum \frac{\pi_i}{n_i} \frac{\partial \pi_i}{\partial \theta_j} = 0, j = 1, 2, \ldots, r$$

(iii) Helinger distance:

If $Y = \cos^{-1} \sum \sqrt{\frac{n_i}{n}} \pi_i$ then solving the equation

$$\frac{\partial Y}{\partial \theta_j} = \frac{1}{2\sqrt{1 - Y^2}} \sum \sqrt{\frac{n_i}{n\pi_i}} \frac{\partial \pi_i}{\partial \theta_j} = 0, j = 1, 2, \ldots, r$$

(iv) Kullback-Leibler separator (K.L.S.):

K.L.S. $= \Sigma \pi_i \log \dfrac{\pi_i}{n_i/n}$ then solving the equation

$$\frac{\partial K.L.S.}{\partial \theta_j} = \sum \frac{\partial \pi_i}{\partial \theta_j} \log \frac{n\pi_i}{n_i} + \sum \frac{n_i}{n\pi_i} \frac{n}{n_i} \frac{\partial \pi_i}{\partial \theta_j} = 0$$

gives

$$\sum_{i=1}^{k} \frac{1}{\pi_i} \frac{\partial \pi_i}{\partial \theta_j} = \sum_{i=1}^{k} \frac{\partial \pi_i}{\partial \theta_j} \log \frac{n_i}{n\pi_i}, \; j = 1, 2, \ldots, r$$

(v) Haldane's discrepancy:

It is defined as

$$D_k = \frac{(n+k)!}{n!} \sum_{i=1}^{k} \frac{n_i! \pi_i^{k+1}}{(n_i + k)!}, \; k \neq 1$$

and $D_{-1} = -\dfrac{1}{n} \Sigma n_i \log \pi_i$ solving the equation $\dfrac{\partial D_k}{\partial \theta_j} = 0$, we have

$$\frac{\partial D_k}{\partial \theta_j} = \frac{(n+k)!}{n!} \sum_{i=1}^{k} \frac{n_i!(k+1)\pi_i^{k+1}}{(n_i + k)!} \frac{\partial \pi_i}{\partial \theta_j} = 0$$

and

$$\frac{\partial D_{-1}}{\partial \theta_j} = -\frac{1}{n} \sum \frac{n_i}{\pi_i} \frac{\partial \pi_i}{\partial \theta_j} = 0$$

which is minimum χ^2 estimator.

All these methods provide reasonably good estimators in the sense that they are consistent and efficient.

Example 4.23

For Poisson distribution with mean θ obtain minimum χ^2 estimator and modified minimum χ^2 estimator of θ.

We have to solve

$$\sum_{i=1}^{k} \frac{n_i}{\frac{e^{-\theta}\theta^{n_i}}{n_i!}} \frac{\partial}{\partial\theta} \frac{e^{-\theta}\theta^{n_i}}{n_i!} = 0$$

$$\Rightarrow \qquad \Sigma n_i \left(\frac{1}{\theta} - 1\right) = 0$$

$$\Rightarrow \qquad \hat{\theta} = \frac{\sum_{i=1}^{k} i n_i}{\Sigma n_i}$$

= mean of the sample

Also modified minimum χ^2 estimator is given by solving the following equation for θ,

$$\sum_{i=1}^{k} \frac{ie^{-2\theta} e^{2i-1}}{n_i (i!)^2} = \sum_{i=1}^{k} \frac{e^{-2\theta} e^{2i}}{n_i (i!)^2}$$

PROBLEMS

4.1 If $X_1, X_2, \ldots, X_n$ is a random sample from p.f. $f(x;\theta)$ then obtain moment estimator $\hat{\theta}$ for each of the following $f(x;\theta)$ and examine whether it is consistent.

(i) $N(\theta, 1)$

(ii) $N(0, \theta^2)$

(iii) $N(\theta, \theta^2)$

(iv) $B(n, \theta)$

(v) $u(0, \theta)$

(vi) $Be(\theta, \theta)$

(vii) Geometric distribution: $f(x;\theta) = \theta(1-\theta)^{x-1}, x = 1, 2, \ldots$

(viii) Discrete uniform: $P(X = x) = \frac{1}{\theta}, x = 1, 2, \ldots, \theta$

(ix) Poisson: $P(\theta)$

(x) Truncated poisson: $P(X=x)=\dfrac{e^{-\theta}\theta^{x}}{x!(1-e^{-\theta})}, x=1,2,\ldots$

(xi) Truncated binomial for $n=3$:

$$P(X=x)=\frac{\binom{3}{x}\theta^{x}(1-\theta)^{3-x}}{1-(1-\theta)^{3}}, x=1,2,3.$$

(xii) $f(x;\theta)=\dfrac{\theta}{x^{2}}, \theta<x<\infty.$

4.2 If $X_1, X_2, \ldots, X_n$ is a random sample from beta distribution with p.d.f.

$$\frac{1}{B(p,q)}x^{p-1}(1-x)^{q-1}, 0\langle x\langle 1, p\rangle 0, q\rangle 0$$

Obtain moment estimators of p, q. Are these consistent?

4.3 If $X_1, X_2, \ldots, X_n$ is a random sample from p.d.f.

$$f(x;\alpha,\lambda)=\frac{\left(\frac{\lambda}{\alpha}\right)^{\lambda}}{\Gamma(\lambda)}e^{-\left(\frac{\lambda}{\alpha}\right)x}x^{\lambda-1}, x>0$$

then obtain moment estimators of α, λ.

4.4 If $X_1, X_2, \ldots, X_n$ is a random sample from Poisson distribution with parameter θ then examine whether UMVU estimators of

(i) θ

(ii) $e^{-\theta}$

(iii) $\theta e^{-\theta}$ are CAN estimators. Find their asymptotic variances.

4.5 If $X_1, X_2, \ldots, X_n$ is a random sample from p.d.f.

$$f(x;\theta)=\frac{1}{\theta}e^{-\frac{x}{\theta}}, x>0, \theta>0$$

then obtain CAN estimator of reliability $P(X>t)$ and find its asymptotic variance.

4.6 Based on a random sample $X_1, X_2, \ldots, X_n$ from p.d.f.

$$f(x;\theta) = \theta x^{\theta-1}, 0 < x \langle 1, \theta \rangle 0$$

show that moment estimator of θ is CAN. Also show that median estimator of θ is CAN.

4.7 If $X_i = \theta + \xi_i, i = 1, 2, \ldots, n$ are i.i.d. with distribution of ξ_i as $f(\xi) = |\xi|, -1 < \xi < 1$. Obtain CAN $\frac{1}{4}$ and $\frac{3}{4}$ quantiles and show that the CAN does not exist at the median.

4.8 If $X_1, X_2, \ldots, X_n$ is a random sample from Weibull distribution with p.d.f.

$$f(x;\theta) = \theta x^{\theta-1} e^{-x^{\theta}}, x > 0, \theta > 0$$

then show that sample p^{th} quantile is CAN.

4.9 If $f(x;\theta) = \frac{1}{2\theta} e^{-\frac{|x|}{\theta}}, -\infty < x \langle \infty, \theta \rangle 0$. Obtain moment estimator and UMVU estimator of θ and show that even though $\bar{X}$ is not consistent but it is CAN.

4.10 If $X_1, X_2, \ldots, X_n$ is i.i.d. from $N(\mu, \sigma^2)$ then obtain CAN estimator of

(i) $\frac{\Sigma X_i}{n}$ and $\frac{\Sigma X_i^2}{n}$

(ii) $\hat{\mu} = \bar{X}$ and $\hat{\sigma}^2 = \frac{1}{n} \Sigma X_i^2 - \bar{X}^2$

4.11 If $f(x, y; \lambda, p) = \binom{x}{y} p^y (1-p)^{x-y} e^{-\lambda} \frac{\lambda^x}{x!}, 0 < p \langle 1, \lambda \rangle 0,\ y = 0, 1, \ldots, x;$ $x = 0, 1, \ldots$

show that $\left(\bar{X}, \frac{\bar{Y}}{\bar{X}}\right)$ is CAN estimator of (λ, p). Obtain its asymptotic covariance matrix.

4.12 For a random sample $X_1, X_2, \ldots, X_n$ from $N(\theta, \theta^2), 0 < \theta < \infty$. Obtain m.l.e. of θ. Is this estimator consistent?

4.13 Let $X_1, X_2, \ldots, X_n$ be a random sample from $N(\mu, \sigma^2)$ and let $Y_1, Y_2, \ldots, Y_m$ be independent random sample from $N(\upsilon, \sigma^2)$. Obtain m.l.e. of

$$\Delta = 2\phi\left(-\frac{\mu - \nu}{2\sigma}\right), \text{ where, } \phi(x) \text{ is c.d.f. of } N(0,\ 1) \text{ distribution.}$$

4.14 If $X_1, X_2, \ldots, X_n$ is a random sample from p.d.f.

$$f(x; \alpha, \lambda) = \left(\frac{\lambda}{\alpha}\right)^{\lambda} \frac{1}{\Gamma(\lambda)} e^{-\frac{\lambda x}{\alpha}} x^{\lambda - 1}, 0 \langle x \langle \infty, \alpha \rangle 0, \lambda \rangle 0$$

Obtain m.l.e. of (α, λ) assuming that $\frac{d}{dx}\log\Gamma(x) \simeq \log x - \frac{1}{2x}$ for large value of x.

4.15 If $X_1, X_2, \ldots, X_n$ is a random sample from Poisson distribution with parameter λ obtain m.l.e. of $P(X=0)+P(X=1)$. Show that this estimator is consistent.

4.16 If $X_1 \sim b(n_1, p)$, $X_2 \sim b(n_2, p)$, $\ldots$, $X_m \sim b(n_m, p)$ are independent obtain m.l.e. of p.

4.17 If $X_1, X_2, \ldots, X_n$ is a random sample from $N(\mu, \sigma^2)$ and if $Y_1, Y_2, \ldots, Y_n$ is an independent random sample from $N(\mu, \lambda\sigma^2), \lambda > 0$. Obtain m.l.e. of λ when

(i) μ and σ^2 are known

(ii) μ and σ^2 are unknown.

4.18 If $X_1, X_2, \ldots, X_n$ is a random sample from $N(\mu, \sigma^2)$ where, both μ and σ^2 are unknown, obtain m.l.e. of $P(X_1 < c)$, where, c is known constant. Show that it is consistent.

4.19 If $X_1, X_2, \ldots, X_n$ is a random sample from geometric distribution $P(X=x) = (1-p)^{x-1}p$ for x = 1, 2, ... find m.l.e. of p.

4.20 Let $X_1, X_2, \ldots, X_n$ be a random sample from beta distribution with parameters (λ, β), here λ is known and α is unknown. Then obtain m.l.e. of α and $\frac{\alpha}{\alpha+1}$ when

(i) $\lambda = 1g$

(ii) $\lambda = 2$.

4.21 Let $X_1, X_2, \ldots, X_n$ be a random sample from Laplace distribution with p.d.f.

$$f(x;\theta) = \frac{1}{2} e^{-|x-\theta|}.$$

Obtain m.l.e. of θ.

4.22 If $X_1, X_2, \ldots, X_n$ is a random sample from $N(\mu, \sigma_1^2)$ and if $Y_1, Y_2, \ldots, Y_m$ is an independent random sample from $N(\mu, \sigma_2^2)$ then obtain m.l.e. of $\mu, \sigma_1^2, \sigma_2^2$ and find the variances of these estimators.

4.23 If $X_1, X_2, \ldots, X_n$ is a random sample from

$$f(x;\theta) = e^{-(x-\theta)}, \theta < x < \infty$$

obtain m.l.e. of θ and if random sample is from

$$f(x;\theta_1,\theta_2) = \frac{1}{\theta_1} e^{-\frac{x-\theta_2}{\theta_1}}, x \geq \theta_2$$

obtain m.l.e. of θ_1, θ_2 and obtain $E(\hat{\theta}_1), E(\hat{\theta}_2)$.

4.24 If $X_1, X_2, \ldots, X_n$ is a random sample from uniform distribution on an interval $(\theta_1 - \theta_2, \theta_1 + \theta_2)$ then obtain m.l.e. of:

(i) θ_1 when θ_2 is known

(ii) θ_1 and θ_2 when both are unknown.

4.25 If $X_1, X_2, \ldots, X_n$ is a random sample from uniform distribution, $u(\theta_1, \theta_2)$ on an interval (θ_1, θ_2) then obtain m.l.e. of:

(i) θ_1 when θ_2 is known

(ii) θ_1 and θ_2 when both are unknown.

4.26 For hyper geometric distribution

$$P(X = x) = \frac{\binom{a}{x}\binom{N-a}{n-x}}{\binom{N}{n}}$$

Find m.l.e. of N when, a, x, n are known.

4.27 If $X_1, X_2, \ldots, X_n$ is a random sample from truncated Laplace distribution with p.d.f.

$$f(x;\theta) = \frac{e^{-|x|}}{2(1-e^{-\theta})}, -\theta < x\langle\theta, \theta\rangle 0$$

Obtain m.l.e. of θ. Is this consistent?

4.28 If $(X_i, Y_i), i = 1, 2, \ldots, n$ is a random sample from bivariate normal distribution with parameters $(\mu_1, \mu_2, \sigma_1^2, \sigma_2^2, \rho)$ obtain m.l.e. of ρ when

(i) $\mu_1, \mu_2, \sigma_1^2, \sigma_2^2$ are known

(ii) $\mu_1, \mu_2, \sigma_1^2, \sigma_2^2$ are unknown.

4.29 If $X_1, X_2, \ldots, X_n$ is a random sample from

$$f(x;\theta) = \frac{\theta}{x^2}, \theta < x\langle\infty, \theta\rangle 0$$

find m.l.e. of

(i) θ

(ii) $P(X < c)$, where, c is known.

4.30 Let $X_1, X_2, \ldots, X_n$ be a random sample from p.d.f.

$$f(x;\alpha,\beta) = \frac{\beta e^{-(\alpha+\beta x)}}{[1+e^{-(\alpha+\beta x)}]^2}, -\infty < x < \infty$$

Obtain m.l.e. of α when β is known.

4.31 If n_1, n_2, n_3 are observed frequencies of three types of twins in a sample of $n = n_1 + n_2 + n_3$ with probability $(\pi + \alpha\phi)$, $2\pi\phi(1 - \alpha)$, $\phi(2\pi + \phi)$ where, $\phi = 1 - \pi$, obtain m.l.e. of α, π.

4.32 If $X_1, X_2, \ldots, X_n$ is a random sample from

$$b(1, p), p \in \left(\frac{1}{3}, \frac{2}{3}\right)$$

then obtain m.l.e. of p and find its asymptotic distribution.

4.33 If n_1, n_2, n_3, n_4 are observed frequencies with probability

$\frac{p}{2}, \frac{q}{2}, \frac{p^2}{2} + pq, \frac{q^2}{2}, p + q = 1$, obtain m.l.e. of $p, \frac{q^2}{2}$.

5

CHAPTER

Decision Theory

5.1 INTRODUCTION

Let $X_1, X_2, \ldots, X_n$ be a random sample from the family of distributions $\mathbb{P} = \{p_\theta, \theta \in \Omega\}$. Then one takes the decision about the parameter θ on the basis of the value $x = (x_1, x_2, \ldots, x_n)$ of $X = (X_1, X_2, \ldots, X_n)$. Let D be a class of decision functions that maps R_n into D and $L(\theta, d)$, $d \in D$ be a function defined on $\Omega \times D$ then $L(\theta, d)$ is called loss function. The average loss function, $L(\theta, d) = EL(\theta, d(X))$, is known as the risk function, $R(\theta, d)$ where average is taken over X. If available decisions are qualitative in nature preference of ordering is done through utility function based on this ordering which we shall discuss later on.

Given the loss function one has to choose $d \in D$ such that $R(\theta, d)$ is minimum for all $\theta \in \Omega$. But unfortunately this is not possible, e.g., if θ is the probability that an item is defective in a lot and if lot is accepted (A) when $\theta \le \theta_0$ and rejected (R) when $\theta > \theta_0$ with loss function

$$L(\theta; A) = \begin{cases} 0 & \text{if } \theta \le \theta_0 \\ 1 & \text{if } \theta > \theta_0 \end{cases}$$

and

$$L(\theta; R) = \begin{cases} 0 & \text{if } \theta > \theta_0 \\ 2 & \text{if } \theta \le \theta_0 \end{cases}$$

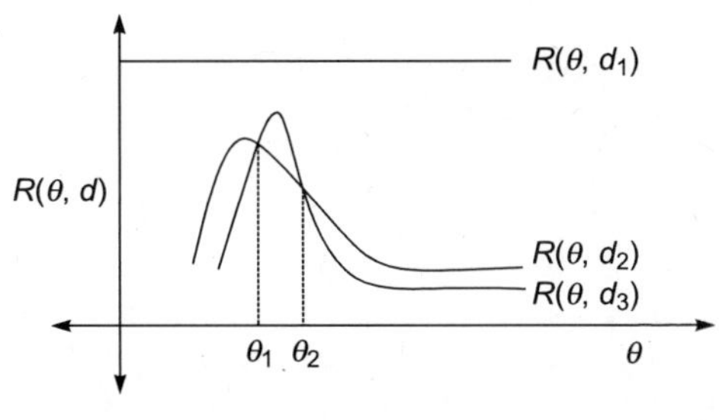

Thus, the choice of decision function depends on θ and hence no uniformly minimum risk decision function is available. In the graph of θ against $R(\theta, d)$ note that d_2 is better in the interval (θ_1, θ_2) otherwise d_3 is better.

To get rid of this difficulty (i) either restrict the available decisions by unbiasedness principle which we have discussed fully in chapter 2 or invariance principle or (ii) ordering principle.

Ordering Principle

There are two ways of ordering principle. One is to take the maximum risk for each decision (d) and then select minimum of them and the second way is to average out the risks for each d and then select the minimum of them. The first principle is called minimax principle and the second one is called Bayes principle.

Definition 5.1

If X has family of distributions $\mathbb{P} = \{p_\theta, \theta \in \Omega\}$ and if D is class of decision functions and if $R(\theta, d)$, $d \in D$ is the risk function then $d_0 \in D$ is called minimax rule if

$$\sup_\theta R(\theta, d_0) \leq \sup_\theta R(\theta, d) \; \forall \; d \in D$$

This principle guards against worst situation and it has limited uses. It is used when error can be tolerated over a range but an error outside this range is catastrophic.

Another ordering principle needs averaging over θ, which needs the probability law for θ, that is θ is considered to be a random variable having prior distribution $\pi(\theta)$. So X has p.f., $p_\theta(x) = p(x/\theta)$ and θ has p.f. $\pi(\theta)$ so that the joint p.d.f., $p(x; \theta)$ of x and θ is

$$p(x;\theta) = p(x/\theta)\pi(\theta) = \pi(\theta/x)m(x)$$

where $\pi(\theta/x)$ is called posterior distribution of θ after x is observed and $m(x)$ is the marginal distribution of X.

Bayes estimation is discussed in section 5.2 whereas section 5.3 is devoted to minimax estimation. The choice of prior distribution is discussed in section 5.4 and some applications of Bayes estimation in sampling are given in section 5.5. Utility function is illustrated in section 5.6.

5.2 BAYES ESTIMATION

As θ has prior distribution, one has to define $R(\theta, d) = EL(\theta, d(x)/\theta)$ the conditional expectation of loss function when θ is held fixed. $E_\pi R(\theta, d)$ is called Bayes mean risk denoted by $R(\pi, d)$.

Definition 5.2

A rule $d^* \in D$ is called Bayes rule if $R(\pi, d^*) \leq R(\pi, d)\ \forall\ d \in D$

Theorem 5.1: If the loss function is $L(g(\theta), d(x)) = [g(\theta) - d(x)]^2$ for $g(\theta)$ and rule $d(x)$ then Bayes rule is $E(g(\theta)/x)$. Since bayes risk is

$$R(\pi, d) = ER(\theta, d)$$

$$= \int R(\theta, d)\pi(\theta)d\theta$$

$$= \int_\Omega \int_{R_1} [d(x) - g(\theta)]^2 \pi(\theta)p(x/\theta)dxd\theta$$

$$= \int_{R_1} \left\{ \int_\Omega [d(x) - g(\theta)]^2 p(x/\theta)\pi(\theta)d\theta \right\} dx$$

minimizing $R(\pi, d)$, that is, minimizing

$$[d(x)]^2 \int p(x/\theta)\pi(\theta)d\theta - [2d(x)]\int g(\theta)p(x/\theta)\pi(\theta)d\theta + \int [g(\theta)]^2 p(x/\theta)\pi(\theta)d\theta$$

which occurs at $d^*(x)$ where

$$d^*(x) = \frac{\int g(\theta)p(x/\theta)\pi(\theta)d\theta}{\int p(x/\theta)\pi(\theta)d\theta}$$

$$= \frac{\int g(\theta)p(x;\theta)d\theta}{\int p(x;\theta)d\theta}$$

$$= \frac{\int g(\theta)p(x;\theta)}{m(x)} d\theta$$

$$= \int g(\theta)\pi(\theta/x)d\theta$$

$$= E[g(\theta)/x]$$

Remark 5.1

If $L[g(\theta), d(x)] = |g(\theta) - d(x)|$ then Bayes rule is the posterior median.

In general, Bayes rule is obtained by minimizing the mean of the posterior loss function, $E[L(g(\theta), d(x))/x]$.

Theorem 5.2: A Bayes rule with constant risk is minimax.

Proof: Let d_0 be Bayes and $\pi(\theta)$ be prior and let d be another rule. Then

$$\sup_{\theta} R(\theta, d) = \sup_{\theta} R(\theta, d)\int \pi(\theta)d\theta$$

$$\geq \int R(\theta, d)\pi(\theta)d\theta \text{ (as average is always less than maximum)}$$

$$\geq \int R(\theta, d_0)\pi(\theta)d\theta,$$

$$= r, \text{ where } r \text{ is constant risk.}$$

$$= \sup_{\theta} R(\theta, d_0)$$

which shows that $R(\theta, d_0)$ is minimax.

Example 5.1

Let $p(x/\theta) = \binom{n}{x}\theta^x(1-\theta)^{n-x}$, $\pi(\theta) = \frac{1}{B(\alpha, \beta)}\theta^{\alpha-1}(1-\theta)^{\beta}$,

$L(\theta, d) = (\theta - d)^2$. Obtain Bayes rule.

The posterior distribution of θ given x is

$$\pi(\theta/x) = \frac{\theta^{x+\alpha-1}(1-\theta)^{n+\beta-x-1}}{\int_0^1 \theta^{x+\alpha-1}(1-\theta)^{n+\beta-x-1}}$$

Bayes rule is

$$d_0(x) = E(\theta/x)$$

$$= \frac{\int_0^1 \theta^{x+\alpha}(1-\theta)^{n+\beta-x-1}d\theta}{\int_0^1 \theta^{x+\alpha-1}(1-\theta)^{n+\beta-x-1}d\theta}$$

$$= \frac{B(x+\alpha+1, n+\beta-x)}{B(x+\alpha, n+\beta-x)}$$

$$= \frac{x+\alpha}{\alpha+\beta+n}$$

Particular Cases:

If $\alpha = \beta = 1$, $\pi(\theta) = 1$ uniform distribution and Bayes rule is $d_0(x) = \dfrac{x+1}{n+2}$ with risk function

$$R(\theta, d) = E\left(\frac{x+1}{n+2} - \theta\right)^2$$

$$= \frac{E[(X-n\theta)-(2\theta-1)]^2}{(n+2)^2}$$

$$= \frac{n\theta(1-\theta)+(2\theta-1)^2}{(n+2)^2}$$

and mean risk is

$$R(\pi, d) = \frac{n}{(n+2)^2}\int_0^1 (\theta-\theta^2)d\theta + \frac{1}{(n+2)^2}\int_0^1 (4\theta^2 - 4\theta + 1)d\theta$$

$$= \frac{1}{6(n+2)}$$

For what loss function Bayes estimator is usual UMVU estimator in this case?

Let loss function be

$$L(\theta, d) = \frac{(d-\theta)^2}{\theta(1-\theta)}.$$

Then for this loss function the Bayes estimator, $d_0(x)$ of θ will be

$$d_0(x) = \frac{B(x+1, n-x)}{B(x, n)}$$

$$= \frac{x}{n}$$

Also UMVU estimator of θ is $\frac{x}{n}$.

Thus, for the above-mentioned loss function Bayes estimator and UMVU estimator are same.

Example 5.2

Let $X_1, X_2, \ldots, X_n$ be a random sample from $N(\mu, 1)$ and let μ have prior distribution $N(0, \sigma^2)$ let $L(\mu, d) = (\mu - d)^2$. Then obtain Bayes estimator of μ.

As $\bar{X}$ is sufficient for μ, instead of $p(x/\mu)$, taking $f(\bar{x}/\mu)$ where, $\bar{X} \sim N\left(\mu, \frac{1}{n}\right)$ and $\mu \sim N(0, \sigma^2)$, the joint density, $h(\bar{x}, \mu)$ of $\bar{x}$ and μ is

$$h(\bar{x}, \mu) = \frac{\sqrt{n}}{2\pi\sigma} e^{-\frac{\mu^2}{2\sigma^2} - \frac{n}{2}(\bar{x}-\mu)^2}$$

$$= \frac{\sqrt{n}}{2\pi\sigma} e^{-\frac{1}{2}\left[\mu^2\left(n+\frac{1}{\sigma^2}\right) - 2n\mu\bar{x} + n\bar{x}^2\right]}$$

$$= \frac{\sqrt{n}}{2\pi\sigma} e^{-\frac{1}{2\left(\frac{1}{n+\frac{1}{\sigma^2}}\right)}\left[\left(\mu^2 - \frac{2n\mu\bar{x}}{\left(n+\frac{1}{\sigma^2}\right)}\right) + \frac{n^2\bar{x}^2}{\left(n+\frac{1}{\sigma^2}\right)} - \frac{n^2\bar{x}^2}{\left(n+\frac{1}{\sigma^2}\right)^2} + n\bar{x}^2\right]}$$

$$= \frac{\sqrt{n}}{2\pi\sigma} e^{-\frac{1}{2\left(\frac{1}{n+\frac{1}{\sigma^2}}\right)}\left[\left(\mu - \frac{n\bar{x}}{\left(n+\frac{1}{\sigma^2}\right)}\right)^2 - \frac{n^2\bar{x}^2}{\left(n+\frac{1}{\sigma^2}\right)^2} + n\bar{x}^2\right]}.$$

$$= \frac{\sqrt{n}}{2\pi\sigma} e^{-\frac{n+\frac{1}{\sigma^2}}{2}\left[\frac{n}{\left(n+\frac{1}{\sigma^2}\right)^2} + 1\right]n\bar{x}^2} e^{-\frac{n+\frac{1}{\sigma^2}}{2}\left(\mu - \frac{n\bar{x}}{\left(n+\frac{1}{\sigma^2}\right)}\right)^2}$$

so that the marginal p.d.f. of $\overline{x}$ is

$$m(\overline{x}) = \frac{\sqrt{n}}{2\pi\sigma}\sqrt{2\pi\frac{1}{\left(n+\frac{1}{\sigma^2}\right)}}\, e^{-\frac{\left(n+\frac{1}{\sigma^2}\right)}{2}\left[\frac{n}{\left(n+\frac{1}{\sigma^2}\right)^2}+1\right]n\overline{x}^2}$$

Therefore,

$$\pi(\mu/\overline{x}) = \frac{h(\overline{x},\mu)}{m(\overline{x})}$$

$$= \frac{1}{\sqrt{2\pi}\sigma\sqrt{\frac{1}{\left(n+\frac{1}{\sigma^2}\right)}}}\, e^{-\frac{1}{2\left(\frac{1}{\left(n+\frac{1}{\sigma^2}\right)}\right)}\left(\mu-\frac{n\overline{x}}{\left(n+\frac{1}{\sigma^2}\right)}\right)^2}$$

which is $N\left(\frac{n\overline{x}}{\left(n+\frac{1}{\sigma^2}\right)}, \frac{1}{\left(n+\frac{1}{\sigma^2}\right)}\right)$

the Bayes rule being posterior mean, it is

$$d^*(x) = \frac{n\overline{x}}{\left(\frac{n\sigma^2+1}{\sigma^2}\right)} = \frac{\sigma^2\overline{x}}{\left(\frac{n\sigma^2+1}{n}\right)} = \frac{\sigma^2\overline{x}}{\sigma^2+\frac{1}{n}}.$$

Example 5.3

If $X \sim N(\theta, \sigma^2)$ and if $\theta \sim N(\mu, \tau^2)$ then for squared error loss function obtain Bayes rule (estimator) of θ when σ^2, μ, τ^2 are known.

Let $p(x/\theta)$, $\pi(\theta)$, $m(x)$, $\pi(\theta/x)$ be p.d.f. of x given θ, prior density of θ marginal of x and posterior density of θ given x respectively. We have to find posterior mean as Bayes rule for squared error loss function is posterior mean.

Here the joint p.d.f. of x and θ is

$$h(x,\theta) = \frac{1}{2\pi\sigma\tau} e^{-\frac{1}{2}\left[\frac{(\theta-\mu)^2}{\tau^2} + \frac{(x-\theta)^2}{\sigma^2}\right]}$$

defining $\rho = \frac{1}{\sigma^2} + \frac{1}{\tau^2} = \frac{\sigma^2+\tau^2}{\sigma^2\tau^2}$ and completing square in the exponent, we get

$$\frac{1}{2}\left[\frac{(\theta-\mu)^2}{\tau^2} + \frac{(x-\theta)^2}{\sigma^2}\right] = \frac{1}{2}\left[\left(\frac{1}{\sigma^2} + \frac{1}{\tau^2}\right)\theta^2 - 2\left(\frac{\mu}{\tau^2} + \frac{x}{\sigma^2}\right)\theta + \left(\frac{\mu^2}{\tau^2} + \frac{x^2}{\sigma^2}\right)\right]$$

$$= \frac{1}{2}\rho\left[\theta^2 - \frac{2}{\rho}\left(\frac{\mu}{\tau^2} + \frac{x}{\sigma^2}\right)\theta\right] + \frac{1}{2}\left(\frac{\mu^2}{\tau^2} + \frac{x^2}{\sigma^2}\right)$$

$$= \frac{1}{2}\rho\left[\theta - \frac{1}{\rho}\left(\frac{\mu}{\tau^2} + \frac{x}{\sigma^2}\right)\right]^2$$

$$-\frac{1}{2\rho}\left(\frac{\mu}{\tau^2} + \frac{x}{\sigma^2}\right)^2 + \frac{1}{2}\left(\frac{\mu^2}{\tau^2} + \frac{x^2}{\sigma^2}\right)$$

$$= \frac{1}{2}\rho\left[\theta - \frac{1}{\rho}\left(\frac{\mu}{\tau^2} + \frac{x}{\sigma^2}\right)\right]^2 + \frac{(\mu-x)^2}{2(\sigma^2+\tau^2)}$$

Hence,

$$h(x,\theta) = \frac{1}{2\pi\sigma\tau} e^{-\frac{\rho}{2}\left[\theta - \frac{1}{\rho}\left(\frac{\mu}{\tau^2} + \frac{x}{\sigma^2}\right)\right]^2} e^{-\frac{1}{2}\frac{(\mu-x)^2}{(\sigma^2+\tau^2)}} \quad \ldots(5.2.1)$$

and marginal p.d.f. x is

$$m(x) = \int_{-\infty}^{\infty} h(x,\theta)d\theta$$

$$= \frac{1}{\sqrt{2\pi\rho}\,\tau\sigma} e^{-\frac{1}{2}\frac{(\mu-x)^2}{(\sigma^2+\tau^2)}}$$

$$= \frac{1}{\sqrt{2\pi}\sqrt{\tau^2+\sigma^2}} e^{-\frac{1}{2}\frac{(\mu-x)^2}{(\sigma^2+\tau^2)}} \qquad \ldots(5.2.2)$$

$$\sim N(\mu, \sigma^2+\tau^2)$$

So that posterior distribution of θ is,

$$\pi(\theta/x) = \frac{h(x,\theta)}{m(x)}$$

$$= \left(\frac{\rho}{2\pi}\right)^{\frac{1}{2}} e^{-\frac{\rho}{2}\left[\theta - \frac{1}{\rho}\left(\frac{\mu}{\tau^2}+\frac{x}{\sigma^2}\right)\right]^2} \sim N\left(\mu(x), \frac{1}{\rho}\right)$$

where

$$\mu(x) = \frac{1}{\rho}\left(\frac{\mu}{\tau^2}+\frac{x}{\sigma^2}\right)$$

$$= \frac{\sigma^2}{\sigma^2+\tau^2}\mu + \frac{\tau^2}{\sigma^2+\tau^2}x$$

$$= x - \frac{\sigma^2}{\sigma^2+\tau^2}(x-\mu).$$

Thus Bayes rule, which is posterior mean, is

$$d^*(x) = x - \frac{\sigma^2}{\sigma^2+\tau^2}(x-\mu).$$

If $X_1, X_2, \ldots, X_n$ is a random sample from $N(\theta, \sigma^2)$ and if prior distribution of θ is $N(\mu, \tau^2)$ then using $\bar{X} \sim N\left(\theta, \frac{\sigma^2}{n}\right)$ in place of X, being sufficient statistic, we get posterior distribution

$$\pi(\theta/x) \sim N\left(\mu(x), \frac{1}{\rho}\right)$$

where
$$\mu(x) = \frac{\frac{\sigma^2}{n}}{\tau^2 + \frac{\sigma^2}{n}}\mu + \frac{\tau^2}{\tau^2 + \frac{\sigma^2}{n}}\bar{x}$$

and ρ being $\frac{n\tau^2 + \sigma^2}{\sigma^2\tau^2}$, posterior variance is $\frac{1}{\rho} = \frac{\sigma^2\tau^2}{n\tau^2 + \sigma^2}$.

Multivariate Extension of the Problem

If $\underline{X} \sim N_p(\underline{\theta}, \Sigma)$ and $\underline{\theta} \sim N_p(\underline{\mu}, A)$ where, $\underline{\mu}$ is known and A, Σ are $p \times p$ known matrices, then posterior distribution of $\underline{\theta}$ given $\underline{X}$ is $N_p(\mu_x^\pi, \rho^{-1})$

where
$$\mu_x^\pi = \rho^{-1}\left(A^{-1}\underline{\mu} + \Sigma^{-1}\underline{x}\right)$$
$$= \underline{x} - \Sigma(A+\Sigma)^{-1}(\underline{x} - \underline{\mu}),$$

using $\quad \rho = \Sigma^{-1} + A^{-1},\ \rho^{-1} = (\Sigma^{-1} + A^{-1})^{-1} = \Sigma - \Sigma(A+\Sigma)^{-1}\Sigma$

This can be shown analogously.

5.3 MINIMAX ESTIMATION

Theorem 5.3: Let $\pi_1, \pi_2, \ldots$ be a sequence of priors and let d_n be Bayes with prior π_n. Let r_n be the Bayes risk of d_n and let d^* be any rule having risk $R(\theta, d^*) = r$. Then d^* is minimax.

Proof: $\sup_{\theta} R(\theta, d) \geq \int R(\theta, d)\pi_n(\theta)d\theta \;\forall\; d \geq \int R(\theta, d_n)\pi_n(\theta)d\theta$

$$= r_n$$
$$= r \text{ as } n \to \infty$$
$$= R(\theta, d^*)$$
$$= \sup R(\theta, d^*)$$

$\Rightarrow$ d^* is minimax.

Remark 5.2

In the above example 5.3 $\pi(\sigma^2) \sim N(0, \sigma^2)$ we have Bayes estimator

$$\frac{\bar{x}\sigma^2}{\left(\frac{1}{n}+\sigma^2\right)} = \frac{\frac{\sigma^2}{n}}{\left(\frac{1}{n}+\sigma^2\right)}\sum x_i \to \frac{\sum x_i}{n} \text{ as } \sigma^2 \to \infty.$$

So we have to find an estimator whose risk is $\frac{1}{n}$. Obviously, it is $\bar{X}$ and risk is $\frac{1}{n}$.

Example 5.4

For $X \sim b(n, p)$, taking Bayes estimator $d(x) = aX + b$ with squared error loss function, obtain minimax estimator.

$$R(p, d) = E(aX + b - p)^2$$
$$= E[a(X - np) + anp + b - p]^2$$
$$= [(an-1)^2 - a^2n]p^2 + [a^2n + 2(an-1)]p + b^2$$

select a, b such that $R(p, d) = \text{constant} = b^2$ which implies

$$\left.\begin{aligned}(an-1)^2 - a^2n = 0\\ a^2n + 2(an-1)b = 0\end{aligned}\right\} \Rightarrow a = \frac{1}{\sqrt{n}(1+\sqrt{n})},\; b = \frac{1}{2(1+\sqrt{n})}$$

Then $$d^*(x) = \frac{X}{\sqrt{n}(1+\sqrt{n})} + \frac{1}{2(1+\sqrt{n})}.$$

What is the prior distribution of p?

Let it be $p^{\alpha-1}(1-p)^{\beta-1}$,

$$h(p/x) = \frac{p^{\alpha+x-1}(1-p)^{n+\beta-x-1}}{\int_0^1 p^{\alpha+x-1}(1-p)^{n+\beta-x-1}\,dp}$$

$$h(p/x) = \frac{x+\alpha}{\alpha+\beta+n}$$

$$\Rightarrow \left.\begin{aligned} \frac{1}{\sqrt{n}(1+\sqrt{n})} &= \frac{1}{\alpha+\beta+n} \\ \frac{1}{2(1+\sqrt{n})} &= \frac{\alpha}{\alpha+\beta+n} \end{aligned}\right\} \Rightarrow \alpha = \beta = \frac{\sqrt{n}}{2}$$

d^* has risk $\dfrac{1}{4(1+\sqrt{n})^2}$.

The UMVU estimator $\dfrac{X}{n} = d$ has risk $\dfrac{p(1-p)}{n}$.

$$\frac{p(1-p)}{n} \le \frac{1}{4(1+\sqrt{n})^2} \quad \text{iff} \quad \left|p-\frac{1}{2}\right| \ge \frac{\sqrt{1+2\sqrt{n}}}{(1+\sqrt{n})}$$

If n is small, prefer minimax otherwise usual estimator is better.

Theorem 5.4: If Y is minimax for θ then $aY + b$ is minimax for $a\theta + b$.

Proof: Since Y is minimax estimator of θ, max $R(\theta, Y) \le \max R(\theta, Z)\ \forall\ Z$. For the squared error loss function

$$\begin{aligned} R(a\theta+b, aY+b) &= E(aY+b-a\theta-b)^2 \\ &= a^2E(Y-\theta)^2 \\ &= a^2R(\theta, Y). \end{aligned}$$

Similarly,

$$R(a\theta+b, aZ+b) = a^2R(\theta, Z)$$

As $a^2 \max R(\theta, Y) \le a^2 \max R(\theta, Z)$ which implies $aY+b$ is minimax.

Theorem 5.5: The sum of two minimax estimators is not minimax estimator.

Proof: Consider an example that $Y_1 \sim b(1, p)$ and $Y_2 \sim b(1, p)$, Y_1 and Y_2 are independently distributed then $Y_1 + Y_2 \sim b(2, p)$ and the minimax estimator of p is $\frac{Y_1 + Y_2}{\sqrt{2}(1-\sqrt{2})} + \frac{1}{\sqrt{2}(1-\sqrt{2})}$. Now as $Y_1 \sim b(1, p)$, the minimax estimator of p is $\frac{Y_1}{2} + \frac{1}{4}$ and that of $\frac{p}{2}$ is $\frac{Y_1}{4} + \frac{1}{8}$. Similarly, as $Y_2 \sim b(1, p)$, minimax estimator of $\frac{p}{2}$ is $\frac{Y_2}{4} + \frac{1}{8}$. Thus, the minimax estimator of $\frac{p}{2} + \frac{p}{2}$ is $\frac{Y_2 + Y_1}{4} + \frac{1}{4}$. This shows that the sum of minimax estimators is not minimax estimator.

5.4 CHOICE OF PRIOR DISTRIBUTION

The Baysian approach for the problem is logical and more realistic but the only difficulty is how to determine prior distribution. Another problem is that for different priors the rules are different. How to select best prior distribution? Some of the methods of determining prior distribution are given below.

(a) Subjective Determination of Prior Density

If Ω is subset of real line, likelihood of various points in Ω are compared and sketch $\pi(\theta)$, e.g., if $\Omega = (0, 1]$ and if $\theta = \frac{3}{4}$ is most likely and $\theta = 0$ is least likely and if $\theta = \frac{3}{4}$ is three times as likely as $\theta = 0$ and $\theta = \frac{1}{2}$ while $\theta = \frac{1}{4}$ is 1.5 times as likely as $\theta = 0$, assigning value 1 to 0 then the graph of likelihoods is drawn in the adjoining figure

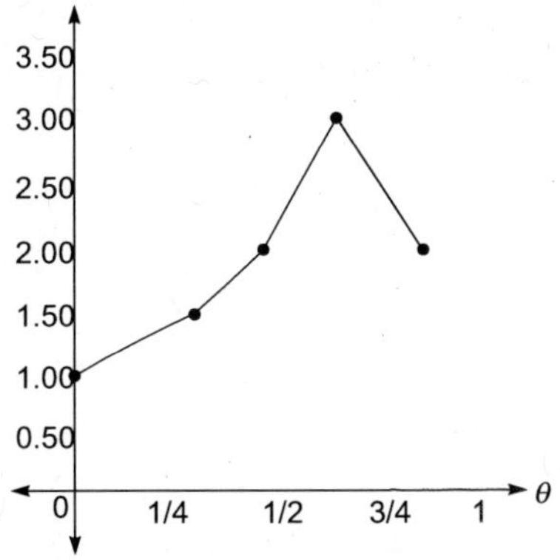

(b) Matching a given Functional Form

In this method one has to match the quantiles of the distribution with subjective quantiles, e.g., if $\Omega = (-\infty, \infty)$ the prior is thought of normal or Cauchy. Let median of prior be zero and $Q_1 = -1, Q_3 = 1$. Then from normal table as $P\left\{Z < -\frac{1}{\sqrt{2.19}}\right\} = \frac{1}{4} \Rightarrow \sigma^2 = 2.19$ and hence, prior $\pi(\theta)$ is $N(\theta, 2.19)$ and for Cauchy distribution $\int_{-\infty}^{-1} \frac{1}{\pi(1+x^2)} dx = \frac{1}{4}$, median is zero. Which match is reasonable? If the difference between two decisions has little effect on Bayes rule then it is reasonable.

(c) Non-informative Prior

A prior which does not contain any information about θ is called non-informative prior, e.g., $\pi(\theta) = 1$, uniform distribution on (0, 1), where, $\int_{-\infty}^{\infty} \pi(\theta) d\theta = \infty$ improper prior with infinite mass. If Ω is finite number of points n then non-informative prior is to assign probability $\frac{1}{n}$ and if Ω is infinite then non-informative prior is uniform $\pi(\theta) = 1$.

For location parameter family of distributions $f(x - \theta)$, the non-informative prior is $\pi(\theta) = 1$ and for scale parameter family of distributions $f\left(\frac{x}{\sigma}\right), \sigma > 0$ non-informative prior is $\pi(\sigma) = \frac{1}{\sigma}$.

For many problems generally informative prior is widely used, such as $\pi(\theta) = [I(\theta)]^{-1}$, where $I(\theta)$ is the Fisher's information, e.g., for $N(\theta, \sigma^2)$ distribution.

$$I(\theta) = -E\begin{bmatrix} \frac{\partial^2}{\partial\theta^2}\left[-\frac{(x-\theta)^2}{2\sigma^2}\right] & \frac{\partial^2}{\partial\theta\partial\sigma}\left[-\frac{(x-\theta)^2}{2\sigma^2}\right] \\ \frac{\partial^2}{\partial\theta\partial\sigma}\left[-\frac{(x-\theta)^2}{2\sigma^2}\right] & \frac{\partial^2}{\partial\sigma^2}\left[-\frac{(x-\theta)^2}{2\sigma^2}\right] \end{bmatrix}$$

$$= \begin{pmatrix} \frac{1}{\sigma^2} & 0 \\ 0 & \frac{3}{\sigma^2} \end{pmatrix}, \text{ so } \pi(\theta) \alpha \frac{1}{\sigma^2}$$

which is improper. Considering independence of θ and σ and multiplying improper non-informative prior of θ and σ as $\pi(\theta, \sigma) = \frac{1}{\sigma}$, which is called Jeffery's prior.

Illustration 5.1: If θ is binomial parameter the possible non-informative priors are

(1) Bayes – $\pi_1(\theta) = 1$

(2) Novik – $\pi_2(\theta) = \dfrac{1}{\theta(1-\theta)}$

(3) Jeffery – $\pi_3(\theta) \alpha [\theta(1-\theta)]^{-\frac{1}{2}}$

(4) Zeltner – $\pi_4(\theta) \alpha \theta^{\theta}(1-\theta)^{1-\theta}$

(d) Conjugate Priors

If the prior distribution, $\pi(\theta)$, is such that the posterior distribution is of the same form as $f(x/\theta)$ then $\pi(\theta)$ is called conjugate prior. Normal prior conjugate to normal distribution because the posterior distribution is again normal. Natural conjugate prior is same functional form as likelihood function, e.g., if $X_1, X_2, \ldots, X_n$ is a random sample from Poisson distribution with parameter λ. Then

$$f(\underline{x}/\theta) = \frac{e^{-n\theta}\theta^{n\bar{x}}}{\prod_{i=1}^{n} x_i!}$$

Which resembles to Gamma density so we take prior as $G(\alpha, \beta)$ with p.d.f. as

$$G(\theta; \alpha, \beta) = \frac{e^{-\frac{\theta}{\beta}}\theta^{\alpha-1}}{\Gamma(\alpha)\beta^{\alpha}}.$$

$$h(x; \theta) = \frac{e^{-n\theta}\theta^{n\bar{x}}}{\prod x_i!} \frac{e^{-\frac{\theta}{\beta}}\theta^{\alpha-1}}{\Gamma(\alpha)\beta^{\alpha}}, 0 < \theta < \infty$$

$$= \frac{e^{-\theta\left(n+\frac{1}{\beta}\right)}\theta^{n\bar{x}+\alpha-1}}{\Gamma(\alpha)\beta^{\alpha}\prod x_i} \sim \text{G}\left(n\bar{x}+\alpha, \left(n+\frac{1}{\beta}\right)^{-1}\right)$$

The conjugate prior of binomial distribution is beta distribution. Some conjugate priors of some distributions are given below:

For	Conjugate prior
Gamma distribution	Gamma distribution
Binomial distribution	Beta distribution
Normal distribution	Normal distribution
Poisson distribution	Gamma distribution
Negative binomial	Beta distribution

5.5 APPLICATIONS OF BAYES ESTIMATION IN SAMPLING PROBLEM

(1) Municipal Bus Problem

If all the municipal buses are numbered 1, 2, ... and if you observe a particular number in some time interval $(t_2 - t_1)$ can you estimate the total number of buses in the city? Such problem occurs also in stock exchange market crises, animal mutation, traffic lights etc.

Particularly, consider bus problem in which we have to estimate total number of buses, N which can take values 1, 2, 3, ... Consider non-informative prior $\pi(N) = \frac{1}{N}$ as N is somewhat scale parameter. If T is the observed bus number it is distributed as

$$f_T(t/N) = P(T) = \frac{1}{N}, \; t = 1, 2, \ldots, N$$

so that posterior distribution of N given T is

$$\pi(N/T) \alpha \frac{1}{N^2}, N \geq T$$

and

$$P_\pi(N \geq n_0/T) = \frac{\sum_{n=n_0}^{\infty} \frac{1}{n^2}}{\sum_{n=T}^{\infty} \frac{1}{n^2}} = \frac{\int_{n_0}^{\infty} \frac{1}{x^2} dx}{\int_{T}^{\infty} \frac{1}{x^2} dx} = \frac{T}{n_0}$$

In this case posterior median is approximated by $N_\pi(T) = 2T$ (n_0 being $\frac{1}{2}$, $\frac{T}{n_0} = 2T$) which is the estimator usually considered in the bus problem. In fact, note that mean of T conditionally on N is $\frac{N-1}{2} \cong \frac{N}{2}$.

(2) Capture-Recapture Problem

When census is impossible (or too costly) it is necessary to provide estimation methods for estimating population size, i.e., N, e.g., number of fishes in a pond, number of dears (tigers, lions, birds) in a particular area. The usual method is to use capture-recapture method.

Suppose that a lake contains unknown number of fishes, N. A random sample of size k fishes is captured and tagged and released in the lake again. After some time a second sample of size n fishes is recaptured and noted that x of them are found tagged ones. Considering that n fishes in the recaptured (second) sample constitute n Bernoulli trials with probability of success (getting tagged fishes), $= \frac{k}{N}$. So the population size $N = \frac{k}{p}$. We have to estimate $\frac{1}{p}$. Reliable estimate of $\frac{1}{p}$ is difficult to obtain if p is close to zero. So it is necessary to take a large sample. The sampling scheme to achieve this is inverse binomial sampling in which sampling continues till one gets specified number of successes, say m. Let $Y + m$ denote the required number of trials. Then Y has negative binomial distribution with probability function

$$P(Y = y) = \binom{m+y-1}{m-1} p^m q^y, \; y = 0, 1, 2, \ldots$$

Since $EY = \frac{mq}{p}$, $\text{var}(Y) = \frac{mq}{p^2}$, $p + q = 1$, $\frac{y+m}{m}$ is unbiased for $\frac{1}{p}$ and since the distribution of Y belongs to exponential family of distribution, Y is complete sufficient for p and $\frac{y+m}{m}$ is UMVU estimator of $\frac{1}{p}$.

Instead of using the above said classical method we use Bayesian method for solving this problem given by Walter who showed that most capture-recapture models can be described by a multinomial distribution for each individual i in the population $(1 \le i \le N)$.

The following table describes the probability of capture

		Sample: 2	
		Captured	Missed
Sample: 1	Captured	p_{11}^i	p_{12}^i
	Missed	p_{21}^i	p_{22}^i

With $p_{11}^i + p_{12}^i + p_{21}^i + p_{22}^i = 1$. For instance, p_{12}^i means the probability that i^{th} individual being captured only in the first sample. After two samples the population is divided into

		Sample: 2	
		Captured	Missed
Sample: 1	Captured	n_{11}	n_{12}
	Missed	n_{21}	n_{22}

with $n_{11} + n_{12} + n_{21} + n_{22} = N$, ($n_{22}$ is unknown)

For the simplest model called uniform, each individual has the same probability, p of being captured in both experiments. So $p_{11} = p^2$,

$$p_{12} = p_{21} = p(1 - p),\ p_{22} = (1 - p)^2.$$

The likelihood can be written as

$$L(N, p/n_{11}, n_{12}, n_{21}) = \frac{N!}{n_{11}!\, n_{12}!\, n_{21}!\, n_{22}!} p^n (1-p)^{2N-n} \qquad \ldots(5.5.1)$$

where $n = 2n_{11} + n_{12} + n_{21}$ is the total number of captured individuals.

Let p have prior distribution $\pi(p)$ as

$$\frac{1}{\beta(\alpha, \gamma)} p^{\alpha-1}(1-p)^{\gamma-1}, 0 < p < 1, \qquad \ldots(5.5.2)$$

and (N, p) have $\Pi(N, p)$ distribution given in (5.5.1). Then the posterior distribution of p is $\Pi(p/N, n_{11}, n_{12}, n_{21}) = \frac{(5.5.1)}{(5.5.2)}$ which is proportional to $p^{\alpha+n-1}(1-p)^{\gamma+2N-n-1}$. i.e. $p/N, n$ has $\beta_e(\alpha+n, \gamma+2N-n)$. Using mean of beta distributions $\beta(\alpha, \gamma)$ as $\frac{\alpha}{\alpha+\gamma}$, we have conditional mean of p given N, n is $\frac{\alpha+n}{\alpha+\gamma+2N}$. Equating this with the observed sample mean $\overline{X}$, we have

$$\overline{X} = \frac{\alpha+n}{\alpha+\gamma+2N} \quad N = \frac{\alpha+n-\alpha\overline{X}-\gamma\overline{X}}{2\overline{X}}.$$

One can use hypergeometric distribution as

$$\binom{Np}{x}\binom{N-pN}{n-x} \Big/ \binom{N}{n}, 0 \le p < 1,\ n < N,\ pN \in N,\ x = 0, 1, \ldots, n$$

to estimate N by using mean $\frac{n^2 p}{N}$ and using two samples $n_1 = n_{11} + n_{12}$ and $n_2 = n_{11} + n_{22}$ (fixed). The classical estimator of N is m.l.e.

$$\hat{N} = \frac{n_1}{\left(\frac{n_{11}}{n_2}\right)} = \frac{n_1 n_2}{n_{11}}$$

which identifies proportion in the population $\left(\frac{n_1}{N}\right)$ and proportion in the sample $\left(\frac{n_{11}}{n_2}\right)$. The drawback is that estimator cannot be defined when $n_{11} = 0$. In this case take a third draw of n_3 elements and repeat the process.

5.6 UTILITY FUNCTION

Sometimes loss function is difficult to obtain, then it is decided by utility function. We shall give some elementary ideas about utility function. Sometimes consequences of certain actions are immeasurable such as prestige, reputation, etc., and even in monetary scale it does not reflect true value, e.g., the value of Rs. 100 to a person (A) having income of Rs. 3000 p.m. and a person (B) having income of Rs. 1,00,000 p.m. are different. For (A) utility of Rs. 100 is for good purpose as it is more valuable while (B) wastes it for bad habits. So there is the theory of utility function in which preferences are given by subjective probability. The following illustration will deal it.

Illustration 5.2

On Sunday afternoon if you have to choose one of the actions (i) a_1 – to see football match (ii) a_2 – to go to movie. There is 40% chance of rain which would ruin the game. Let θ_1 be the state of raining and θ_2 be the state of not raining and let their corresponding probabilities be $\pi(\theta_1) = 0.4$, $\pi(\theta_2) = 0.6$. In order to get loss matrix we shall obtain utility matrix.

	a_1	a_2
θ_1	r_1	r_3
θ_2	r_2	r_4

where r_i denotes return (gain) of ith situation. Clearly preference ordering is

$$r_1 \prec r_4 \prec r_3 \prec r_2, [a \prec b \text{ means } b \text{ is prefered to } a]$$

In order to construct utility function. We assign value zero to worst choice and value one to the best choice. So $U(r_1) = 0, U(r_2) = 1$ and utility of r_4 is given by

$$U(r_4) = 0.4, U(r_1) + 0.6, U(r_2) = 0.6$$

by gambling subjective probabilities are chosen as 0.4 and 0.6.

$$U(r_3) = 0.3 \text{ and } U(r_1) + 0.7, U(r_2) = 0.7$$

by gambling subjective probability as 0.3 and 0.7. Thus, for action a_1

$$E(U(r)) = \pi(\theta_1)4U(r_1) + \pi(\theta_2)7U(r_2) = 0.4 * 0 + 0.6 * 1 = 0.6$$

and for action a_2 is

$$E(U(r)) = \pi(\theta_1)U(r_3) + \pi(\theta_2)U(r_4) = 0.4 * 0.7 + 0.6 * 0.6 = 0.64.$$

Action a_2 preferred, that is, go to movie.

Here loss function is given by using $L(\theta, a) = -U(\theta, a)$

	a_1	a_2
θ_1	1	−0.7
θ_2	0	−0.6

PROBLEMS

5.1 Given $X \sim N(\theta, 1)$, $L(\theta, d) = (cx - \theta)^2$. Obtain

(i) Risk function $R(\theta, d)$ for $\pi(\theta) = \frac{1}{2}$ and $\pi(\theta) = 1$
(ii) Mean risk
(iii) Bayes solution
(iv) Minimax rule.

5.2 An investor wants to buy risky bonds. At maturity the net gain in bonds is Rs. 500 but when default occurs he loses Rs. 1000. If he puts his money as fixed deposit he gets Rs. 300 as net gain given the prior information that probability of no default is 0.9 and that of default is 0.1. Obtain Bayes and minimax rules.

5.3 Given $X \sim b(n, \theta)$, $L(\theta, d) = \left(\frac{X}{n} - \theta\right)^2$, $\pi(\theta) = 1, 0 < \theta < 1$, obtain Bayes rule. What will be Bayes rule if $L(\theta, d) = \dfrac{\left(\frac{X}{n} - \theta\right)^2}{\theta(1-\theta)}$?

5.4 Obtain posterior distribution in the following cases and state Bayes rules for squares error loss function.

(i) $f(x/\sigma) \sim N(\theta, \sigma^2)$, $\frac{1}{\sigma^2} \sim G(1, 2)$

(ii) $f(x/\lambda) \sim P(\lambda)$, $\lambda \sim G(2, 1)$

(iii) $f(x/p) \sim$ negative binomial $(10, p)$, $p \sim B_e\left(\frac{1}{2}, \frac{1}{2}\right)$.

5.5 Let $X_1, X_2, \ldots, X_n$ be a random sample from $N(\theta, \sigma^2)$ and

$$\pi(\theta, \sigma^2) = \frac{1}{\sigma^2(\alpha+1)} e^{-\frac{s_0^2}{2\sigma^2}}, \; s_0^2 = \frac{1}{n}\sum(x_i - \bar{x})^2$$

Then obtain posterior distribution $\pi(\theta, \sigma^2/x_1, x_2, \ldots, x_n)$ and derive posterior expectation $E_\pi(\theta/x_1, x_2, \ldots, x_n)$.

5.6 Let $x \sim P(\lambda)$ and $\pi(\lambda) = e^{-\lambda}$, $L(\lambda, d) = (d_c - \lambda)^2$. Taking $d_c = cX$ compute $R(d_c, \lambda)$ and show that d_c is not admissible for $c > 1$.

[An estimator d_0 is admissible if it dominates every other estimator i.e. $R(\theta, d_0) \le R(\theta, d) \; \forall \; d \in D, \forall \; \theta \in \Omega$

$$R(\theta_0, d_0) \le R(\theta_0, d) \text{ for at least one } \theta_0 \; \forall \; d \in D]$$

Compute Bayes mean risk $r(\pi, d_c)$ and obtain optimal value of c. Also obtain the best estimator d_c for the minimax criterion. Solve the above questions for the loss function

$$L'(\lambda, d) = \left(\frac{d}{\lambda} - 1\right)^2$$

5.7 If $X \sim N(\theta, 1), \theta \sim N(0, 1)$ and loss function $L(\theta, d)' = 1$ if $d < \theta$ then show that in this case there is no Bayes estimator.

5.8 If $X_1, X_2, \ldots, X_n$ is a random sample from $N(0, \sigma^2)$ and $\frac{1}{2\sigma^2} \sim G(\alpha, p)$, obtain Bayes estimator of $\frac{1}{2\sigma^2}$.

5.9 If $X \sim b(n, p)$ and $p \sim B_e(a, b)$ then obtain Bayes estimator of pq.

5.10 If $X_1, X_2, \ldots, X_n$ is a random sample from $P(\lambda)$ distribution and if $\lambda \sim G(\alpha, p)$, for squared error loss function, obtain Bayes estimator of $e^{-\lambda}$.

5.11 Let $X_1, X_2, \ldots, X_n$ be a random sample from $N(\theta, \sigma^2)$, σ^2 is known. Then for squared error loss function show that $\bar{X}$ is minimax for θ.

6

CHAPTER

Basic Principles of Testing of Hypotheses

6.1 INTRODUCTION

In every-day life one may be confronted with problems like (i) whether new variety of wheat is superior to old one in the production; (ii) whether a candidate will win in the election; (iii) whether a girl does not study after marriage; (iv) whether the country will have to import groundnut oil, etc. These are the problems in which one has to check the truthfulness of the given statement. We consider that the experiment is random and the experiment is associated with a random variable X having family of distribution $\mathbb{P} = \{P_\theta, \theta \in \Omega\}$ if 'A' is any statement regarding P_θ, we call it the hypothesis. If the statement is regarding general form of P_θ then it is called *non-parametric hypothesis* and if the statement is regarding θ then it is called *parametric hypothesis*. The parametric space (all possible values of θ) is completely known. Let $\Omega_H \subset \Omega, \Omega_K \subset \Omega$ such that $\Omega_H \cup \Omega_K = \Omega, \Omega_H \cap \Omega_K = \phi$. Then $\theta \in \Omega_H$ is called the null hypothesis H, that is $\theta \in \Omega_H \equiv H$ is true and $\theta \in \Omega_K \equiv$ the alternative hypothesis, K is true. We utilize information available from the random sample $X = (X_1, X_2, \ldots, X_n)$ to decide the truthfulness of H or K, that is, whether the information from X supports H or K, is decided. Let $\mathfrak{x} = \{(x_1, x_2, \ldots, x_n)\}$ be the sample space of all possible observations $x = (x_1, x_2, \ldots, x_n)$ and let $S \subset \mathfrak{x}$ and let us follow the rule to reject H when $x \in S$, otherwise accept H.

Choice of S is called the test for testing H against K, S is called the critical or rejection region and its compliment (S^c) is called the acceptance region. In describing the rule for rejection or acceptance of H one commits two types of errors: rejecting H when it is true, that is, $x \in S$ when $\theta \in \Omega_H$, is type-I error and accepting H when K, is true, that is, $x \in S^c$ when $\theta \in \Omega_K$, is type-II error. Then probabilities of errors are

$$P\{\text{type} - \text{I error}\} = P_\theta\{X \in S/H\} \qquad \ldots(6.1.1)$$

$$P\{\text{type} - \text{II error}\} = P_\theta\{X \in S^c/K\} \qquad \ldots(6.1.2)$$

Obviously, one would like to choose S such that both the types of errors are simultaneously small but this is not possible because minimizing (6.1.1) means taking S small and minimizing (6.1.2) means S^c must be small, that is, S is large and thus they behave oppositely. If we take type-I error as more serious and set the null hypothesis accordingly then we keep it at a fixed level as we desire and then minimize type-II error, which is Neyman's theory. How to set the null hypothesis we illustrate it. The error made by declaring that the medicine is not poisonous when it is really poisonous is more serious than to declare it that it is poisonous when it is not really poisonous and so that null hypothesis must be set as H: medicine is poisonous.

Thus, one has to decide a number $0 \le \alpha \le 1$ to represent the seriousness of type-I error well in advance. This number α is called level of significance and α is determined by the experts of the field and not by the statistician. The critical region is one which satisfies

$P_\theta\{X \in S/H\} \le \alpha$ relates to the level-α test

or $P_\theta\{X \in S/H\} = \alpha$ relates to the size-α test.

The function, $\beta(\theta) = P_\theta\{X \in S/K\}$ of θ is called the power function of the test which is the probability of rejecting H when it is false. Thus, $\beta(\theta)$ is the probability of the complimentary event of type-II error.

The Neyman's problem of minimizing probability of type-II error amongst all level-α tests is same as maximizing $\beta(\theta)$ amongst all level-α tests, that is to maximize $P_\theta\{X \in S/K\}$ when $P_\theta\{X \in S/H\} \le \alpha$ and obtain S. Then S is called the critical region of the most powerful (MP) test. $\sup_{\theta \in \Omega_H} P_\theta(X \in S)$ is called the size of the test.

Definition 6.1

If $\{S_\alpha\}$ is the class of level-α tests for testing $H:\theta = \theta_0$ vs $K:\theta = \theta_1$ in the family $\mathbb{P} = \{P_\theta, \theta \in \Omega\}$ of distributions of X and if $S_0 \in \{S_\alpha\}$ such that $P_{\theta_1}\{S_0\} \ge P_{\theta_1}\{S\}, \forall\, S \in \{S_\alpha\}$ then S_0 is called the critical region of the Most Powerful (MP) test.

The following illustration shows that the power function increases as the level of significance increases.

Illustration

Let X be a random variable such that, $P_\theta\{X=1\}=\theta$, $P_\theta\{X=0\}=1-\theta$.

To test $H:\theta=\frac{1}{2}$ vs $K:\theta>\frac{1}{2}$ at level α. $\Omega=\left[\frac{1}{2},1\right)$. Here $\mathfrak{X}$ has only two points $\{0\}$, $\{1\}$. Let us compute the probability of type-I error and power function for different critical regions.

Critical region	ϕ	$\{0\}$	$\{1\}$	$\mathfrak{X}$
$P_\theta\{X\in S/H\}$	0	$\frac{1}{2}$	$\frac{1}{2}$	1
$P_\theta\{X\in S/K\}$	0	$1-\theta$	θ	1

Case 1 Let $\alpha<\frac{1}{2}$. Then $S=\phi$ is the only critical region of level-α test. H is always accepted. The power is 0.

Case 2 Let $\frac{1}{2}\le\alpha<1$ Then there are three level α critical regions $S_1=\phi$, $S_2=\{0\}$, $S_3=\{1\}$. Since $\theta>1-\theta$, S_3 is the best critical region.

Case 3 Let $\alpha=1$. All four regions are critical regions of level-α tests. $\mathfrak{X}$ has maximum power but the type-I error is also very large. Always rejection of H is not a nice situation at all. The main drawback in the above set-up is the dimensionality of the sample space when n is large. To avoid large dimension one defines a function on $X_1, X_2, \ldots, X_n$, say statistic $T=u(X_1, X_2, \ldots, X_n)$ on $(X_1, X_2, \ldots, X_n)$ and then considers the partitioning of the sample space generated by T, say Γ of T. If $\Gamma_1\subset\Gamma$ then selects Γ_1 such that max $P_{\theta_1}\{t\in\Gamma_1\}$ subject to $P_{\theta_0}\{t\in\Gamma_1\}\le\alpha$.

The need of randomized test is illustrated by example. Randomized test is defined and many examples have been worked out in section 6.2.

6.2 RANDOMIZED TESTS

For introducing a randomized test consider the following illustration.

Let X_1, X_2, X_3, X_4 be an independent observations with $P(X_i=0)=1-\theta=1-$

$P(X_i = 1)$, $i = 1, 2, 3, 4$ and let $X = \sum_{i=1}^{4} X_i$. Then X has $b(4, \theta)$ distribution.

Let us reject $H:\theta = \frac{1}{2}$ vs $K:\theta = \frac{3}{4}$ when $X \geq c$, where c is such that $P_{\theta=\frac{1}{2}}\{X \geq c\} = \alpha = \frac{3}{32}$, say.

We prepare probability table to determine c

$X = x$	0	1	2	3	4
$P_{\theta=\frac{1}{2}}\{X = x\}$	$\frac{1}{16}$	$\frac{4}{16}$	$\frac{6}{18}$	$\frac{4}{16}$	$\frac{1}{16}$

Obviously,

$$\frac{1}{16} = P(X \geq 4) < \frac{3}{32} < P(X \geq 3) = P(X = 4) + P(X = 3) = \frac{1}{16} + \frac{4}{16} = \frac{5}{16},$$

so c is neither 4 nor 3. But we can include $x = 4$ in the critical region and we should search for a part of $x = 3$ to be included in the critical region, so that total probability of rejection is $\frac{3}{32}$. We need a part of $x = 3$ to compensate the probability $\frac{1}{32}\left(= \frac{3}{32} - \frac{1}{16}\right)$. As $\frac{4}{16}$ is the probability for the whole of $x = 3$, $\frac{1}{32}$th part of $x = 3$ will require $\frac{1}{8}\left(= \frac{1}{32} \Big/ \frac{4}{16}\right)^{\text{th}}$ part of $x = 3$, that is, the probability of including $x = 3$ in the critical region must be $\frac{1}{8}$. How shall we do this? We construct an extraneous Bernoullian experiment with regions R_1, R_2 such that R_1 has probability of rejection $P(R_1) = \frac{1}{8}$.

For doing this let us take eight cards numbered 1 to 8 and draw one card randomly.

Let $R_1 = \{1\}$ and $R_2 = \{2, 3, \ldots, 8\}$. If the card drawn has number 1 on the face then we place $x = 3$ in the critical region and if $x \in R_2 (= \{2, 3, \ldots, 8\})$ we keep $x = 3$ in the acceptance region.

In non-randomized test we either include the sample point in the critical region or in the acceptance region, that is, either a point is included in critical region with probability 1 or with probability zero. Instead of this situation a more general case is randomized test. We assign probability to each point to include in critical region, that is, we define a function on sample space mapping onto [0, 1].

Definition 6.2

A statistical test function $\phi(X_1, X_2, \ldots, X_n)$ is a function defined on the sample space, $\mathfrak{X}$ of $X_1, X_2, \ldots, X_n$ and which takes values on the interval [0, 1] such that $E(\phi(X_1, X_2, \ldots, X_n)) = \propto$, when H is true, that is, $\phi(X) = \phi(X_1, X_2, \ldots, X_n)$ is a Borel measurable function defined from $R_n \to [0, 1]$.

Illustration

(i) $\phi(x) = 0$ if $x \in \mathfrak{X}$.

(ii) $\phi(x) = 1$ if $x \in \mathfrak{X}$.

(iii) $\phi(x) = \alpha$ if $x \in \mathfrak{X}$.

(iv) $\phi(x) = 1$ if $x \in s$, $\phi(x) = 0$ if $x \in s^c$.

(v) $\phi(x) = \begin{cases} 1 \text{ if } x \in s_1 \\ \gamma \text{ if } x \in s_2, \text{ where } s_1 \cup s_2 \cup s_3 = \mathfrak{X}. \\ 0 \text{ if } x \in s_3 \end{cases}$

Remarks

(i) $\phi(x)$ is interpreted as conditional probability of rejecting H whenever $X = x$ is observed.

(ii) $E_\theta \phi(x)$ represents power function of the test, defined as the probability of ejecting H for $\theta \in \Omega_k$

$$E_\theta \phi(x) = \int \phi(x) p_\theta(x) dx$$

$$= \int P_\theta(\text{rejecting } H/X = x) P(X = x) dx$$

$$= \int P_\theta(\text{rejecting } H, X = x)dx$$
$$= P_\theta(\text{rejecting } H)$$
$$= \text{power function of the test}$$

Definition 6.3

$$\underset{\theta \in \Omega_H}{\text{Sup}}\ E_\theta \phi(x) = \text{size}$$

if size $\leq \alpha$, then the test is called level-α test, and if size $= \alpha$, the test is said to be size-α test.

Definition 6.4

Let $\{\phi_\alpha\}$ be class of level-α tests for testing $\theta = \Omega_H$ against $\theta = \Omega_K$ in $\mathbb{P} = \{P_\theta, \theta \in \Omega\}$ family of distributions. Then $\phi_0 \in \{\phi_\alpha\}$ is MP test if $E_K\phi_0(X) \geq E_K\phi(X), \forall\ \phi(X) \in \{\phi_\alpha\}$ and is Uniformly Most Powerful (UMP) test if $E_\theta\phi_0(X) \geq E_\theta\phi(X),\ \forall\ \phi(X) \in \{\phi_\alpha\}, \forall\ \theta \in \Omega_K$ for testing $H{:}\theta \in \Omega_H$ vs $K{:}\theta \in \Omega_K$.

Theorem 6.1: If S is sufficient for P_θ and if $\phi(X)$ and $\Psi(S)$ are test functions defined on X and on support of S respectively then $E_\theta\phi(X) = E_\theta\Psi(S)$.

Proof: Let $\Psi(S) = E(\phi(X)/S = s)$. Then $E_\theta\Psi(S) = E_\theta E(\phi(X)/S = s) = E_\theta\phi(X)$.

Example 6.1

Let $X \sim b(2, p)$ distribution and let $x = 1$ or 2 be critical region for testing $H{:}p = \frac{1}{2}$ vs $K{:}p = \frac{3}{4}$. Then obtain size and power of the test.

The given test function is $\phi(x) = \begin{cases} 1 & \text{if } x = 1 \text{ or } 2 \\ 0 & \text{otherwise} \end{cases}$.

$$\text{size} = \alpha = E_{p=\frac{1}{2}}\phi(X) = P_H(X = 1) + P_H(X = 2)$$

$$= \binom{2}{1}\frac{1}{4} + \binom{2}{2}\frac{1}{4}$$

$$= \frac{2}{4} + \frac{1}{4} = \frac{3}{4}$$

$$\beta = E_{p=\frac{3}{4}}\phi(X) = \binom{2}{1}\left(\frac{3}{4}\right)\left(\frac{1}{4}\right) + \binom{2}{2}\left(\frac{3}{4}\right)^2$$

$$= \frac{6}{16} + \frac{9}{16} = \frac{15}{16}.$$

Example 6.2

Let X_1, X_2 be a random sample from $Be(\theta, 1)$ distribution and let

$$\phi(X_1, X_2) = \begin{cases} 1 & \text{if } \frac{3}{x_1} \leq 4x_2 \\ 0 & \text{otherwise} \end{cases},$$

be the test function for testing $H:\theta = 1$ vs $K:\theta = 2$, obtain size and power of the test.

Here p.d.f. of X is

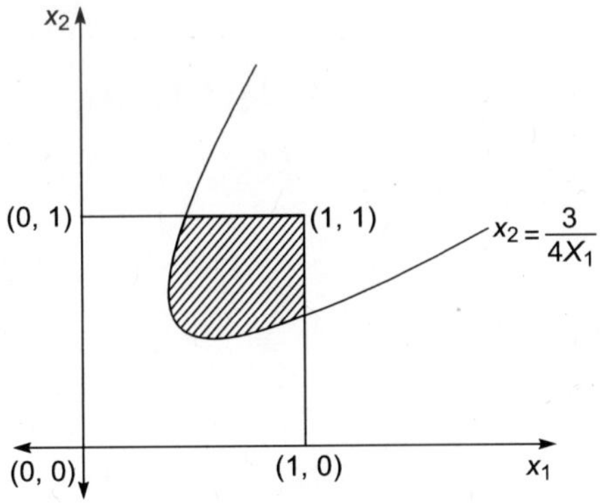

$$f_X(x;\theta) = \begin{cases} \theta x^{\theta-1}, & 0 \leq x \leq 1 \\ 0 & \text{otherwise} \end{cases},$$

$$\text{size} = \alpha = E_H[\phi(X_1, X_2)]$$

$$= P_H\left\{X_2 \geq \frac{3}{4X_1}\right\}$$

$$= \int_{\frac{3}{4}}^{1} \int_{\frac{3}{4X_1}}^{1} dx_2 dx_1$$

$$= \int_{\frac{3}{4}}^{1} \left(1 - \frac{3}{4x_1}\right) dx_1$$

$$= \left[x - \frac{3}{4}\log x_1\right]_{3/4}^{1}$$

$$= \frac{1}{4} + \frac{3}{4}\log\frac{3}{4}$$

$$\text{power} = \beta = E_K[\phi(X_1, X_2)]$$

$$= P_K\left\{X_2 \geq \frac{3}{4X_1}\right\}$$

$$= \int_{\frac{3}{4}}^{1} \int_{\frac{3}{4X_1}}^{1} 4x_2 x_1 dx_2 dx_1$$

$$= \int_{\frac{3}{4}}^{1} 4x_1 \left[\frac{x_2^2}{2}\right]_{\frac{3}{4X_1}}^{1} dx_1$$

$$= \int_{\frac{3}{4}}^{1} 2x_1 \left(1 - \frac{9}{16x_1^2}\right) dx_1$$

$$= 1 - \frac{9}{16} + \frac{9}{8}\log\frac{3}{4}$$

$$= \frac{7}{16} + \frac{9}{8}\log\frac{3}{4}$$

Example 6.3

A random sample, $X_1, X_2, \dots, X_{10}$, is taken from Poisson distribution with parameter λ, for testing $H:\lambda = 0.1$ vs $\lambda > 0.1$. Obtain size and power function for the test function

$$\phi(x) = \begin{cases} 1 & \text{if } \Sigma x_i > 1 \\ 0 & \text{if } \Sigma x_i < 1 \\ \dfrac{1}{2} & \text{if } \Sigma x_i = 1. \end{cases}$$

Here, $Y = \sum_{i=1}^{10} X_i$ has Poisson distribution with parameter 10λ. Under null hypothesis H, $Y \sim P(1)$

$$\text{size} = \propto = P_H(Y > 1) + \frac{1}{2} P_H(Y = 1)$$

$$= 1 - [P_H(Y = 0) + P_H(Y = 1)] + \frac{1}{2} P_H(Y = 1)$$

$$= 1 - (e^{-1} + e^{-1}) + \frac{1}{2}(e^{-1})$$

$$= 1 - 2e^{-1} + \frac{1}{2} e^{-1}$$

$$= 1 - \frac{3}{2e}.$$

$$\text{Power function} = \beta(\phi) = P_K(Y > 1) + \frac{1}{2} P_K(Y = 1)$$

$$= 1 - P_K(Y \leq 1) + \frac{1}{2} P_K(Y = 1)$$

$$= 1 - [e^{-10\lambda} + 10\lambda e^{-10\lambda}] + \frac{1}{2} 10\lambda e^{-10\lambda}$$

$$= 1 - e^{-10\lambda}[1 + 10\lambda - 5\lambda]$$

$$= 1 - e^{-10\lambda}(1 + 5\lambda)$$

Example 6.4

If X has binomial distribution, $b(5, \theta)$ and if the test function for testing

$H:\theta = \frac{1}{2}$ vs $K:\theta = \frac{3}{4}$ with size $\alpha = 0.25$ is

$$\phi(x) = \begin{cases} 1 & \text{if } x > 3 \\ \gamma & \text{if } x = 3 \\ 0 & \text{if } x < 3 \end{cases}$$

then find γ and power of the test.

$$P(X = x) = \binom{5}{x} \theta^x (1-\theta)^{5-x}, x = 0, 1, 2, 3, 4, 5$$

$$E\phi(X) = P(X = 4) + P(X = 5) + \gamma P(X = 3)$$

$$\alpha = \frac{1}{4} = E_H \phi(X) = P_H(X = 4) + P_H(X = 5) + \gamma P_H(X = 3)$$

$$\Rightarrow \qquad \frac{1}{4} = \frac{5}{32} + \frac{1}{32} + \gamma^* \frac{10}{32}$$

$$\Rightarrow \qquad \gamma = \frac{1}{5} = 0.2$$

Now power $\beta(\phi) = E_K \phi(X) = P_K(X = 4) + P_K(X = 5) + \frac{1}{5} P_K(X = 3)$

$$= \frac{5 * 3^4}{4^5} + \frac{3^5}{4^5} + \frac{10 * 3^3}{5 * 4^5} = \frac{702}{1024}.$$

Example 6.5

For testing H against K, probabilities are given in the following table

Table 6.1

x	0	1
H	0.5	0.5
K	0.4	0.6

Determine the class of all test functions of size 0.01 by considering randomized test functions. Let $\phi_1(x)$ be a test function such as

$$\phi_1(x) = \begin{cases} \gamma_1 & \text{if } x = 0 \\ 0 & \text{if } x = 1 \end{cases}$$

Then $$\alpha = E_H\phi(X) = \gamma_1 P(X = 0) = \gamma_1 * 0.5 = 0.01$$

$\Rightarrow$ $$\gamma_1 = 0.02$$

Thus, $\phi_1(x) = \begin{cases} 0.02 & \text{if } x = 0 \\ 0 & \text{if } x = 1 \end{cases}$ has size 0.01

Similarly, $\phi_2(x) = \begin{cases} 0 & \text{if } x = 0 \\ 0.02 & \text{if } x = 1 \end{cases}$ has size 0.01

Let $\phi(x) = \begin{cases} \gamma_1 & \text{if } x = 0 \\ \gamma_2 & \text{if } x = 1 \end{cases}$ be of size 0.01.

Then $0.01 = E_H\phi(X) = \gamma_1 * 0.5 + \gamma_2 * 0.5 \Rightarrow \gamma_1 + \gamma_2 = 0.02$

Different choices of γ_1, γ_2 give different tests of size 0.01.

Example 6.6

Given the following Table 6.2 for probabilities under H and K, obtain MP test for testing H vs K with size $\alpha = 0.02$ by using graphical method of linear programming (LP).

Table 6.2

x	0	1
H	0.2	0.8
K	0.7	0.3
K	0.1	0.9

Neyman's problem can be looked upon as LP problem if linear functions are involved in test function $\phi(x)$. Maximize $E_K\phi(X)$, subject to $E_H\phi(X) \le \alpha$.

In this example let $\phi(x)$ be

$$\phi(x)=\begin{cases}\gamma_1 & \text{if } x=0\\ \gamma_2 & \text{if } x=1\end{cases}$$

Then one has to obtain γ_1, γ_2 such as to maximize $0.8\gamma_1 + 1.2\gamma_2$, subject to $0.2\gamma_1 + 0.8\gamma_2 = 0.02$, $0 \le \gamma_1, \gamma_2 \le 1$.

(γ_1, γ_2)	$\beta(\phi)$
(0, 0)	0
(0.1, 0)	0.08
$\left(0, \frac{1}{40}\right)$	0.03

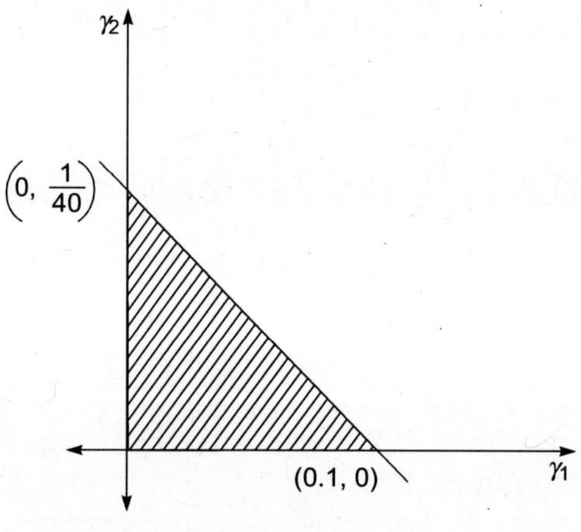

So there are two tests of size $\propto = 0.02$,

$$\phi_1(x)=\begin{cases}0.1 & \text{if } x=0\\ 0 & \text{if } x=1\end{cases} \quad \text{and} \quad \phi_2(x)=\begin{cases}0 & \text{if } x=0\\ \dfrac{1}{40} & \text{if } x=1\end{cases}$$

The power of the test ϕ_1 is $\beta_{\phi_1} = 0.08$ and that of ϕ_2 is $\beta_{\phi_2} = 0.03$. Since $\beta_{\phi_1} > \beta_{\phi_2}$, ϕ_1 is MP test.

PROBLEMS

6.1 Draw graphs of the error probabilities for the following hypotheses.

A. $H: X \in G(1, 1) \quad K: X \in G\left(\frac{1}{2}, 1\right)$

B. $H: X \in U(0, 1) \quad K: X \in U\left(\frac{1}{2}, \frac{3}{2}\right)$

C. $H: X \in b\left(2, \frac{1}{2}\right) \quad K: X \in b\left(2, \frac{2}{3}\right)$

D. $H: X \in P(1) \quad K: X \in P(2)$

where, G, U, b, P denote gamma, uniform, binomial and Poisson distributions, respectively.

6.2 An urn contains 10 marbles of which N are white and $10 - N$ are black. To test $N = 5$ against $N = 6$ one draws 3 marbles from the urn:

(i) without replacement, and

(ii) with replacement.

The null hypothesis is rejected if the sample contains 2 or 3 white marbles. Find the size and power of the test.

6.3 Given a random sample of size n from $N(\theta, 1)$ and given the following two size $\propto$ test for testing $\theta = \theta_0$ against $\theta = \theta_1 (> \theta_0)$, obtain the values of k_1 and k_2. Further examine which one has larger power

$$\phi(x) = \begin{cases} 1 & \text{if } \bar{X} > k_1 \\ 0 & \text{otherwise} \end{cases}$$

$$\phi(x) = \begin{cases} 1 & \text{if } \bar{X} > k_2 \\ 0 & \text{otherwise.} \end{cases}$$

6.4 Obtain power functions for testing $\theta = 0$ against $\theta \neq 0$ in $N(\theta, 1)$ distribution for which the following tests have been suggested:

$$\phi_1(x) = \begin{cases} 1 & \text{if } \bar{X} \geq \frac{Z_\alpha}{\sqrt{n}} \\ 0 & \text{otherwise} \end{cases}$$

$$\phi_2(x) = \begin{cases} 1 & \text{if } \bar{X} \leq -\frac{Z_\alpha}{\sqrt{n}} \\ 0 & \text{otherwise} \end{cases}$$

$$\phi_3(x) = \begin{cases} 1 & \text{if } |\bar{X}| \leq \frac{Z_{\alpha/2}}{\sqrt{n}} \\ 0 & \text{otherwise} \end{cases}$$

where, Z_c is such that $P(Z > Z_c) = C$ and $Z \sim N(0, 1)$

6.5 Given a random sample of size n from $N(0, \theta)$ distribution the following test is suggested for testing $\theta = \theta_0$ against $\theta \neq \theta_0$

$$\phi(x) = \begin{cases} 1 & \text{if } \Sigma X_i^2 \langle C_1 \text{ or } \Sigma X_i^2 \rangle C_2 \\ 0 & \text{if } C_1 \leq \Sigma X_i^2 \leq C_2 \end{cases}$$

Find C_1 and C_2 for a test function ϕ to be of size $\propto$ test. What is power function of this test?

6.6 Let (X_1, X_2, X_3, X_4) be a random sample from $U(0, \theta)$ distribution and let Y_4 be the largest order statistic. Obtain power function of the test for testing $H: \theta = 1$ vs $K: \theta \neq 1$, where,

$$\phi(x) = \begin{cases} 1 & \text{if } Y_4 \leq \frac{1}{2} \text{ or } Y_4 \geq 1 \\ 0 & \text{otherwise} \end{cases}$$

6.7 A box contains 4 balls which are either red or green. Let H: 2 red and 2 green K: 1 red and 3 green. We take randomly one ball from the box. Calculate the size and the power of the test procedure: Accept H if ball is red and accept K with probability 0.2, if ball is green.

6.8 Let $P_\theta(X=1)=\theta=1-P_\theta(X=0)$. We reject $H:\theta=\frac{1}{3}$ when X = 1 and when X = 0 reject it if two heads occur in tossing two unbiased coins. Find the size of the test. Also compute power function of the test when $K:\theta>\frac{1}{3}$.

6.9 Given the following test based on a random sample of size 10 from Poisson distribution with parameter θ, for testing $H:\theta=0.1$ against $\theta>0.1$. Compute power function of the test

$$\phi(x)=\begin{cases}1 & \text{if } \Sigma x_i>6\\ 0 & \text{if } \Sigma x_i<6\\ 0.43 & \text{if } \Sigma x_i=6\end{cases}$$

Find size of the test and power at $\theta = 0.3$

6.10 Let ϕ be MP test of size α for testing $\theta=\theta_0$ against $\theta=\theta_1$. Show that $E_{\theta_1}\phi(x)>\alpha$ unless $P_{\theta_1}=P_{\theta_0}$.

6.11 Let $X\sim b(5,p)$ distribution for testing $H:p=\frac{1}{4}$ vs $p=\frac{1}{2}$ and let test function be

$$\phi(x)=\begin{cases}0.3 & \text{if} \quad x=0\\ 0.2 & \text{if} \quad x=1\\ 0 & \text{otherwise}\end{cases}$$

Obtain α and power β of the test.

6.12 For a random sample of size one from the distribution with p.d.f.

$$f_X(x;\theta)=\frac{\theta}{x^2},\ \theta<x\langle\infty,\theta\rangle 0$$

$$=0 \text{ otherwise}$$

the critical region, S = $\{x: x<c\}$ for testing $H:\theta=\theta_0$ vs $K:\theta=\theta_1$ $(\neq\theta_0)$ is given. Obtain α and β

6.13 For testing $H:X\sim p_0$ vs $K\sim p_1$ with level $\propto=0.02$, the p.m.f. p_0 and p_1 are given below:

x	0	1	
H	0.4	0.6	p_0
K	0.2	0.8	p_1

Let $\phi(x)=\begin{cases}\gamma_1 & \text{if } x=0\\ \gamma_2 & \text{if } x=1\end{cases}$, $0\le\gamma_1,\gamma_2\le 1$. Obtain MP test.

6.14 For the following probability distributions

x	p_0	p_1	p_2
0	0.2	0.3	0.4
1	0.8	0.7	0.6

Obtain MP test for testing $H:X\sim p_0$ or p_1 vs $K:X\sim p_2$, at level 0.01 when test function is $\phi(x)=\begin{cases}\gamma_1 & \text{if } x=0\\ \gamma_2 & \text{if } x=1\end{cases}$, $0\le\gamma_1,\gamma_2\le 1$. Use simplex method to solve LPP.

6.15 Let

x	p_0	p_1	p_2	p_3
–1	0.2	0.3	0.4	0.7
1	0.8	0.7	0.6	0.3

and test function $\phi(x)=\begin{cases}\gamma_1 & \text{if } x=0\\ \gamma_2 & \text{if } x=1\end{cases}$, $0\le\gamma_1,\gamma_2\le 1$. Obtain MP test at level α = 0.01 by simplex method of LPP for testing $H:X\sim p_0$ or p_1 vs $K:X\sim p_2$ or p_3.

6.16 Let $X\sim b(3,\theta)$ obtain MP test at level $\alpha=0.05$ test by using simplex method for testing $H:\theta\in(0.1,0.3,0.5)$ vs $K:\theta\in(0.6,0.7)$.

7

CHAPTER

Neyman Theory

7.1 INTRODUCTION

A housewife wants to purchase maximum number of vegetables say $V_1, V_2, V_3, V_4, V_5, V_6, V_7, V_8$ with her limited budget of Rs. 50. Let cost of vegetables be

Vegetable	V_1	V_2	V_3	V_4	V_5	V_6	V_7	V_8
Vegetable quantity	Q_1	Q_2	Q_3	Q_4	Q_5	Q_6	Q_7	Q_8
Cost/kg in Rs.	c_1	c_2	c_3	c_4	c_5	c_6	c_7	c_8
Quantity per 1 Re.	$\frac{Q_1}{c_1}$	$\frac{Q_2}{c_2}$	$\frac{Q_3}{c_3}$	$\frac{Q_4}{c_4}$	$\frac{Q_5}{c_5}$	$\frac{Q_6}{c_6}$	$\frac{Q_7}{c_7}$	$\frac{Q_8}{c_8}$

Suppose

$$\frac{Q_1}{c_1} = \frac{Q_5}{c_5} = 600 \text{ gm} \Rightarrow \text{select vegetable } (V_1, V_5)$$

$$\frac{Q_2}{c_2} = \frac{Q_3}{c_3} = \frac{Q_8}{c_8} = 400 \text{ gm} \Rightarrow \text{select vegetable } (V_2, V_3, V_8)$$

$$\frac{Q_4}{c_4} = \frac{Q_7}{c_7} = 250 \text{ gm} \Rightarrow \text{select vegetable } (V_4, V_7).$$

Then maximum quantity which she gets is vegetable (V_1, V_5) subject to $\sum_{i=1}^{8} c_i = 50$.

In the same way Neyman theory selects that test which maximizes power of the test (p_1 = proability of rejecting H when it is false) subject to level of the test is fixed at (p_0 = proability of rejecting H when it is true = α). Thus, in the MP test one collects those points of x which satisfy $\frac{p_1(x)}{p_0(x)} \geq k$ for the formation of critical region of MP test. Thus, naturally ratio occurs when some quantity is fixed and the rate is taken that is by rule of three one gets ratio. We give below Neyman-Pearson lemma for randomized and non-randomized tests for most general form of the distribution.

Neyman-Pearson lemma for randomized and non-randomized versions have been treated in section 7.2. Defining monotone class of distributions UMP test for testing one sided hypothesis for this class has been obtained in section 7.3 whereas section 7.4 deals with the testing of one sided hypothesis with loss function. Section 7.5 discusses the locally MP tests and section 7.6 is devoted to least favourable distributions.

7.2 (A) NEYMAN-PEARSON LEMMA (NON-RANDOMIZED VERSION)

Theorem 7.1: Let X have p.f. $p_\theta(x)$ and sample space be $\mathfrak{X}$, $S \subset \mathfrak{X}$.

Then for testing $H:\theta = \theta_0$ vs $K:\theta = \theta_1$ a test having critical region S as

$$S = \{x: p_{\theta_1}(x) \geq kp_{\theta_0}(x)\},\ S^c = \{x: p_{\theta_1}(x) < kp_{\theta_0}(x)\} \qquad \ldots(7.2.1)$$

is MP test of level α where $\alpha = p_{\theta_0}(S)$.

Proof: Let S_1 be any other critical region $p_{\theta_0}(S_1) \leq \alpha$. Now

$$p_\theta(S) = p_\theta(S \cap S_1) + p_\theta(S \cap S_1^c),$$

$$p_\theta(S_1) = p_\theta(S \cap S_1) + p_\theta(S_1 \cap S^c)$$

$$\Rightarrow \qquad p_\theta(S) - p_\theta(S_1) = p_\theta(S \cap S_1^c) - p_\theta(S_1 \cap S^c) \qquad \ldots(7.2.2)$$

If $E \subset S, F \subset S^c$ then from (7.2.1) we conclude that $p_{\theta_1}(E) \geq p_{\theta_0}(E)$, $p_{\theta_1}(F) < p_{\theta_0}(F)$. Taking $E = S \cap S_1^c$, $F = S_1 \cap S^c$, we have

$$p_{\theta_1}(S \cap S_1^c) \geq kp_{\theta_0}(S \cap S_1^c),\ p_{\theta_1}(S_1 \cap S^c) \leq kp_{\theta_0}(S_1 \cap S^c) \qquad \ldots(7.2.3)$$

As $$p_{\theta_0}(S) = \alpha \geq p_{\theta_0}(S_1),$$

$$p_{\theta_0}(S \cap S_1^c) - p_{\theta_0}(S_1 \cap S^c) \geq 0 \qquad \ldots(7.2.4)$$

using (7.1.2). Further from (7.2.2)

$$p_{\theta_1}(S) - p_{\theta_1}(S_1) = p_{\theta_1}(S \cap S_1^c) - p_{\theta_1}(S_1 \cap S^c)$$

$$\geq k[p_{\theta_0}(S \cap S_1^c) - p_{\theta_0}(S_1 \cap S^c)] \text{ using (7.2.3)}$$

$$\geq 0, \text{ using (7.2.4)}$$

Theorem 7.2: Let $\{p_\theta,\ \theta \in \Omega\}$ be the family of probability functions of X. Then the test having critical region, S satisfying

$$x \in S \Rightarrow p_{\theta_1}(x) \geq p_{\theta_0}(x),$$

$$x \in S^c \Rightarrow p_{\theta_1}(x) \leq p_{\theta_0}(x)$$

minimizing the sum of error probabilities for testing

$$H: \theta = \theta_0 \text{ vs } K: \theta = \theta_1.$$

Proof: Let S_1 be any other critical region. If $E \in S$ and $F \in S^c$ then by the given condition, $p_{\theta_1}(E) \geq p_{\theta_0}(E)$, $p_{\theta_1}(F) < p_{\theta_0}(F)$. Taking $E = S \cap S_1^c$, $F = S^c \cap S_1$,

$$p_{\theta_0}(S \cap S_1^c) \leq p_{\theta_1}(S \cap S_1^c),\ p_{\theta_1}(S_1 \cap S^c) \leq p_{\theta_0}(S_1 \cap S^c)$$

$$\Rightarrow \quad p_{\theta_0}(S \cap S_1^c) + p_{\theta_1}(S_1 \cap S^c) \leq p_{\theta_1}(S \cap S_1^c) + p_{\theta_0}(S_1 \cap S^c)$$

$$\Rightarrow \quad [p_{\theta_0}(S \cap S_1^c) + p_{\theta_0}(S \cap S_1)] + [p_{\theta_1}(S^c \cap S_1) + p_{\theta_1}(S^c \cap S_1^c)]$$

$$\leq [p_{\theta_0}(S_1 \cap S^c) + p_{\theta_0}(S \cap S_1)] + [p_{\theta_1}(S_1^c \cap S) + p_{\theta_1}(S_1^c \cap S^c)]$$

$$\Rightarrow \quad p_{\theta_0}(S) + p_{\theta_1}(S^c) \leq p_{\theta_0}(S_1) + p_{\theta_1}(S_1^c)$$

i.e., sum of error probabilities with critical region $S \leq$ sum of error probabilities with critical region S_1

7.2 (B) NEYMAN-PEARSON THEOREM (RANDOMIZED VERSION)

Theorem 7.3: Let $\mathbb{P} = \{p_\theta, \theta \in \Omega\}$ be the family of probability functions of $X = (X_1, X_2, \ldots, X_n)$ with $\Omega = \{\theta_0, \theta_1\}$ and let us denote p_1, p_0 for $p_{\theta_1}(x), p_{\theta_0}(x)$, respectively. Then

(a) Existence: For testing $H: p = p_0$ against the alternative $K: p = p_1$, a test function $\phi_0(x)$ is of the form for $0 < \alpha \le 1$,

$$\phi_0(x) = \begin{cases} 1 & \text{if} \quad p_1 > kp_0 \\ \gamma(x) & \text{if} \quad p_1 = kp_0 \\ 0 & \text{if} \quad p_1 < kp_0 \end{cases} \qquad \ldots(7.2.5)$$

and for $\alpha = 0$, the test is of the form

$$\phi_0(x) = \begin{cases} 1 & \text{if} \quad p_0 = 0 \\ 0 & \text{if} \quad p_0 > 0 \end{cases} \qquad \ldots(7.2.6)$$

where, k, γ are such that $E_{\theta_0}\phi_0(X) = \alpha$.

(b) The test is MP iff it is of the form (7.2.5) or (7.2.6). (Necessary and sufficient condition.)

Proof:

(a) Existence:

(i) Let $\alpha = 0$. Then $\phi_0(x)$ defined in (7.2.6) is of size zero, because $E_{\theta_0}\phi_0(X) = P\{p_0 = 0\} = 0$.

(ii) Let $0 < \alpha \le 1$ and let $\gamma(x) = \gamma$. Then

$$\begin{aligned} \alpha &= E_{\theta_0}\phi_0(X) \\ &= P_{\theta_0}\{p_1 > kp_0\} + \gamma P_{\theta_0}\{p_1 = kp_0\} \\ &= 1 - P_{\theta_0}\{p_1 \le kp_0\} + \gamma P_{\theta_0}\{p_1 = kp_0\} \end{aligned}$$

$$\Rightarrow P_{\theta_0}\{Y \le k\} - \gamma P_{\theta_0}\{Y = k\} = 1 - \alpha$$

where $Y = \dfrac{p_1}{p_0}$. We note that $P_{\theta_0}\{Y \le k\}$ is a distribution function so that it is non-decreasing and right continuous function of k. If there exists k_0 such that the equation

$$P_{\theta_0}\{Y \le k_0\} = 1 - \alpha$$

is satisfied, then we choose $\gamma = 0, k = k_0$. If not, then there exists k_0 such that

$$P_{\theta_0}\{Y < k_0\} < 1 - \alpha < \{Y \le k_0\}$$

and in this case we choose $k = k_0$ and

$$\gamma = \frac{P\{Y \le k_0\} - (1-\alpha)}{P_{\theta_0}\{Y = k_0\}}$$

(b) Necessary Condition (if part)

(i) Let $\alpha = 0$. Consider test function $\phi_0(x)$ defined in (7.2.6). Let $\phi(x)$ be any other function which vanishes on the set $\{x : p_0 > 0\}$ and of size zero. Then ϕ_0 is of size zero, as

$$E_{\theta_0}\phi_0(x) = P(p_0 = 0) + P\{p_0 > 0\} = 0$$

Now

$$E_\theta\phi_0(X) - E_\theta\phi(X) = \int_{\{x:p_0=0\}} [1 - \phi(X)]pdx \ge 0$$

$\Rightarrow \quad \phi_0$ is MP

(ii) Let $0 < \alpha \le 1$. Let us consider test function $\phi_0(x)$ defined in (7.2.5). Let $\phi(x)$ be any other test of size α, that is, $E_{\theta_0}\phi(X) = \alpha$.

Let

$$S^+ = \{x : p_1 > kp_0\}$$

$$S^- = \{x : p_1 < kp_0\}$$

$$S^0 = \{x : p_1 = kp_0\}$$

such that $S^+ \cap S^- \cup S^0 = \mathfrak{X}$, sample space of X. Then

$$\phi_0 - \phi \ge 0 \text{ in } S^+$$

$$\phi_0 - \phi \le 0 \text{ in } S^-$$

$$\phi_0 - \phi = \text{arbitrary value in } S^0$$

and

$$(\phi_0 - \phi)(p_1 - kp_0) \geq 0 \text{ in } \mathscr{X}$$

so that

$$\int_{\mathscr{X}} (\phi_0 - \phi)(p_1 - kp_0)dx \geq 0$$

i.e.,

$$\int (\phi_0 - \phi)p_1 dx \geq k\left[\int \phi_0 p_0 dx - \int \phi p_0 dx\right]$$

$$= k(\alpha - \alpha)$$

$$= 0$$

$$\Rightarrow \qquad E_{\theta_1}\phi_0(X) - E_{\theta_1}\phi(X) \geq 0$$

which means that ϕ_0 is MP test of size α.

(c) *Sufficient Condition (only if part) (uniqueness)*

(i) Let $\alpha = 0$. $\phi_0(x)$ defined in (7.2.6) is given to be MP test. Let $\phi(x)$ be another MP test of size zero which vanishes on the set $\{x : p_0 > 0\}$. Since ϕ_0 is MP, $E_{\theta_1}\phi_0(X) \geq E_{\theta_1}\phi(X)$

i.e.,

$$\int_{\{x:p_0=0\}} (\phi_0 - \phi)p_1 dx \geq 0$$

i.e.,

$$\int_{\{x:p_0=0\}} (1 - \phi)p_1 dx \geq 0$$

Similarly, if ϕ is MP,

$$\int_{\{x:p_0=0\}} (\phi - 1)p_1 dx \geq 0$$

or

$$\int_{\{x:p_0=0\}} (1 - \phi)p_1 dx \leq 0$$

which implies that

$$\int_{\{x:p_0=0\}} (1-\phi)p_1 dx = 0$$

$\Rightarrow \phi = 1$ a.e. on the set $\{x:p_0 = 0\}$. Thus $\phi_0 = \phi$. So any test which is MP must be of the form (7.2.6).

(ii) Let $0 < \alpha \le 1$. $\phi_0(x)$ defined in (7.2.5) is given to be MP of size α. Let $\phi(x)$ be any other MP test of size α. Then

$$E_{\theta_0}\phi_0(X) = E_{\theta_0}\phi(X) = \alpha.$$

Since ϕ_0 is MP,

$$E_{\theta_1}\phi_0(X) - E_{\theta_1}\phi(X) \ge 0$$

i.e.,
$$\int(\phi_0 - \phi)p_1 dx \ge 0$$

i.e.,
$$\int(\phi_0 - \phi)(p_1 - kp_0)dx \ge 0.$$

Similarly, as ϕ is MP test of size α,

$$\int(\phi - \phi_0)(p_1 - kp_0)dx \ge 0$$

i.e.,

$$\int(\phi_0 - \phi)(p_1 - kp_0)dx \le 0$$

$\Rightarrow$
$$\int(\phi_0 - \phi)(p_1 - kp_0)dx = 0,$$

as the integral is non-negative

$\phi_0 = \phi$ except for set $\{p_1 = kp_0\}$.

Thus, the MP test must be of the form (7.2.5).

7.3 MONOTONIC CLASS OF TESTS

For one-sided hypothesis like $H:\theta \le \theta_0$ vs $K:\theta \ge \theta_0$, sometimes UMP test does not exist. In order that such test exists we restrict class of distributions, such as distribution possessing monotonicity.

Definition 7.1

A real parameter family of probability functions, $\{p_\theta, \theta \in \Omega\}$ is said to have monotone likelihood ratio (MLR) in $T(x)$ if for any $\theta_1 < \theta_2 \in \Omega$, whenever $p_{\theta_1} \neq p_{\theta_2}$, the ratio $\dfrac{p_{\theta_2}}{p_{\theta_1}}$ is non-decreasing in $T(x)$.

Illustration 7.1

(1) A one parameter exponential family with density

$$p_\theta(x) = e^{Q(\theta)T(x) + S(x) + D(\theta)}$$

where, $Q(\theta)$ is non-decreasing has MLR in $T(x)$ because for $\theta_1 < \theta_2$,

$$\frac{p_{\theta_2}(x)}{p_{\theta_1}(x)} = e^{\{(Q(\theta_2) - Q(\theta_1))T(x) + (D(\theta_2) - D(\theta_1))\}}$$

is non-deceasing function of $T(x)$. The non-decreasingness of $Q(\theta)$ can be obtained by reparametrization.

(2) Let X have p.d.f.

$$f(x;\theta) = \frac{1}{\pi[1+(x-\theta)^2]}, \quad -\infty < x < \infty.$$

Then for $\theta_1 < \theta_2$,

$$\frac{p_{\theta_2}(x)}{p_{\theta_1}(x)} = \frac{1+(x-\theta_1)^2}{1+(x-\theta_2)^2} \to 1 \text{ as } x \to \pm\infty$$

and hence it does not possess MLR in X.

Theorem 7.4: If $X = (X_1, X_2, \ldots, X_n)$ has family of p.f. $\{p_\theta(x), \theta \in \Omega\}$ and if it has MLR in $T(x)$ then size α test $\phi_0(x)$ of the form

$$\phi_0(x) = \begin{cases} 1 & \text{if } T > c \\ \gamma & \text{if } T = c \\ 0 & \text{if } T < c \end{cases} \qquad \ldots(7.3.1)$$

exists and is UMP for testing $H:\theta \le \theta_0$ vs $K:\theta > \theta_0$, where, c and γ are such that

$$E_{\theta_0}\phi_0(X) = \alpha.$$

Proof: Let $\theta_1, \theta_2 \in \Omega, \theta_1 < \theta_2$. By N-P lemma there exists a MP test $\phi_1(x)$ of the form

$$\phi_1(x) = \begin{cases} 1 & \text{if } \dfrac{p_{\theta_2}}{p_{\theta_1}} > k \\ \gamma & \text{if } \dfrac{p_{\theta_2}}{p_{\theta_1}} = k \\ 0 & \text{if } \dfrac{p_{\theta_2}}{p_{\theta_1}} < k \end{cases}$$

for testing $H_1:\theta = \theta_1$ vs $K_1:\theta = \theta_2$ of size α, such that

$$E_{\theta_1}\phi_1(X) = \alpha, \text{ for } 0 < \alpha \le 1$$

and a test $\phi_1(x)$ of the form

$$\phi_1(x) = \begin{cases} 1 & \text{if } p_{\theta_1} = 0 \\ 0 & \text{if } p_{\theta_1} > 0 \end{cases}$$

exists and is MP test of size zero when $\alpha = 0$. As $p_\theta(x)$ possesses MLR in $T(x)$,

$$\frac{p_{\theta_2}(x)}{p_{\theta_1}(x)} = u(T),$$

where u is non-decreasing function of $T(x)$

$$\Rightarrow \qquad \frac{p_{\theta_2}}{p_{\theta_1}} > k$$

$$\Rightarrow \qquad u(T) > k$$

$$\Rightarrow \qquad T > u^{-1}(k)$$

$$\Rightarrow \qquad T > c$$

Thus, test ϕ_1 is same as test ϕ_0 defined in (7.3.1). The trivial test $\phi^*(x) \equiv \alpha$ has size and power α and hence ϕ_0 being MP,

$$E_{\theta_2}\phi_0(X) \geq E_{\theta_2}\phi^*(X) = \alpha = E_{\theta_1}\phi_0(X)$$

which shows that power function of the test ϕ_0 is non-decreasing.

Let $\theta_1 = \theta_0$. Then due to above argument ϕ_0 is MP for testing

$$H_2 : \theta = \theta_1 \text{ vs } K_2 : \theta = \theta_2 (> \theta_0)$$

such that

$$E_{\theta_0}\phi_0(X) = \alpha \qquad \text{...(7.3.2)}$$

As the test determined by (7.3.2) does not involve alternative hypothesis ϕ_0 is UMP test for testing

$$H_3 : \theta = \theta_0 \text{ vs } K_3 : \theta > \theta_0$$

satisfying (7.3.2). Further, as power function of ϕ_0 is non-decreasing

$$E_{\theta}\phi_0(X) \leq E_{\theta_0}\phi_0(X) = \alpha \text{ for } \theta \leq \theta_0. \qquad \text{...(7.3.3)}$$

Now as the class of tests satisfying (7.3.3) is included in the class of tests satisfying (7.3.2) it follows that ϕ_0 which is UMP test for testing H_3 in the larger class must also be UMP test in smaller class for testing $H_1 : \theta \leq \theta_0$.

Remark 7.1

By changing inequalities in (7.3.1), one gets UMP test for testing

$$H : \theta > \theta_0 \text{ vs } K : \theta \leq \theta_0 .$$

Corollary 7.1

If the distribution of X has MLR in $T(x)$ and if $\phi_0(x)$ is a test function of size α of the form

$$\phi_0(x) = \begin{cases} 1 & \text{if} \quad T(x) > c \\ \gamma & \text{if} \quad T(x) = c \\ 0 & \text{if} \quad T(x) < c \end{cases}$$

for testing $H : \theta \le \theta_0$ vs $K : \theta > \theta_0$ then

$$E_\theta \phi_0(X) \le E_\theta \phi(X) \text{ for } \theta \le \theta_0$$

$$E_\theta \phi_0(X) \ge E_\theta \phi(X) \text{ for } \theta > \theta_0$$

for every test $\phi(x)$ of size α.

Proof: Since ϕ_0 and ϕ are of size α,

$$E_{\theta_0} \phi_0(X) = E_\theta \phi(X) = \alpha$$

and since $\phi_0(x)$ is UMP test,

$$E_\theta \phi_0(X) \ge E_\theta \phi(X) \text{ for } \theta \ge \theta_0 \ \forall \ \phi(x).$$

Further, due to symmetry of the theorem 7.4 it implies that $1 - \phi_0(x)$ is UMP of size α for testing

$$H_1 : \theta \ge \theta_0 \text{ vs } K_1 : \theta < \theta_0$$

and hence,

$$E_\theta [1 - \phi_0(X)] \ge E_\theta [1 - \phi(X)] \text{ for } \theta \le \theta_0$$

$$\Rightarrow \qquad E_\theta \phi_0(X) \le E_\theta \phi(X) \text{ for } \theta \le \theta_0$$

7.4 PROBLEM OF TESTING OF HYPOTHESIS WITH LOSS FUNCTION

A more general problem is to include loss function while considering two types of errors. Let us recall the preliminaries.

A loss function, $L(\theta, d)$, is a real valued function defined on $\Omega \times D$, where, Ω is parametric space, D is decision space $\{d_0, d_1\}$, and the expected loss function is risk function, $R(\theta, d)$,

$$R(\theta, d) = E_\theta L(\theta, d).$$

Let $\phi(x)$ be a test function which interprets as probability of rejecting H when x is observed. Then

$$R(\theta, d) = L(\theta, d_0)[1 - E_\theta\phi(X)] + L(\theta, d_1)E_\theta\phi(X)$$

$$= L(\theta, d_0) + [L(\theta, d_1) - L(\theta, d_0)]E_\theta\phi(X)$$

where d_0 is the decision to accept H when H is true and d_1 is the decision to reject H when H is true.

Natural assumptions on loss function are

$$L(\theta, d_0) = 0 \text{ if } \theta \in \Omega_H,\ L(\theta, d_0) = a \text{ for } \theta \in \Omega_K$$

$$L(\theta, d_1) = 0 \text{ if } \theta \in \Omega_K,\ L(\theta, d_1) = b \text{ for } \theta \in \Omega_H$$

Then for $\theta \in \Omega_K$,

$$\min R(\theta, \phi) = \min\{1 - E_\theta\phi(X)\}a,\ a > 0$$

$$= \max E_\theta\phi(X)$$

$$= \text{max of the power function.}$$

We recall some definitions from decision theory.

Definition 7.2

A rule d_0 is as good as a rule d_1 if $R(\theta, d_0) \le R(\theta, d_1)$ for all $\theta \in \Omega$.

Definition 7.3

A class c of decision rules is essentially complete if for every rule d outside c there exists a rule d_0 in c which is as good as d.

Theorem 7.5: If the loss function satisfies

$$L(\theta, d_1) - L(\theta, d_0) > 0 \text{ for } \theta < \theta_0$$

$$L(\theta, d_1) - L(\theta, d_0) < 0 \text{ for } \theta \ge \theta_0$$

and if the distribution of X has MLR in $T(x)$ for testing

$$H: \theta \le \theta_0 \text{ vs } K: \theta > \theta_0,$$

the class of tests $\phi_0(x)$ of the form

$$\phi_0(x) = \begin{cases} 1 & \text{if } T(x) > c \\ \gamma & \text{if } T(x) = c \\ 0 & \text{if } T(x) < c \end{cases} \qquad \ldots(7.4.1)$$

is essentially complete.

Proof: Let $E_{\theta_0}\phi_0(X) = \alpha$ and let $\phi(x)$ be any other test of size α, $E_{\theta_0}\phi(X) = \alpha$.

For showing ϕ_0 to be essentially complete we have to show

$$R(\theta, \phi_0) \le R(\theta, \phi), \forall\, \theta.$$

As

$$R(\theta, \phi) = L(\theta, d_0) + E_\theta\phi(X)[L(\theta, d_1) - L(\theta, d_0)]$$

$$R(\theta, \phi) - R(\theta, \phi_0) = [L(\theta, d_1) - L(\theta, d_0)][E_\theta\phi(X) - E_\theta\phi_0(X)].$$

Due to the assumption in the enunciation and due to the Corollary 7.1,

$$[L(\theta, d_1) - L(\theta, d_0)][E_\theta\phi(X) - E_\theta\phi_0(X)] \ge 0$$

for $\theta < \theta_0$ or $\theta \ge \theta_0$. Thus,

$$R(\theta, \phi) - R(\theta, \phi_0) \ge 0 \text{ for } \theta \in \Omega,$$

which proves the theorem.

Example 7.1

A random sample of size 4 is taken from the probability function $P_\theta(X_i = 1) = \theta$ and $P_\theta(X_i = 0) = 1 - \theta$, for $i = 1, 2, 3, 4$. Obtain MP test for testing: $\theta = \frac{1}{3}$ vs $K: \theta = \frac{1}{4}$, with size $\alpha = \frac{8}{27}$.

Let $X = \sum_{i=1}^{n} X_i$. Then X has $b(4, \theta)$ distribution.

As

$$p_{\theta_0}(x) = \binom{4}{x}\left(\frac{1}{3}\right)^x\left(\frac{2}{3}\right)^{4-x}$$

$$= \binom{4}{x}\frac{2^{4-x}}{81}$$

and under hypothesis K the probability function being

$$p_{\theta_1}(x) = \binom{4}{x}\left(\frac{1}{4}\right)^x\left(\frac{3}{4}\right)^{4-x}$$

$$= \binom{4}{x}\frac{3^{4-x}}{256},$$

$$\frac{p_{\theta_1}(x)}{p_{\theta_0}(x)} > k_1 \Rightarrow \left(\frac{3}{2}\right)^{4-x}\frac{81}{256} > k_1$$

$$\Rightarrow \quad 4 - x > k_2$$

$$\Rightarrow \quad x < k.$$

Then by N-P lemma the MP test function $\phi(x)$ will be

$$\phi(x) = \begin{cases} 1 & \text{if } x < k \\ \gamma & \text{if } x = k. \\ 0 & \text{if } x > k \end{cases}$$

We have to determine k and γ such that $E_{\theta=\frac{1}{3}}\phi(x) = \alpha$. The probability table under H and K will be

x	0	1	2	3	4
$p_{\theta_0}(x)$	16/81	32/81	24/81	8/81	1/81
$p_{\theta_1}(x)$	8/256	108/256	54/256	12/256	1/256

Taking $k = 1$, $E_{\theta=\frac{1}{3}}[\phi(x)] = P_{\theta=\frac{1}{3}}(X=0) + \gamma P_{\theta=\frac{1}{3}}(X=1) = \frac{24}{81}$

$$\Rightarrow \quad \frac{16}{81} + \gamma\frac{32}{81} = \frac{24}{81}$$

$$\Rightarrow \qquad \gamma = \frac{1}{4}.$$

Thus MP test is of the form

$$\phi(x) = \begin{cases} 1 & \text{if } x = 0 \\ \frac{1}{4} & \text{if } x = 1 \\ 0 & \text{otherwise} \end{cases}$$

$$\beta = \frac{35}{256}$$

Example 7.2

Let X have truncated Poisson distribution with parameter λ of the form

$$P(X = x) = \frac{e^{-\lambda}\lambda^x}{x!\,k(\lambda)},\ x = 3, 4, \ldots;\ \lambda > 0$$

where $k(\lambda) = 1 - \sum_{x=0}^{2} \frac{e^{-\lambda}\lambda^x}{x!}$. Obtain MP test of size $\alpha = 0.3$ for testing

$$H: \lambda = 1 \text{ vs } K: \lambda = 2.$$

Here, $$p_\lambda(X = x) = p_\lambda(x)$$

$$= \frac{e^{-\lambda}\lambda^x}{x!} \Bigg/ \left(1 - \left[e^{-\lambda} + \lambda e^{-\lambda} + \frac{\lambda^2}{2} e^{-\lambda}\right]\right)$$

$$= \frac{e^{-\lambda}\lambda^x}{x!\left[1 - e^{-\lambda}\left(1 + \lambda + \frac{\lambda^2}{2}\right)\right]},\ x = 3, 4, 5, \ldots$$

and so $$p_0(x) = P_\lambda(X = x/H) = \frac{e^{-1}}{x![1 - 2.5e^{-1}]},\ x = 3, 4, 5, \ldots$$

$$p_1(x) = P_\lambda(X = x/H) = \frac{2^x e^{-2}}{x![1-5e^{-2}]}, x = 3, 4, 5, \ldots$$

So

$$\frac{p_1(x)}{p_0(x)} = \frac{2^x e^{-1}[1-2.5e^{-1}]}{[1-5e^{-2}]} > c_1 \Rightarrow 2^x > c_2 \Rightarrow x > c$$

By Neymann-Pearson Lemma, the MP test function is

$$\phi(x) = \begin{cases} 1 & \text{if} \quad x > c \\ \gamma & \text{if} \quad x = c \\ 0 & \text{if} \quad \text{otherwise} \end{cases}$$

where (c, γ) will be determined by using

$X =$	3	> 3
$p_0(x)$	0.7641	0.2359
$p_1(x)$	0.5576	0.4424

$$\alpha = E(\phi(x)/H)$$

$\Rightarrow$ $$0.3 = P_{\lambda=1}(X > 3) + \gamma P_{\lambda=1}(X = 3), \text{ letting } c = 3$$

$\Rightarrow$ $$P_{\lambda=1}(X > 3) + \gamma P_{\lambda=1}(X = 3)$$

$\Rightarrow$ $$0.3 = 0.2359 + 0.7641\gamma$$

$\Rightarrow$ $$\gamma = 0.084.$$

Thus MP test of size $\alpha = 0.3$ is

$$\phi(x) = \begin{cases} 1 & \text{if} \quad x > 3 \\ 0.0840 & \text{if} \quad x = 3 \\ 0 & \text{Otherwise} \end{cases}$$

and power $\beta_\phi(\lambda)$ is

$$\beta_\phi(\lambda = 2) = P_{\lambda=2}(X > 3) + 0.0840 P_{\lambda=2}(X = 3)$$

$$= 0.4424 + 0.0840\ (0.5576)$$

$$= 0.4870$$

Example 7.3

Let $X_1, X_2, \ldots, X_n$ be a random sample from geometric distribution with parameter θ. Obtain UMP test for testing

$$H:\theta = \theta_0 \text{ vs } K:\theta > \theta_0 \text{ at level } \alpha.$$

$$\begin{aligned} p_\theta(x) &= \prod_{i=1}^{n} [\theta(1-\theta)^{x_i}] \\ &= \theta^n (1-\theta)^{\Sigma x_i} \\ &= \theta^n e^{[\log(1-\theta)]\Sigma x_i} \\ &= c(\theta) e^{q(\theta)t(x)} h(x) \end{aligned}$$

where $c(\theta) = \theta^n$, $q(\theta) = \log(1-\theta)$, $t(x) = \Sigma x_i$, which has exponential family of distributions. As

$$\frac{d}{d\theta} q(\theta) = -\frac{1}{1-\theta} < 0 \Rightarrow q(\theta) \downarrow \theta(\geq \theta_0)$$

or

$$\frac{p_{\theta_1}(x)}{p_{\theta_0}(x)} = \left(\frac{\theta_1}{\theta_0}\right)^n \left(\frac{1-\theta_1}{1-\theta_0}\right)^{\Sigma x_i}$$

as $\theta_1 > \theta_0, 1-\theta_1 \leq 1-\theta_0 \Rightarrow \dfrac{1-\theta_1}{1-\theta_0} \leq 1$, which is a decreasing function of $T(x) = \Sigma x_i$.

So the UMP test is

$$\phi(x) = \begin{cases} 1 & \text{if } \ t < c \\ \gamma & \text{if } \ t = c \\ 0 & \text{if } \ \text{otherwise} \end{cases}$$

where c and γ are determined by solving $E_{\theta_0}\phi(x) = \alpha$,

$$P_{\theta_0}\{T(x) < c\} + \gamma P_{\theta_0}\{T(x) = c\} = \alpha$$

where $T(x) = \Sigma x_i$ follow negative binomial distribution and its p.m.f. is

$$P_\theta(T=t)=\binom{n+t-1}{t}\theta^n(1-\theta)^t, t=0,1,2,\ldots.$$

Thus, c and γ can be determined by solving

$$P_{\theta_0}\{T<c\}+\gamma P_{\theta_0}\{T=c\}=\alpha$$

that is, by solving

$$\sum_{t=0}^{c-1}\binom{n+t-1}{t}\theta_0^n(1-\theta_0)^t+\gamma\binom{n+c-1}{c}\theta_0^n(1-\theta_0)^c=\alpha$$

and power function of the test is

$$E_\theta(\phi(x))=\sum_{t=0}^{c-1}\binom{n+t-1}{t}\theta^n(1-\theta)^t+\gamma\binom{n+c-1}{c}\theta^n(1-\theta)^c$$

Example 7.4

From a lot containing N items of the manufactured product having M defective items a random sample of n items is taken without replacement. Obtain UMP test of size α for testing

$$H:M\le M_0 \text{ vs } K:M>M_0.$$

For $M_0 = 2$, $N = 20$, $n = 5$, $\alpha = 0.05$, obtain UMP test and power function for $M = 5$.

Here the p.m.f. of the product, X is

$$P_M(X=x)=\frac{\binom{M}{x}\binom{N-M}{n-x}}{\binom{N}{n}}, x=0,1,2,\ldots,\min(n,M)$$

where N is population size, n is the sample size, M is the number of defectives in the population which is parameter of this hypergeometric distribution.

Since

$$\frac{P_{M+1}(X=x)}{P_M(X=x)} = \frac{(M+1)}{(N-M)}\frac{N-M-n+x}{M+1-x} \uparrow x,$$

$P_M(x)$ has MLR in X. So for testing H vs K, the UMP test $\phi(x)$ is given by

$$\phi(x) = \begin{cases} 1 & \text{if } x > k \\ \gamma & \text{if } x = k \\ o & \text{if } x < k \end{cases}$$

where k, γ are such that

$$E_H \phi(x) = \alpha.$$

For $N = 20$, $M_0 = 2$, $n = 5$, $\alpha = 0.05$, let us take $k = 2$. Then the UMP test size $\alpha = 0.05$ is given by

$$\phi(x) = \begin{cases} 1 & \text{if } x > 2 \\ \gamma & \text{if } x = 2 \\ o & \text{if } x < 2 \end{cases}$$

where γ is determined by solving $E_H\phi(x) = 0.05$, i.e.,

$$1 \times P_H(X > 2) + \gamma P_H(X = 2) = 0.05.$$

As the largest value of $x = \min(5, 2) = 2$, x takes the values 0, 1, 2. $P_H(X > 2) = 0$,

$$P_H(X = 2, M = 2) = \frac{\binom{18}{3}}{\binom{20}{5}} = \frac{1}{19} \Rightarrow \gamma = 19 \times 0.05 = 0.95.$$

Thus UMP test is

$$\phi(x) = \begin{cases} 1 & \text{if } x > 2 \\ 0.95 & \text{if } x = 2. \\ 0 & \text{if } x < 2 \end{cases}$$

The power β at $M = 5$ is

$$\beta = 0.95 \times P_{M=5}(X=2)$$

$$= 0.95 \times \binom{5}{2}\frac{\binom{15}{3}}{\binom{20}{5}}$$

$$= 0.28.$$

Example 7.5

Let $X_1, X_2, \ldots, X_n$ be a random sample from exponential distribution with p.d.f.

$$p_\theta(x) = \theta e^{-\theta x}, x > 0, \theta > 0.$$

Then obtain UMP test of size α for testing

$$H: \theta \le \theta_0 \text{ vs } K: \theta > \theta_0$$

and obtain power function.

Since $p_\theta(x_1, x_2, \ldots, x_n) = \theta^n e^{-\theta \Sigma x_i} = \theta^n e^{-\theta t}$, where, $T(x) = \sum_{i=1}^{n} x_i$.

Let $\theta_1 > \theta_0$. Then

$$\frac{p_{\theta_1}(x_1, x_2, \ldots, x_n)}{p_{\theta_0}(x_1, x_2, \ldots, x_n)} = \left(\frac{\theta_1}{\theta_0}\right)^n e^{-(\theta_1-\theta_0)t} \downarrow t(x) \text{ as } \theta_1 - \theta_0 > 0.$$

So size α UMP test will be

$$\phi(t) = \begin{cases} 1 & \text{if } \; t < k \\ 0 & \text{otherwise} \end{cases}$$

i.e.,

$$\phi(t) = \begin{cases} 1 & \text{if } \; 2\theta t < 2\theta k \\ 0 & \text{otherwise} \end{cases}.$$

If $Y = 2\theta t$, then $Y \sim \chi^2_{2n}$, as $t \sim G(\theta, n)$ distribution, k is determined by solving the equation

$$E_{\theta_0}\phi(Y) = \alpha$$

i.e., $$p_{\theta_0}(Y < 2\theta_0 k) = \alpha$$

i.e., $$p_{\theta_0}(Y \geq 2\theta_0 k) = 1 - \alpha.$$

If $P_{\theta_0}(Y \geq \chi^2_{2n,1-\alpha}) = 1 - \alpha$, where, $\chi^2_{n,\alpha}$ is upper α percentage point of χ^2_n distribution. Then

$$2\theta_0 k = \chi^2_{2n,1-\alpha}$$

$\Rightarrow$ $$k = \frac{\chi^2_{2n,1-\alpha}}{2\theta_0} = k_0, \text{ say.}$$

Now power function of the test is

$$\beta(\theta) = E_K\phi(Y)$$

$$= P_\theta(Y \leq 2\theta k_0)$$

$$= \int_0^{k_1} \frac{1}{2^n \Gamma(n)} e^{-\frac{Y}{2}} Y^{n-1} dY$$

where $$k_1 = \frac{\theta}{\theta_0}\chi^2_{2n,1-\alpha}.$$

For different values of $\theta, (\geq \theta_0), k_1$ is calculated and hence powers at different values of θ will be determined.

Example 7.6

Given a random sample of size n from the distribution of X with p.d.f.

$$f_X(x;\theta) = e^{-(x-\theta)}, \theta < x < \infty.$$

Obtain UMP tests of size α for testing

(i) $H_1 : \theta \leq \theta_0$ vs $K_1 : \theta > \theta_0$

(ii) $H_2 : \theta = \theta_0$ vs $K_2 : \theta \neq \theta_0$

Further determine power functions for the tests.

Let $Y = \min X_i, i = 1, 2, \ldots, n$. Then the p.d.f. of Y will be

$$g_Y(y;\theta) = \begin{cases} ne^{-n(y-\theta)} & \text{if } \theta \le y \\ 0 & \text{if } \theta > y \end{cases}$$

obtained by using

$$F_X(x) = 1 - e^{-(x-\theta)} \quad \text{in} \quad g_Y(y) = n[1 - F_X(y)]^{n-1} F_X(y).$$

Since $P(Y \le \theta_0) = 0$, the set $A = \{y : y \le \theta_0\}$ is included in the c.r. If we include A in the critical region, we get MP test of size 0 and if we exclude A from c.r. the MP test has size 1. Both the situations are unacceptable as we need size α test. Thus, there is a need for randomization by taking a subset of a sample space. Let us take one of the subsets as $B = \{y : Y \ge k\}$, such that

$$\alpha = \int_k^\infty ne^{-n(y-\theta_0)} dy$$

$$\Rightarrow \qquad k = \theta_0 - \frac{\log \alpha}{n}$$

$$\Rightarrow \qquad B = \left\{ y : Y \ge \theta_0 - \frac{\log \alpha}{n} \right\}.$$

Thus, UMP test $\phi(y)$ has critical region $A \cup B$ as test does not depend on alternatives,

$$\phi(y) = \begin{cases} 1 & \text{if } \; y \le \theta_0 \text{ or } y \ge \theta_0 - \dfrac{\log \alpha}{n} \\ 0 & \text{if } \; \text{otherwise} \end{cases}.$$

The power function for three different cases will be due to θ lying in three different intervals such as

(i) (θ, θ_0)

(ii) $\left[\theta_0, \theta_0 - \dfrac{\log \alpha}{n}\right)$

(iii) $\left[\theta_0 - \dfrac{\log \alpha}{n}, \infty\right)$

Case 1 $\theta > \theta_0$:

$$\beta_\phi(\theta) = \int_{\theta}^{\theta_0} ne^{-n(y-\theta)}dy + \int_{\theta_0 - \frac{\log\alpha}{n}}^{\infty} ne^{-n(y-\theta)}dy$$

$$= 1-(1-\alpha)e^{-n(\theta_0-\theta)}$$

Case 2 $\theta_0 < \theta < \theta_0 - \dfrac{\log\alpha}{n}$:

$$\beta_\phi(\theta) = \int_{\theta_0 - \frac{\log\alpha}{n}}^{\infty} ne^{-n(y-\theta)}dy$$

$$= \alpha e^{-n(\theta_0-\theta)}$$

Case 3 $\theta < \theta_0$:

$$\beta_\phi(\theta) = 0 + \int_{\theta}^{\infty} ne^{-n(y-\theta)}dy = 1.$$

Example 7.7

Obtain MP test of size α for testing

$$H: X \sim f(x) \text{ vs } K: X \sim g(x)$$

where

$$f(x) = \frac{e^{-\frac{x^2}{2}}}{\sqrt{2\pi}}, -\infty < x < \infty \text{ and } g(x) = \frac{1}{2}e^{-|x|}, -\infty < x < \infty.$$

By Neyman-Pearson lemma the MP test is of the form

$$\phi(x) = \begin{cases} 1 & \text{if } \dfrac{g(x)}{f(x)} > k \\ 0 & \text{otherwise} \end{cases}.$$

Now

$$\frac{g(x)}{f(x)} = \frac{\frac{1}{2}e^{-|x|}}{\frac{e^{-\frac{x^2}{2}}}{\sqrt{2\pi}}} > k \Rightarrow |x|^2 - 2|x| - k > 0$$

where k is such that

$$E_H\phi(X) = \alpha$$

i.e., $$P_H\{|X|^2 - 2|x| - k > 0\} = \alpha$$

i.e., $$P_H\{|X| > 1 + \sqrt{1+k} > 0\} = \alpha$$

i.e., $$P_H\{X < 1 - \sqrt{1+k} \text{ or } X > 1 + \sqrt{1+k}\} = \alpha$$

i.e., $$2P_H\{X > 1 + \sqrt{1+k}\} = \alpha, \text{ as } X \text{ is symmetric}$$

i.e., $$P_H\{X > 1 + \sqrt{1+k}\} = \frac{\alpha}{2}.$$

If $P\left\{X > -Z_{\frac{\alpha}{2}}\right\} = \frac{\alpha}{2}$, where, $Z_{\frac{\alpha}{2}}$ is upper $\frac{\alpha}{2}$ percentage point of $N(0, 1)$. Then

$$1 + \sqrt{1+k} = -Z_{\frac{\alpha}{2}} \Rightarrow k = \left(1 + Z_{\frac{\alpha}{2}}\right)^2 - 1$$

Particularly if $\alpha = 0.05$, $Z_{\frac{\alpha}{2}} = 1.96$,

$$k = (2.96)^2 - 1 \Rightarrow k = 7.92.$$

So in this case UMP test is

$$\phi(x) = \begin{cases} 1 & \text{if } \quad x \langle -7.92 \text{ or } x \rangle 7.92 \\ 0 & \text{otherwise} \end{cases}.$$

Example 7.8

Let $X_1, X_2, \dots, X_n$ be a random sample from $N(\theta, 1)$. Then obtain UMP test of size α for testing

$$H_1: \theta = \theta_0 \text{ vs } K_1: \theta > \theta_0.$$

Also obtain power function of the test. Now obtain sample size n for MP test of size α and of power $(1 - \beta(\theta))$ for testing

$$H_1: \theta = \theta_0 \text{ vs } K_1: \theta = \theta_1.$$

As

$$p_\theta(x) = \left(\frac{1}{\sqrt{2\pi}}\right)^n e^{-\frac{1}{2}\sum_1^n (x_i - \theta)^2}$$

$$= \left(\frac{1}{\sqrt{2\pi}}\right)^n e^{-\frac{\sum x_i^2}{2}} e^{-\frac{n\theta^2}{2}} e^{n\theta\bar{x}}$$

$$= \left(\frac{1}{\sqrt{2\pi}}\right)^n e^{-\frac{\sum x_i^2}{2}} e^{-\frac{n\theta^2}{2}} e^{n\theta\bar{x}}$$

$$= c(\theta) h(x) e^{Q(\theta)T(x)}$$

where $c(\theta) = e^{-\frac{n\theta^2}{2}}$, $h(x) = \left(\frac{1}{\sqrt{2\pi}}\right)^n e^{-\frac{\sum x_i^2}{2}}$, $Q(\theta) = n\theta$, $T(x) = \bar{x}$, which is the exponential family of distributions. As $n\,\theta \uparrow$ for $\theta > \theta_0$, the UMP test of size α is

$$\phi(x) = \begin{cases} 1 & \text{if } \bar{x} > k \\ 0 & \text{otherwise} \end{cases}$$

where k is determined by

$$\alpha = E_H \phi(X)$$

$$= P_{\theta_0}(\bar{X} > k)$$

$$= P_{\theta_0}\{(\bar{X}-\theta_0)\sqrt{n} > (k-\theta_0)\sqrt{n}\}$$

$$= P_{\theta_0}\{Z > (k-\theta_0)\sqrt{n}\}$$

where $Z \sim N(0,1)$ under H.

If $P(Z > Z_\alpha) = \alpha$, where, Z_α is upper α percentage value of $N(0, 1)$ distribution then

$$(k-\theta_0)\sqrt{n} = Z_\alpha$$

$$\Rightarrow \qquad k = \theta_0 + \frac{Z_\alpha}{\sqrt{n}}$$

Thus, UMP size α test is

$$\phi(x) = \begin{cases} 1 & \text{if } \bar{x} > \theta_0 + \dfrac{Z_\alpha}{\sqrt{n}}. \\ 0 & \text{otherwise} \end{cases}$$

The power function of the test is

$$\beta(\theta) = E_\theta \phi(X), \theta > \theta_0$$

$$= P_\theta\left\{\bar{X} > \theta_0 + \frac{Z_\alpha}{\sqrt{n}}\right\}$$

$$= P_\theta\{(\bar{X}-\theta)\sqrt{n} > Z_\alpha - (\theta-\theta_0)\sqrt{n}\}$$

$$= P_\theta\{Z > Z_\alpha - (\theta-\theta_0)\sqrt{n}\}$$

$$1-\beta(\theta_1) = 1-\phi(Z_\alpha + (\theta_1-\theta_0)\sqrt{n})$$

$$= \phi((\theta_1-\theta_0)\sqrt{n} - Z_\alpha)$$

using $\qquad \phi(-x) = 1-\phi(x)$.

As $\qquad \phi(Z_\beta) = 1-\beta$,

$$(\theta_1-\theta_0)\sqrt{n} - Z_\alpha = Z_\beta$$

$$\Rightarrow \qquad n = \frac{(Z_\alpha + Z_\beta)^2}{(\theta_1-\theta_0)^2}.$$

7.5 LOCALLY MOST POWERFUL TEST

Whenever UMP test does not exist one would like to find a MP test in the neighbourhood of null hypothesis. If null hypothesis $H:\theta=\theta_0$ then we have to obtain a test $\phi(x)$ such that its size $E_{\theta_0}\phi(X)=\alpha$ and it maximizes power $E_\theta\phi(X)=\beta_\phi(\theta)$ in the interval $(\theta_0-\in, \theta_0+\in)$. To do this let $\beta_\phi(x)$ admit Taylor expansion about θ_0 (taking $\beta_\phi(\theta_0)=\alpha$)

$$\beta_\phi(\theta)=\alpha+(\theta-\theta_0)\frac{\partial}{\partial\theta_0}\beta_\phi(\theta)+\frac{(\theta-\theta_0)^2}{2!}\frac{\partial^2}{\partial\theta_0^2}\beta_\phi(\theta)+\ldots$$

If the alternative hypothesis $K:\theta>\theta_0$, we have to maximize

$$\frac{\partial}{\partial\theta_0}\beta_\phi(\theta)=\int\phi(x)\frac{\partial}{\partial\theta_0}p_\theta(x)dx,$$

in which we have assumed differentiation under integral sign such that

$$\beta_\phi(\theta_0)=\int\phi(x)p_{\theta_0}(\theta)dx=\alpha.$$

By Neyman-Pearson lemma LMP test $\phi_0(x)$ is

$$\phi_0(x)=\begin{cases}1 \text{ if } & \dfrac{\partial}{\partial\theta_0}p_\theta(x)>kp_{\theta_0}(x)\\ \gamma \text{ if } & \dfrac{\partial}{\partial\theta_0}p_\theta(x)=kp_{\theta_0}(x).\\ 0 \text{ if } & \dfrac{\partial}{\partial\theta_0}p_\theta(x)<kp_{\theta_0}(x)\end{cases}$$

where k and γ are determined by $E_{\theta_0}\phi_0(x)=\alpha$, $\phi_0(x)$ is also written as

$$\phi_0(x)=\begin{cases}1 \text{ if } & \dfrac{\partial}{\partial\theta_0}\log p(x)>k\\ \gamma \text{ if } & \dfrac{\partial}{\partial\theta_0}\log p(x)=k\\ 0 \text{ if } & \dfrac{\partial}{\partial\theta_0}\log p(x)<k\end{cases}$$

If the alternatives are $\theta < \theta_0$ LMP test is

$$\phi_1(x) = \begin{cases} 1 & \text{if } \dfrac{\partial}{\partial\theta_0}\log p(x) < k \\ \gamma & \text{if } \dfrac{\partial}{\partial\theta_0}\log p(x) = k. \\ 0 & \text{if } \dfrac{\partial}{\partial\theta_0}\log p(x) > k \end{cases}$$

Example 7.9

Obtain locally most powerful level α test for testing

$$H:\mu = \mu_0 \text{ vs } K:\mu > \mu_0$$

on the basis of n independent observations drawn from $N(\mu, 1)$ distribution.
As

$$\log p(x_1, x_2, \ldots, x_n) = -\frac{n}{2}\log(2\pi) - \frac{1}{2}\sum_{i=1}^{n}(x_i - \mu)^2$$

and

$$\frac{\partial \log p(x_1, x_2, \ldots, x_n)}{\partial \mu_0} = \sum_{i=1}^{n}(x_i - \mu_0),$$

LMP level α test is given by

$$\phi_0(x) = \begin{cases} 1 & \text{if } \sum_{i=1}^{n}(x_i - \mu_0) > k \\ 0 & \text{otherwise} \end{cases}$$

$$\Rightarrow \qquad \phi_0(x) = \begin{cases} 1 & \text{if } \bar{x} > c \\ 0 & \text{otherwise} \end{cases}$$

where c is determined by

$$E_{\mu_0}\phi_0(x) = \alpha$$

$$\text{i.e.,}\quad P_{\mu_0}\{(\bar{X}-\mu_0)\sqrt{n} > (c-\mu_0)\sqrt{n}\} = \alpha$$

$$\text{i.e.,}\qquad P_{\mu_0}\{Z > (c-\mu_0)\sqrt{n}\} = \alpha, \text{ where } Z \sim N(0,1).$$

If $P\{Z > Z_\alpha\} = \alpha$ then

$$Z_\alpha = (c-\mu_0)\sqrt{n}$$

$$c = \mu_0 + \frac{Z_\alpha}{\sqrt{n}}.$$

Example 7.10

Obtain LMP level α test for testing

$$H:\theta \le 0 \text{ vs } K:\theta > 0$$

where θ is the parameter of Cauchy distribution. Take $n = 1$.

As

$$\log p_\theta(x) = -\log(\pi) - \log[1+(x-\theta)^2]$$

$$\left.\frac{\partial \log p_\theta(x)}{\partial\theta}\right|_{\theta=0} = \left.\frac{2(x-\theta)}{1+(x-\theta)^2}\right|_{\theta=0} = \frac{2x}{1+x^2}$$

LMP test is given by

$$\phi_0(x) = \begin{cases} 1 & \text{if } \dfrac{2x}{1+x^2} > k \\ 0 & \text{otherwise} \end{cases}$$

where k is determined by

$$E_{\theta=0}\phi_0(x) = \alpha.$$

7.6 LEAST FAVOURABLE DISTRIBUTIONS

For testing composite hypothesis

$$H:X \sim p_\theta(x), \theta \in \omega \text{ vs } K:X \sim g(x)$$

One may reduce the composite hypothesis into simple hypothesis by using undetermined multipliers but this will involve considerable difficulty.

Another way is to reduce the composite hypothesis into a simple hypothesis. This is done by considering weighted averages of the distributions of H. The composite hypothesis (H) is reduced by the simple H_λ such that

$$H_\lambda(x) = \int_{\theta \in \omega} p_\theta(x)\lambda(\theta)d\theta$$

where $\lambda(\theta)$ is the p.f. over ω and then testing

$$H_\lambda : X \sim h_\lambda(x) \text{ vs } K : X \sim g(x)$$

by N-P lemma. But the difficulty is to select the distribution of λ. The choice of λ is made easier if that gives least power.

Example of least power distribution, $p_\theta(x) = \theta^x(1-\theta)^{1-x}$,

$$H : \theta \le \frac{1}{2} \text{ vs } K : \theta = \frac{3}{4}$$

the hardest value of θ is $\frac{1}{2}$ and hence, $P(\lambda = 1) = \frac{1}{2}$.

Definition 7.4

If ϕ_λ is MP level α test for testing

$$H_\lambda : X \sim h_\lambda(x) \text{ vs } K : X \sim g(x)$$

and β_λ is the power of test then λ is said to be least favourable if for any other distribution $\lambda^*(\theta)$ the inequality $\beta_\lambda \le \beta_{\lambda^*}$ holds.

Theorem 7.6: Let ϕ_λ be MP level α test for testing H_λ vs g and let it be of size $\le \alpha$ with respect to the original H. Then

(i) The test ϕ_λ is MP for testing H vs g

(ii) If ϕ_λ is unique MP level α test for testing H_λ vs $K = g(x)$ it is also unique MP test of H vs g.

(iii) The distribution of λ is least favourable.

Proof: Note that h_λ is a density function, since by the theorem on change of order of integration,

$$\int h_\lambda(x)dx = \int_\omega \left[\int_\omega p_\theta(x)\lambda(\theta)d\theta \right] dx$$

$$= \int_\omega \lambda(\theta) \left[\int p_\theta(x)dx \right] d\theta$$

$$= \int_\omega \lambda(\theta)d\theta = 1.$$

Suppose that ϕ_λ is a level α test for testing H and let ϕ^* be any other level α test. Since $E_\theta \phi^*(x) \le \alpha \;\forall\; \theta \in \omega$. We have

$$\int \phi^*(x) h_\lambda(x)dx = \int \phi^*(x) \left[\int_\omega p_\theta(x)\lambda(\theta)d\theta \right] dx$$

$$= \int_\omega \left[\int \phi^*(x) p_\theta(x)dx \right] \lambda(\theta)d\theta$$

$$= \int_\omega E\phi^*(x)\lambda(\theta)d\theta \le \alpha$$

which shows that $\phi^*(x)$ is a level α test for testing H_λ vs K and its power cannot exceed that of ϕ_λ and hence,

$$\int \phi_\lambda(x)g(x)dx \ge \int \phi^*(x)g(x)dx$$

$\Rightarrow$ ϕ_λ is MP for testing H vs K and if ϕ_λ is unique MP for testing H_λ vs $K: g(x)$ it is true for H vs K.

Let $\lambda^*(\theta)$ be any other distribution of $\theta \in \omega$. Then as

$$\int \phi_\lambda(x) p_\theta(x)dx \le \alpha \Rightarrow \int_\omega \left[\int \phi_\lambda(x) p_\theta(x)dx \right] h_{\lambda^*}(\theta)d\theta \le \alpha \;\forall\; \lambda^*$$

$$\Rightarrow \qquad \int \phi_\lambda(x) \left[\int p_\theta(x)\lambda^*(\theta)d\theta \right] dx \le \alpha \Rightarrow \int \phi_\lambda(x) h_{\lambda^*}(x)dx \le \alpha$$

$\Rightarrow \phi_\lambda$, is level α test for testing h_{λ^*} vs g.

Since ϕ_λ is MP test for h_{λ^*} vs g, $\beta_{\lambda^*} \geq \beta_\lambda$, i.e., λ is least favourable. This can be rewritten in the following corollary.

Corollary 7.2

Suppose $\lambda(w)$ is p.f. over w and w' is a subset of w with $\lambda(w')=1$. Let ϕ_λ be a test such as

$$\phi_\lambda(x)=\begin{cases}1 & \text{if} \quad g(x)>k\int p_\theta(x)\lambda(\theta)d\theta \\ 0 & \text{if} \quad g(x)<k\int p_\theta(x)\lambda(\theta)d\theta\end{cases}.$$

Then ϕ_λ is an MP level α test for testing H vs g provided

$$E_{\theta'}\phi_\lambda(x)=\sup_{\theta\in\Omega} E_\theta\phi_\lambda(x)=\alpha \;\forall\; \theta' \in w'.$$

Choice of λ

If λ is a step function, we have $\int \phi_\lambda p_\theta(x)dx=\alpha$. If

$$\lambda(\theta)=\int \phi_\lambda(\theta)p_\theta(x)dx$$

is analytic then try simple step function or try to find λ which makes ϕ_λ the MP α-similar test.

Example of Composite Hypothesis vs Simple Hypothesis

Example 7.11

Let $X=(X_1, X_2,\ldots, X_n)$ where, X_i's are independent $N(\xi, 1)$. To test

$$H:\xi \leq \xi_0 \text{ vs } K:\xi=\xi_1(>\xi_0).$$

Try

$$\lambda(\xi)=P_\lambda(\xi=\xi_0)=1.$$

Then

$$h_\lambda(x) = \frac{1}{(2\pi)^{\frac{n}{2}}} e^{-\frac{1}{2}\sum(x_i - \xi_0)^2}$$

$$g(x) = \frac{1}{(2\pi)^{\frac{n}{2}}} e^{-\frac{1}{2}\sum(x_i - \xi_1)^2}.$$

The MP test of h_λ is to reject h_λ whenever $\bar{X} > c$ where, c is determined by the equation

$$P_{\xi_0}\{\bar{X} > c\} = \alpha.$$

By the theorem ϕ_λ is the MP test of H vs K provided one has to show that

$$P_\xi\{\bar{X} > c\} \le \alpha \ \forall\ \xi \le \xi_0.$$

Since $\bar{X} \sim N\left(\xi, \frac{1}{n}\right)$. We see that $P_\xi\{\bar{X} > c\}$ increases as ξ increases.

Hence,

$$P_{\xi_0}\{\bar{X} > c\} = \alpha \Rightarrow P_\xi\{\bar{X} > c\} \le \alpha \ \forall\ \xi \le \xi_0.$$

We include that

$$\phi_\lambda(x) = 1 \quad \text{when} \quad \bar{X} > c$$

is unique MP test of H vs K.

Since ϕ_λ does not depend on ξ_1, this is UMP for testing

$$H: \xi \le 0 \text{ vs } K: \xi > 0.$$

Example 7.12

Let $X = (X_1, X_2, \ldots, X_n)$ be independently distributed as $N(\xi, \sigma^2)$,

$$H: \sigma = \sigma_0 \text{ vs } K :: \sigma = \sigma_1, \xi = \xi_1, (\sigma_1 < \sigma_0)$$

We take least favourable distribution as

$$P(\xi = \xi_1) = 1$$

$$h_\lambda(x) = \left(\frac{1}{\sqrt{2\pi}\sigma_0}\right)^n e^{-\frac{1}{2\sigma_0^2}\Sigma(x_i-\xi_1)^2}$$

$$g(x) = \left(\frac{1}{\sqrt{2\pi}\sigma_1}\right)^n e^{-\frac{1}{2\sigma_1^2}\Sigma(x_i-\xi_1)^2}.$$

The MP size α test for testing h_λ vs g is

$$\phi(x) = \frac{e^{-\frac{1}{2\sigma_1^2}\Sigma(x_i-\xi_1)^2}}{e^{-\frac{1}{2\sigma_0^2}\Sigma(x_i-\xi_1)^2}} > c_1$$

$$\Rightarrow \qquad \left(\frac{1}{\sigma_0^2} - \frac{1}{\sigma_1^2}\right)\Sigma(x_i - \xi_1)^2 > c_2$$

$$\Rightarrow \qquad \Sigma(x_i - \xi_1)^2 < c$$

where c is determined from

$$P_{\xi_0,\sigma_0^2}\{\Sigma(x_i - \xi_1)^2 < c\} = \alpha.$$

We obtain c by referring χ^2 table on n degree of freedom. It remains to show that

$$P_{\xi_1,\sigma_0^2}\{\Sigma(x_i - \xi_1)^2 < c\} \leq \alpha$$

$$P_{\xi,\sigma_0^2}\{\Sigma(x_i - \xi_1)^2 < c\} \leq P_{\xi_1,\sigma_0^2}\{\Sigma(y_i)^2 < c\},$$

where $y_i = x_i - \xi_1$, that is, to show that

$$P_{\xi,\sigma_0^2}\{\Sigma y_i^2 < c\} \leq P_{\xi_1,\sigma_0^2}\{\Sigma y_i^2 < c\}.$$

This is geometrically obvious because normal density decreases as distance from the origin increases. This is not UMP since alternative ξ_1 is involved in the test.

Example 7.13

Obtain MP test for testing

$$H:\sigma=\sigma_0 \text{ vs } K:\sigma>\sigma_1,\ \xi=\xi_1,\ (\sigma_1>\sigma_0)$$

Given a random sample of size n from $N(\xi,\sigma^2)$ distribution. Let $\lambda(\xi)$ be least favourable distribution

$$\phi_\lambda(x)=1 \text{ if } \begin{cases} \dfrac{g(x)}{\int p_\theta(x)\lambda(\xi)}>c & \text{or} \\[2ex] \dfrac{e^{-\frac{1}{2\sigma_1^2}\Sigma(x_i-\xi_1)^2}}{\int_{-\infty}^{\infty} e^{-\frac{1}{2\sigma_0^2}\Sigma(x_i-\xi)^2}\,d\lambda(\xi)}>c & \text{or} \\[2ex] \left\{\dfrac{e^{-\frac{1}{2\sigma_1^2}\Sigma(x_i-\bar{x})^2}}{e^{-\frac{1}{2\sigma_0^2}\Sigma(x_i-\bar{x})^2}}\right\}\left\{\dfrac{e^{-\frac{n}{2\sigma_1^2}\Sigma(\bar{x}-\xi_1)^2}}{\int_{-\infty}^{\infty} e^{-\frac{n}{2\sigma_0^2}\Sigma(\bar{x}-\xi)^2}\,d\lambda(\xi)}\right\}>c \end{cases}$$

Observe that above inequality is satisfied if second bracket is constant. So $\lambda(\xi)$ be chosen such as

$$\int_{-\infty}^{\infty} e^{-\frac{n}{2\sigma_0^2}\Sigma(\bar{x}-\xi)^2}\,d\lambda(\xi)=ce^{-\frac{n}{2\sigma_1^2}\Sigma(\bar{x}-\xi_1)^2}.$$

By Laplace transform theory $d\lambda(\xi)$ should be

$$d\lambda(\xi)=e^{-\frac{n}{2(\sigma_1^2-\sigma_0^2)}(\xi-\xi_1)^2}\,d\xi$$

The test is to reject H if

$$\frac{e^{-\frac{1}{2\sigma_1^2}\Sigma(x_i-\bar{x})^2}}{e^{-\frac{1}{2\sigma_0^2}\Sigma(x_i-\bar{x})^2}} > c \Rightarrow \Sigma(x_i-\bar{x})^2 > c.$$

PROBLEMS

7.1 Let $f(x)$ and $g(x)$ be

(i) $f(x)=\begin{cases}1 & \text{if } 0<x<1\\ 0 & \text{otherwise}\end{cases}$

$$g(x)=\begin{cases}4x & \text{if } \quad 0<x\le\frac{1}{2}\\ 4(1-x), & \text{if } \quad \frac{1}{2}\le x<1\\ 0 & \text{if } \quad \text{otherwise}\end{cases}$$

(ii) $f(x)=\dfrac{e^{-2}2^x}{x!}$, $x=0,1,2,\ldots..$

$$g(x)=\frac{1}{3}\left(\frac{2}{3}\right)^x, x=0,1,2,\ldots..$$

(iii) $f(x)=\begin{cases}4x & \text{if } \quad 0<x<\frac{1}{2}\\ 4(1-x) & \text{if } \quad \frac{1}{2}\le x<1\end{cases}$

$$g(x)=\begin{cases}1 & \text{if } 0<x<1\\ 0 & \text{if otherwise}\end{cases}$$

(iv) $f(x)=\begin{cases}1 & \text{if } 0<x<1\\ 0 & \text{if otherwise}\end{cases}$

$$g(x) = \begin{cases} 6x & \text{if } 0 < x < \frac{1}{3} \\ 3(1-x) & \text{if } \frac{1}{3} \le 3x < 1 \\ 0 & \text{if otherwise} \end{cases}$$

then obtain MP test of size α for testing H : X has probability function f against K : X has probability function g. Compute its power.

7.2 For the following family of p.m.f./p.d.f. $f(x)$, find MP test of size α for testing $H:\theta < \theta_0$ vs $K:\theta \ge \theta_0$, based on a random sample of size n. Also obtain power function in each case.

(i) $f(x, \theta) = \theta^x (1-\theta)^{1-x}, x = 0, 1.0 < \theta < 1$

(ii) $f(x, \theta) = \frac{1}{\sqrt{2\pi\theta}} e^{-\frac{x^2}{2\theta^2}}, -\infty < x \langle \infty, \theta \rangle 0.$

(iii) $f(x, \theta) = \frac{1}{\theta} e^{-\frac{x}{\theta}}, x > 0, \theta > 0$

(iv) $f(x, \theta) = \frac{1}{\Gamma(\theta)} x^{\theta-1} e^{-x}, x > 0, \theta > 0$

(v) $f(x, \theta) = \theta x^{\theta-1}, 0 < x < 1, \theta > 0$

(vi) $f(x, \theta) = \frac{\theta}{x^2}, 0 > \theta \le x < \infty$

7.3 Let X have Cauchy distribution with density function

$$[\pi(1 + (x-\theta)^2)]^{-1}, -\infty < x < \infty.$$

Then show that an MP test for testing $H:\theta = 0$ vs $K:\theta = 1$ is to reject H if $1 < x < 3$.

7.4 If X has Cauchy distribution with density

$$\frac{\theta}{\pi(\theta^2 + x^2)}, -\infty < x \langle \infty, \theta \rangle 0$$

then show that for testing $H:\theta = \theta_0$ vs $K:\theta = \theta_1(> \theta_0)$, a test which rejects H when x^2 is too large is MP. If $K:\theta > \theta_0$ then is the test uniform? obtain power function of the test.

7.5 If X has density

(i) $f(x, \theta) = \frac{2}{\theta^2}(\theta - x), 0 < x < \theta$

(ii) $f(x, \theta) = 2\theta x + 2(1-\theta)(1-x)$.

Obtain MP test of size α for testing $H:\theta = \theta_0$ vs $K:\theta = \theta_1(< \theta_0)$.

7.6 Let $z_1, z_2, \ldots, z_k$ have multinomial distribution

$$\frac{n!\, \theta_1^{z_1}\, \theta_2^{z_2}\, \theta_3^{z_3} \ldots \theta_k^{k}}{z_1!\, z_2! \ldots z_k!}, z_i \geq 0, i = 1, 2, \ldots, k.$$

$\Sigma z_i = n$, where, $\theta_i > 0$, and $\Sigma \theta_i = 1$. Find the form of MP test for testing $H:\theta_i = \theta_i^0, i = 1, 2, \ldots, k$ vs $K:\theta_i = \theta_i^1, i = 1, 2, \ldots, k$, for $k = 4$, $\theta_1^0 = 0.55$, $\theta_2^0 = 0.20$, $\theta_3^0 = 0.15$, $\theta_4^0 = 0.10$, and $\theta_1^1 = 0.10$, $\theta_2^1 = 0.40$, $\theta_3^1 = 0.30$, $\theta_4^1 = 0.20$. Show how the probability of error may be calculated from the binomial distribution.

7.7 Let $\underline{X} \sim N_p(\mu, \Sigma)$ distribution. Obtain MP tests for testing

(i) $H:\underline{\mu} = \underline{0}$ vs $K:\underline{\mu} = \underline{\mu}_1$ when $\Sigma = I$

(ii) $H:\Sigma = I$ vs $K:\Sigma = 2I$ when $\underline{\mu} = \underline{0}$

7.8 If $(X_1, X_2, \ldots, X_n)$ is a random sample of size n from $N(\mu, \sigma^2)$ then obtain MP test for testing $H:\mu = \mu_0, \sigma = \sigma_0$ vs $K:\mu = \mu_1, \sigma = \sigma_1$, where,

(i) $\mu_1 > \mu_0, \sigma_1 > \sigma_0$

(ii) $\mu_1 > \mu_0, \sigma_1 < \sigma_0$

(iii) $\mu_1 < \mu_0, \sigma_1 > \sigma_0$

(iv) $\mu_1 < \mu_0, \sigma_1 < \sigma_0$.

7.9 For the probability function $\frac{pe^{-\frac{x^2}{2}}}{\sqrt{2\pi}} + \frac{qe^{-\frac{1}{2}(x-\mu_0)_1}}{\sqrt{2\pi}}$, $p + q = 1$, $0 < p < 1$, μ_0 is known, $-\infty < x < \infty$, obtain MP test for testing $p = 1$ vs $p = p_0$ (specified) < 1. Is the test UMP against the alternative $p \neq 1$?

7.10 Let $X_i \sim IN(\theta_i, 1), i = 1, 2, 3 \ldots, n$. Then obtain MP test of size α for testing

$$H: \theta_i = 0, i = 1, 2, 3, \ldots, n \text{ vs } K: \begin{cases} \theta_i = \frac{1}{2}, i = 1, 2, 3, \ldots, r \\ \theta_i = -\frac{1}{2}, i = r+1, r+2, \ldots, n \end{cases}.$$

How large n should be to ensure that the power of 0.05 level test is at least 0.90?

7.11 If X has p.d.f.

$$f(x, \theta) = \frac{(x(1-x))^{\theta-1}}{B(\theta, \theta)}, 0 < x < 1,$$

then obtain UMP size α test for testing $\theta = 1$ vs $\theta > 1$. Compute power at $\theta = 2$.

7.12 Based on a random sample of size n from $U(0, \theta)$ distributions, obtain UMP tests of size α for testing

(i) $\theta = \theta_0$ vs $\theta \neq \theta_0$

(ii) $\theta \leq \theta_0$ vs $\theta > \theta_0$.

Determine power functions of the tests.

7.13 Based on random sample of size n from $U(\theta, \theta + 1)$ distribution. Obtain UMP test of size α for testing $\theta \leq 0$ vs $\theta > 0$. Further obtain the power function of the test.

7.14 Obtain UMP test of size α for testing

$$a = a_0, b = b_0 \text{ vs } a > a_0, b < b_0,$$

based on a random sample of size n from

$$f(x, a, b) = a \exp(-a(x - b)), b < x < \infty.$$

7.15 Based on first r ordered observations from the distribution having p.d.f. as

$$f(x,\theta)=(2\theta)^{-1}\exp\left(-\frac{x}{2\theta}\right), x\geq 0.$$

Obtain UMP test of size α for testing $\theta>\theta_0$ vs $\theta\leq\theta_0$.

(i) Determine c.r. for r = 4, θ_0 = 1000 and power function for θ_1 = 500

(ii) Find r when $\beta\geq 0.95$.

7.16 Based on a random sample of size n obtain UMP test of size α for testing $\theta\leq\theta_0$ vs $\theta>\theta_0$ in $u(2\theta, 3\theta)$ distribution. Find power function of the test.

7.17 Show that the following family does not possess MLR

$$f(x;\theta)=\begin{cases} 2(1-\theta) & \text{if} \quad \theta<x<\dfrac{1}{2} \\ 2\theta & \text{if} \quad \dfrac{1}{2}\leq x<1 \end{cases}, 0<\theta<1.$$

7.18 For a random sample of size n from $N(\mu, 1)$ show that UMP tests of size α for testing

(i) $H_1:\theta<\theta_0$ vs $K_1:\theta\geq\theta_0$

(ii) $H_2:\theta>\theta_0$ vs $K_2:\theta\leq\theta_0$

exist but show that there does not exist UMP test for testing

$$H_3:\theta=\theta_0 \text{ vs } K_3:\theta\neq\theta_0.$$

7.19 Construct locally the most powerful level α test for testing

$$H:\sigma=\sigma_0 \text{ vs } K:\sigma>\sigma_0,$$

based on a random sample from $N(0,\sigma^2)$.

7.20 If $X_1, X_2, \ldots, X_n$ is a random sample from

$$f(x;\theta)=\theta\frac{e^{-\frac{(x-1)^2}{2}}}{\sqrt{2\pi}}+(1-\theta)\frac{e^{-\frac{(x)^2}{2}}}{\sqrt{2\pi}}, 0<\theta<1, x\in R,$$

then construct LMP test for testing

$$H:\theta=0 \text{ vs } K:\theta>0.$$

8

CHAPTER

Unbiased Tests

8.1 INTRODUCTION

We obtained UMP test for testing one sided hypothesis of distributions possessing monotone likelihood in Chapter 7. In this chapter we will test the following two-sided hypotheses.

(i) H_1: $\theta \leq \theta_1$ or $\theta \geq \theta_2$ vs K_1: $\theta_1 < \theta < \theta_2$

(ii) H_2: $\theta_1 < \theta < \theta_2$ vs K_2: $\theta \leq \theta_1$ or $\theta \geq \theta_2$

(iii) H_3: $\theta = \theta_0$ vs K_3: $\theta \neq \theta_0$

For testing these types of two sided hypotheses, we need generalized Neyman-Pearson lemma, which is given in section 8.2 and the hypothesis H_1 is tested by the use of this lemma. In real life such hypothesis exists when one wants to know whether measuring instrument is not balanced (H_1: $\theta \leq \theta_1$ or $\theta \geq \theta_2$) or balanced (K_1: $\theta_1 < \theta < \theta_2$). After defining unbiased test, uniformly most powerful unbiased (UMPU) tests for testing H_2 and H_3 have been obtained for exponential family of distributions in section 8.3 whereas section 8.4 deals with locally most powerful unbiased tests.

8.2 GENERALIZED N-P LEMMA

Theorem 8.1: Generalized Neyman-Pearson lemma

Let $p_1, p_2, \ldots, p_{m+1}$ be integrable functions of x and let ϕ_0, ϕ be point functions, $0 < \phi_0 < 1$, $0 < \phi < 1$, such that

$$\int \phi_0 \, p_j(x)dx = \int \phi \, p_{m+1}(x)dx = c_j, \; j = 1, 2, \ldots, m+1$$

then

$$\int \phi_0\, p_{m+1}(x)dx \geq \int \phi\, p_j(x)dx$$

if and only if

$$\phi_0(x) = \begin{cases} 1 & \text{if} \quad p_{m+1} > \sum_{j=1}^{m} k_j p_j \\ \gamma & \text{if} \quad p_{m+1} = \sum_{j=1}^{m} k_j p_j, \\ 0 & \text{if} \quad p_{m+1} < \sum_{j=1}^{m} k_j p_j \end{cases}$$

where k_j, $j = 1, 2, \ldots, m$ are non-negative real numbers and $0 < \gamma < 1$.

Proof being same as *N-P* lemma it is left.

One-Sided Alternative Hypothesis and Two-Sided Null Hypothesis

For two-sided hypothesis H: $\theta \leq \theta_1$ or $\theta \geq \theta_2$ vs K: $\theta_1 < \theta < \theta_2$ UMP test exists.

Theorem 8.2: For testing the hypothesis

$$H: \theta \leq \theta_1 \text{ or } \theta \geq \theta_2 (\theta_1 < \theta_2) \text{ vs } K: \theta_1 < \theta < \theta_2,$$

in one parameter exponential family

$$p_\theta(x) = c(\theta)h(x)e^{\theta t},$$

a test $\phi_0(x)$ of the form

$$\phi_0(x) = \begin{cases} 1 & \text{if} \quad c_1 < T < c_2 \\ \gamma_i & \text{if} \quad T = c_i, i = 1, 2 \\ 0 & \text{if} \quad T < c_1 \text{ or } T > c_2 \end{cases}$$

where c's and γ's are such that

$$E\theta_1\phi_0(X) = E\theta_2\phi_0(X) = \alpha,$$

is the UMP test.

Proof: By generalized Neyman-Pearson lemma a test $\phi^*(x)$ which maximizes $E_{\theta'}\phi(X)$, $\theta_1 < \theta' < \theta_2$ subject to $E_{\theta_1}\phi^*(X) = E_{\theta_2}\phi(X) = \alpha$ is of the form

$$\phi^*(x) = \begin{cases} 1 & \text{if} \quad p'_\theta(x) > k_1 p_{\theta_1}(x) + k_2 p_{\theta_2}(x) \\ \gamma & \text{if} \quad p'_\theta(x) = k_1 p_{\theta_1}(x) + k_2 p_{\theta_2}(x) \\ 0 & \text{if} \quad p'_\theta(x) < k_1 p_{\theta_1}(x) + k_2 p_{\theta_2}(x) \end{cases} \qquad \ldots(8.2.1)$$

Now $p'_\theta(x) > k_1 p_{\theta_1}(x) + k_2 p_{\theta_2}(x) \Rightarrow a_1 e^{b_1 t} + a_2 e^{b_2 t} < 1$, $b_1 < 0 < b_2$. Here both a's cannot be negative since then the test would always reject. If one of the a's is negative and the other a's is positive then the left hand side is strictly monotone and the test cannot satisfy $E_{\theta_1}\phi^*(X) = E_{\theta_2}\phi^*(X) = \alpha$. Therefore, both a's must be positive.

[Alternative proof for both a's to be positive. Consider auxiliary equations for $t = c_1$ and $t = c_2 (c_2 < c_1)$, $a_1 e^{b_1 c_1} + a_2 e^{b_2 c_1} = 1$, $a_1 e^{b_1 c_2} + a_2 e^{b_2 c_2} = 1$. Solving these for a_1, a_2 we get $a_1 = (e^{b_2 c_2} - e^{b_2 c_1})/(e^{b_1 c_1 + b_2 c_2} - e^{b_1 c_2 + b_2 c_1})$ and

$a_2 = (e^{b_1 c_1} - e^{b_1 c_2})/(e^{b_1 c_1 + b_2 c_2} - e^{b_1 c_2 + b_2 c_1})$ Since $b_1 < 0$, $b_2 > 0$, $c_1 < c_2$,

$b_2 c_1 < b_2 c_2 \Rightarrow e^{b_2 c_2} - e^{b_2 c_1} > 0$, $b_1 c_1 > b_2 c_2 \Rightarrow e^{b_1 c_1} - e^{b_1 c_2} > 0$,

$b_1 c_1 + b_2 c_2 > b_2 c_1 + b_1 c_2 \Rightarrow e^{b_1 c_1 + b_2 c_2} - e^{b_1 c_2 + b_2 c_1} > 0 \Rightarrow a_1 > 0$, $a_2 > 0$.]

As $a_1 > 0$, $a_2 > 0$, $b_1 < 0$, $b_2 > 0$, $e^{b_1 t}$ and $e^{b_2 t}$ are monotonic and $a_1 e^{b_1 c_1} + a_2 e^{b_2 c_1} \to \infty$ as $t \to \pm\infty$ which does not satisfy (8.2.1), it must be bounded below as well as above. Thus, $a_1 e^{b_1 c_1} + a_2 e^{b_2 c_1} < 1 \Rightarrow c_1 < t < c_2$ and taking $\phi^* \equiv \phi_0$. As ϕ^* is MP, ϕ_0 is too. Since the determination of test does not involve alternative, it is also UMP.

Two-Sided Alternative Hypotheses

Theorem 8.3: For a simple hypothesis H: $\theta = \theta_0$ no UMP test exists in general for testing H: $\theta = \theta_0$ vs K: $\theta \neq \theta_0$ or $\theta_1 - \theta_0 \geq 0$, under regularity conditions of continuity of $p_\theta(x)$, the probability function of X.

Proof: We expand $p_{\theta_1}(x)$ about θ_0 by Taylor series as

$$p_{\theta_1}(x) = p_{\theta_0}(x) + (\theta_1 - \theta_0) p'_{\theta^*}(x), \quad \theta^* \in (\theta_1, \theta_0).$$

For MP test we must have critical region satisfying

$$\frac{p_{\theta_1}(x)}{p_{\theta_0}(x)} = 1 + \frac{(\theta_1 - \theta_0)p'_{\theta^*}(x)}{p_{\theta_0}(x)} \geq k(\theta_1)$$

$$\Rightarrow \qquad \frac{p'_{\theta^*}(x)}{p_{\theta_0}(x)} \begin{cases} \geq a_\alpha & \text{for} \quad \theta_1 > \theta_0 \\ \leq b_\alpha & \text{for} \quad \theta_1 < \theta_0 \end{cases}$$

As $\theta_1 \to \theta_0$, $\theta^* \to \theta_0$ and

$$\Rightarrow \qquad \frac{p'_{\theta^*}(x)}{p_{\theta_0}(x)} = \left[\frac{\partial \log p_\theta(x)}{\partial \theta}\right]_{\theta=\theta_0} \begin{cases} \geq a_\alpha & \text{for} \quad \theta > \theta_0 \quad \ldots(8.2.2) \\ \leq b_\alpha & \text{for} \quad \theta < \theta_0 \quad \ldots(8.2.3) \end{cases}$$

Thus, in immediate neighbourhood of θ_0, one-sided tests based on $\left[\frac{\partial \log p_\theta(x)}{\partial \theta}\right]_{\theta=0}$ are UMP. Now if we consider an interval of alternatives including positive and negative values of $(\theta_1 - \theta_0)$, (8.2.2) and (8.2.3) both cannot hold and hence no UMP test exists unless

$$E\left[\frac{\partial \log p_\theta(x)}{\partial \theta}\right]_{\theta=\theta_0} = \text{constant} \qquad \ldots(8.2.4)$$

Thus, (8.2.4) is an essential condition for the existence of two sided UMP test. It cannot be satisfied for distributions with range independent of θ unless the constant is zero, that is, the condition is

$$\left[\frac{\partial \log p_\theta(x)}{\partial \theta}\right]_{\theta=\theta_0} = 0 \Rightarrow \left[\frac{\partial \log p_\theta(x)}{\partial \theta}\right]_{\theta=0} = 0$$

We have examples for two-sided UMP tests existing for continuous distributions, e.g., $u(0, \theta)$, $e^{-(x-\theta)}$ as $\left[\frac{\partial \log p_\theta(x)}{\partial \theta}\right]_{\theta=\theta_0} = \text{constant} \neq 0$ is satisfied.

8.3 UNBIASED TESTS

Whenever UMP test does not exist, one would like to obtain a test whose power is at least as large as its size. Such a class of tests where restriction is on the class of tests is called unbiased.

Definition 8.1

For testing H: $\theta \in \Omega_H$ vs K: $\theta \in \Omega_K$ in the family $\{P_\theta, \theta \in \Omega\}$ of X a test $\phi(x)$ satisfying

$$E_\theta \phi(X) \geq \alpha \quad \text{for} \quad \theta \in \Omega_K$$
$$E_\theta \phi(X) \leq \alpha \quad \text{for} \quad \theta \in \Omega_H$$

is said to be unbiased test. Unbiasdness is reasonable because the probability of rejecting H when it is false is at least the probability of rejecting H when it is true.

Definition 8.2

Let $\{\phi_\alpha\}$ be class of unbiased tests of size α and let $\phi_0 \in \{\phi_\alpha\}$. Then ϕ_0 is UMP unbiased if

$$E_\theta \phi_0(X) \geq E_\theta \phi(X) \ \ \forall \ \ \theta \in \Omega_K \ \text{ and } \ \forall \ \ \phi \in \{\phi_\alpha\}.$$

Example 8.1

Show that the UMP level α test is unbiased.

Let ϕ_0 be the UMP level α test for testing H: $\theta \in \Omega_H$ vs K: $\theta \in \Omega_K$ and let $\phi(x) \equiv \alpha$. Then as ϕ_0 is the UMP level α test then

$$\beta = E_K \phi_0(X) \geq E_K \phi(X) = \alpha \Rightarrow \phi_0 \text{ is unbiased.}$$

Example 8.2

Let p.d.f. of X be

$$f(x;\theta) = \begin{cases} 2(1-\theta) & \text{if} \quad 0 < x < \dfrac{1}{2} \\ 2\theta & \text{if} \quad \dfrac{1}{2} \leq x < 1 \\ 0 & \text{otherwise} \end{cases}$$

and let $H: \theta \le \frac{1}{2}$ vs $K: \theta > \frac{1}{2}$ be rejected with size α = 0.1, when $x > c$ and when single observation is taken. Examine whether the test is unbiased. The test being

$$\phi(x) = \begin{cases} 1 & \text{if} \quad x > c \\ 0 & \text{otherwise} \end{cases}$$

$$\alpha = E_H \phi(X)$$

$$= \int_c^1 dx$$

$$= 1 - c.$$

As $a = 0.1$, $c = 0.9$ and

$$\phi(x) = \begin{cases} 1 & \text{if} \quad x > 0.9 \\ 0 & \text{otherwise} \end{cases}$$

Power function,

$$\beta(\theta) = E_K \phi(X)$$

$$= 2\theta \int_{0.9}^1 dx$$

$$= 0.2\theta$$

$$> 0.1 \; \forall \; \theta \ge \frac{1}{2}.$$

Thus, the test is unbiased.

Example 8.3

Let a random variable X have p.d.f.

$$f(x; \theta) = \theta e^{-\theta x}, \theta > 0, x > 0$$

and let a test based on a single observation for testing $H\colon \theta \le 1$ vs $K\colon \theta > 1$ be to reject H when $x < c$ with size α. Examine whether the test is unbiased.

Here the test function is

$$\phi(x) = \begin{cases} 1 & \text{if} \quad x < c \\ 0 & \text{otherwise} \end{cases}$$

where c is such that

$$E_H \phi(X) = \alpha$$

that is,

$$\alpha = P_H(X < c)$$

$$= \int_0^c \theta e^{-\theta x} dx$$

$$= 1 - e^{-c}$$

$$\Rightarrow \qquad c = -\log(1 - \alpha)$$

Thus, the test is

$$\phi(x) = \begin{cases} 1 & \text{if} \quad x < -\log(1-\alpha) \\ 0 & \quad \text{otherwise} \end{cases}.$$

The power function of the test is

$$\beta(\theta) = \int_0^{-\log(1-\alpha)} \theta e^{-\theta x}\, dx,\ \theta > 1$$

$$= 1 - e^{\theta \log(1 - \alpha)}$$

$$\frac{d\beta(\theta)}{d\theta} = -\log(1-\alpha) e^{\theta \log(1-\alpha)} > 0 \quad \text{for} \quad \theta > 1$$

$\Rightarrow$ $\beta(\theta)$ is non-decreasing function of θ and as $\beta(\theta) = \alpha$ for $\theta = 1$. Hence, $\beta(\theta) > \alpha$ for $\theta > 1$ which shows that the test is unbiased.

8.4 UNIFORMLY MOST POWERFUL UNBIASED (UMPU) TESTS FOR TWO-SIDED ALTERNATIVES

Theorem 8.4: For an exponential family of distributions,

$$p_\theta(x) = c(\theta)h(x)e^{\theta t(x)}$$

where $X = (X_1, X_2, \ldots, X_n)$ any two-sided test ϕ_0 of the form

$$\phi_0(x) = \begin{cases} 1 & \text{if} \quad T < c_1 \text{ or } T > c_2 \\ \gamma_i & \text{if} \quad T = c_i, \quad i = 1, 2 \\ 0 & \text{if} \quad c_1 < T < c_2 \end{cases} \qquad \ldots(8.4.1)$$

where $c_1, c_2, \gamma_i, i = 1, 2$ are determined by

$$E_{\theta_1}\phi_0(X) = \alpha, i = 1, 2. \qquad \ldots(8.4.2)$$

is UMP unbiased test for testing

$$H: \theta_1 < \theta < \theta_2 \text{ vs } K: \theta \leq \theta_1 \text{ or } \theta \geq \theta_2.$$

Proof: We shall first show that

$$E_\theta\phi_0(X) \geq E_\theta\phi(X) \text{ for } \theta \leq \theta_1 \text{ or } \theta \geq \theta_2 \qquad \ldots(8.4.3)$$

$$E_\theta\phi_0(X) \geq E_\theta\phi(X) \text{ for } \theta_1 < \theta < \theta_2 \qquad \ldots(8.4.4)$$

for any test $\phi(x)$ satisfying

$$E_{\theta i}\phi_0(X) = \alpha, i = 1, 2.$$

Case I Let $\theta < \theta_1$.

By generalized Neyman-Pearson lemma any test $\phi^*(x)$ of the form

$$\phi^*(x) = \begin{cases} 1 & \text{if} \quad p_\theta(x) > k_1 p_{\theta_1}(x) + k_2 p_{\theta_2}(x) \\ \gamma & \text{if} \quad p_\theta(x) = k_1 p_{\theta_1}(x) + k_2 p_{\theta_2}(x) \\ 0 & \text{if} \quad p_\theta(x) < k_1 p_{\theta_1}(x) + k_2 p_{\theta_2}(x) \end{cases} \qquad \ldots(8.4.5)$$

maximizes the quantity $E_{\theta}\phi(X)$ out of all tests $\phi(x)$ satisfying

$$E_{\theta}\phi(X) = \alpha,\ i = 1, 2.$$

Consider the lower inequality of (8.4.5)

$$c(\theta)h(x)e^{\theta t(x)} < k_1 c(\theta_1)h(x)e^{\theta_1 t(x)} + k_2 c(\theta_2)h(x)e^{\theta_2 t(x)}$$

$$\Rightarrow \qquad a_1 e^{b_1 t} + a_2 e^{b_2 t} > 1 \qquad \text{...(8.4.6)}$$

[Here both a's cannot be negative since then the test would always reject and if both are positive the test would always accept. Therefore, $a_1 > 0,\ a_2 < 0$ or $a_1 < 0,\ a_2 > 0$.]

Alternative proof for a_1, a_2: Consider auxiliary simultaneous equations in a_1, a_2 at $t = c_1,\ t = c_2, > c_1$ such as

$$a_1 e^{b_1 c_1} + a_2 e^{b_2 c_1} = 1$$

$$a_1 e^{b_1 c_2} + a_2 e^{b_2 c_2} = 1$$

and solving for a_1, a_2 we have

$$a_1 = e^{b_2 c_1} - \frac{e^{b_2 c_2}}{\Delta}$$

$$a_2 = e^{b_1 c_2} - \frac{e^{b_1 c_1}}{\Delta}$$

where $\Delta = e^{b_1 c_1 + b_2 c_2} - e^{b_1 c_2 + b_2 c_1}$

$$e^{b_2 c_1} - e^{b_2 c_2} < 0,\ e^{b_1 c_2} - e^{b_1 c_1} > 0$$

$$e^{b_1 c_1 + b_2 c_2} - e^{b_1 c_2 + b_2 c_1} \gtrless 0$$

$$\Rightarrow \qquad a_1 < 0,\ a_2 > 0 \text{ or } a_1 > 0,\ a_2 < 0.$$

Since $e^{b_1 t}, e^{b_2 t}$ are functions of t, as $t \to \infty$,

$$a_1 e^{b_1 t} + a_2 e^{b_2 t} \to \infty \text{ or } -\infty,$$

and as $t \to -\infty$,

$$a_1 e^{b_1 t} + a_2 e^{b_2 t} \to 0$$

so that for $t \to \pm\infty$, $a_1 e^{b_1 t} + a_2 e^{b_2 t} > 1$ is not satisfied

Thus, for satisfying (8.4.6) t must be bounded above as well as below, that is, $c_1 < t < c_2$

$$\Rightarrow \qquad \phi^* \equiv \phi_0$$

Case II $\theta > \theta_2$. Similar case

Case III $\theta_1 < \theta < \theta_2$.

By generalized Neyman-Pearson lemma $\phi^{**} = 1 - \phi$ of the form

$$\phi^{**} = \begin{cases} 0 & \text{if} \quad p_\theta(x) > k_1 p_{\theta_1}(x) + k_2 p_{\theta_2}(x) \\ \gamma & \text{if} \quad p_\theta(x) = k_1 p_{\theta_1}(x) + k_2 p_{\theta_2}(x) \\ 1 & \text{if} \quad p_\theta(x) < k_1 p_{\theta_1}(x) + k_2 p_{\theta_2}(x) \end{cases} \qquad \ldots(8.4.7)$$

minimizes $E_\theta \phi(X)$ among all tests ϕ satisfying

$$E_{\theta_i} \phi(X) = \alpha, \; i = 1, 2.$$

The upper inequality of (8.4.7) is equivalent to

$$a_1 e^{b_1 t} + a_2 e^{b_2 t} < 1 \qquad \ldots(8.4.8)$$

$b_1 < 0$, $b_2 > 0$, considering auxiliary equations we get $a_1 > 0$, $a_2 > 0$

$$a_1 e^{b_1 t} + a_2 e^{b_2 t} \to \infty \text{ as } t \to \pm\infty$$

and hence, (8.4.8) is not satisfied, which implies

$$c_1 < t < c_2 \Rightarrow \phi^{**} \equiv \phi_0$$

ϕ_0 is unbiased by comparing with $\phi(x) \equiv \alpha$. As the test does not depend on alternatives hypothesis, it is UMP unbiased.

Theorem 8.5: For testing H: $\theta = \theta_0$ vs K: $\theta \neq \theta_0$ for one parameter exponential family of distribution $c(\theta)h(x)e^{\theta t}$, any two-sided test $\phi_0(x)$ of the form

$$\phi_0(x) = \begin{cases} 1 & \text{if} \quad T < c_1 \text{ or } T > c_2 \\ \gamma_i & \text{if} \quad T = c_i, i = 1, 2 \\ 0 & \text{if} \quad c_1 < t < c_2 \end{cases} \qquad \ldots(8.4.9)$$

to be UMP unbiased of size α iff

$$E_{\theta_0}\phi_0(X) = \alpha \qquad \ldots(8.4.10)$$

$$E_{\theta_0}T\phi_0(X) = \alpha E_{\theta_0}T \qquad \ldots(8.4.11)$$

Proof: Let us suppose that X has esponentail family of distributions with p.f.

$$c(\theta)h(x)e^{\theta T(x)}.$$

Then the statistic $T(x)$ has exponential family of distribution with

$$c(\theta)h(x)e^{\theta t}.$$

Necessary condition: Since the test ϕ_0 is unbiased of size α,

$$E_{\theta_0}\phi_0(X) \geq \alpha \text{ for } \theta < \theta_0 \text{ or } \theta > \theta_0$$

$$E_{\theta_0}\phi_0(X) \leq \alpha \text{ for } \theta < \theta_0$$

$\Rightarrow E_{\theta_0}\phi_0(X) = \alpha$ for $\theta < \theta_0$, which is (8.4.10)

Further the test ϕ_0 being unbiased it implies that power function

$$\beta(\theta) = E_\theta\phi_0(X) = E_\theta\Psi(T)$$

must have minimum at $\theta = \theta_0$. Assuming that the derivative with respect to θ can go under integral sign we have

$$\frac{d}{d\theta}\beta_\Psi(\theta) = \int t\Psi(t)c(\theta)h(t)e^{\theta t}dt + \frac{c'(\theta)}{c(\theta)}\int \Psi(t)c(\theta)h(t)e^{\theta t}dt$$

$$= E_\theta T\Psi(T) + \frac{c'(\theta)}{c(\theta)}E_\theta\Psi(T) \qquad \ldots(8.4.12)$$

Let $\Psi(T) \equiv \alpha$ in (8.4.12). Then

$$0 = \alpha E_\theta(T) + \frac{c'(\theta)}{c(\theta)}\alpha \Rightarrow \frac{c'(\theta)}{c(\theta)} = -E_\theta(T)$$

and hence (8.4.12) at $\theta = \theta_0$ gives

$$0 = \frac{d}{d\theta_0}\beta(\theta) = E_{\theta_0}T\Psi(T) - E_{\theta_0}\Psi(T)E_{\theta_0}T$$

$$\Rightarrow \qquad E_{\theta_0} T\phi_0(X) = \alpha E_{\theta_0} T$$

which is the necessary condition given in (8.4.11)

Sufficient condition

Case: I Let $\theta < \theta_0$

By generalized Neyman-Pearson lemma any test ϕ^* of the form

$$\phi^*(x) = \begin{cases} 1 & \text{if} \quad p_\theta(x) > k_1 p_{\theta_0}(x) + k_2 \dfrac{\partial}{\partial \theta_0} p_{\theta_0}(x) \\ \gamma & \text{if} \quad p_\theta(x) = k_1 p_{\theta_0}(x) + k_2 \dfrac{\partial}{\partial \theta_0} p_{\theta_0}(x) \\ 0 & \text{if} \quad p_\theta(x) < k_1 p_{\theta_0}(x) + k_2 \dfrac{\partial}{\partial \theta_0} p_{\theta_0}(x) \end{cases} \qquad \text{...(8.4.13)}$$

maximizes the quantity $E_\theta \phi(X)$ out of all tests satisfying

$$E_{\theta_0} \phi(X) = \alpha$$

$$\frac{\partial}{\partial \theta_0} E_\theta \phi(X) = 0$$

considering the lower inequality of (8.4.13) which is

$$c(\theta)h(t)e^{\theta t} < k_1 c(\theta_0)h(t)e^{\theta_0 t} + k_2[c(\theta_0)h(t)te^{\theta_0 t} + c'(\theta_0)h(t)e^{\theta_0 t}]$$

$$\Rightarrow \qquad a_1 + a_2 t > e^{(\theta - \theta_0)t} \qquad \text{...(8.4.14)}$$

Taking auxiliary equations in a_1, a_2 at $t = c_1$, $t = c_2$, $c_2 > c_1$

$$a_1 + a_2 c_1 = e^{(\theta - \theta_0)c_1}$$

$$a_1 + a_2 c_2 = e^{(\theta - \theta_0)c_2}$$

and solving them for a_1, a_2

$$a_1 = \frac{(c_2 e^{(\theta - \theta_0)c_1} - c_1 e^{(\theta - \theta_0)c_2})}{c_2 - c_1} > 0$$

$$a_2 = \frac{(e^{(\theta-\theta_0)c_1} - e^{(\theta-\theta_0)c_2})}{c_2 - c_1} < 0$$

As $t \to \infty$, (8.4.14) gives absurd result $-\infty > 0$ and as $t \to -\infty$, (8.4.14) gives another absurd result $\infty > 0$. So T is bounded above and below in order to satisfy (8.4.14).

Thus, (8.4.14) $\Rightarrow c_1 < T < c_2$ and hence $\phi^* \equiv \phi_0$.

Case II Let $\theta > \theta_0$, $\phi^* = \phi_0$:

Similarly, for $\theta > \theta_0$, $\phi^* = \phi_0$. Let $\phi(x) \equiv \alpha$. Then as $E_{\theta_0}\phi_0(X) = \alpha$, it is unbiased. It is also seen that ϕ_0 is UMP unbiased since the class of tests satisfying (8.4.10), (8.4.11) includes the class of unbiased tests.

Remark 8.1

If the distribution of T is symmetric about a p.f. (a) that is $P_{\theta_0}(T < a < u) = P_{\theta_0}(T > a + u)\ \forall\ u$. Then any test which is symmetric about a and which satisfies (8.4.10) must also satisfy (8.4.11) because

$$E_{\theta_0}T\Psi(T) = E_{\theta_0}[(T-a)\Psi(T) + aE_{\theta_0}\Psi(T)] = a\alpha = \alpha E_{\theta_0}T$$

If we take $a = E_{\theta_0}T$,

$$\begin{aligned} E(T-a)\Psi(T) &= E(T - E_{\theta_0}T)\Psi(T) \\ &= ET\Psi(T) - E_{\theta_0}TE\Psi(T) \\ &= 0 \text{ due to symmetry} \end{aligned}$$

or

$$\begin{aligned} E_{\theta_0}(T-a)\Psi(T) &= -uP_{\theta_0}\{(T-a)\Psi(T) < -u\} + uP_{\theta_0}\{(T-a)\Psi(T) > u\} \\ &= -uP_{\theta_0}\{(T-a)\Psi(T) > u\} + uP_{\theta_0}\{(T-a)\Psi(T) > u\} \\ &= 0 \end{aligned}$$

Example 8.4

Let $X_1, X_2, \ldots, X_n$ be a random sample from $N(\theta, 1)$. Obtain the UMPU test for testing H: $\theta = \theta_0$ vs K: $\theta \neq \theta_0$.

The distribution of X being member of exponential family of distributions, the UMPU test rejects H whenever $\bar{X} < c_1$ or $\bar{X} > c_2$ and the distribution of $\bar{X}$ being symmetric only single condition will determine c_2 such as

$$P_{\theta_0}\{\bar{X} > c_2\} = \frac{\alpha}{2}$$

$$\Rightarrow \qquad P_{\theta_0}\{Z > (c_2 - \theta_0)\sqrt{n}\} = \frac{\alpha}{2}$$

where $Z = (\bar{X} - \theta_0)\sqrt{n} \sim N(0,1)$. Let $\phi\left(Z_{\frac{\alpha}{2}}\right) = 1 - \frac{\alpha}{2}$.

Then $(c_2 - \theta_0)\sqrt{n} = Z_{\frac{\alpha}{2}}$

$$\Rightarrow \qquad c_2 = \frac{Z_{\frac{\alpha}{2}}}{\sqrt{n}} + \theta_0.$$

The point of symmetry being θ_0, $\theta_0 - c_1 = c_2 - \theta_0 \Rightarrow c_1 = 2\theta_0 - c_2$.

For testing H: $\theta_1 < \theta < \theta_2$ vs $\theta < \theta_1$ or $\theta < \theta_2$ the same test works only the determination of c_1 will be $c_1 = \theta_1 + \theta_2 - c_2$ because the point of symmetry is $\frac{\theta_1 + \theta_2}{2}$.

Example 8.5

Let $X_1, X_2, \ldots, X_n$ be a random sample from $N(0, \sigma^2)$. To test

$$\sigma^2 = \sigma_0^2 \text{ vs } \sigma^2 \neq \sigma_0^2.$$

$T = \Sigma X_i^2$ being sufficient statistic for the exponential family of distributions the UMPU test will accept H whenever

$$c_1 < Y = \frac{\sum X_i^2}{\sigma_0^2} < c_2$$

with

$$\int_{c_1}^{c_2} f_n(y)dy = 1 - \alpha \qquad \ldots(8.4.15)$$

where $Y \sim \chi^2_n$, $f_n(y) = \dfrac{\left(e^{-\frac{Y}{2}} Y^{\frac{n}{2}-1}\right)}{2^{\frac{n}{2}} \Gamma\left(\frac{n}{2}\right)}$

and

$$\int_{c_1}^{c_2} y f_n(y) = (1-\alpha) E_{\sigma_0^2} Y = (1-\alpha) n \qquad \text{...(8.4.16)}$$

Because $E_{\theta_0}[T(1 - \phi(x)] = (1 - \alpha)E_{\theta_0} T$

One may use $yf_n(y) = nf_{n+2}(y)$ and (8.4.16) give

$$\int_{c_1}^{c_2} f_{n+2}(y)\, dy = 1-\alpha.$$

Alternatively, the equation (8.4.15) will be

$$\int_{c_1}^{c_2} \frac{\left(e^{-\frac{y}{2}} y^{\frac{n}{2}-1}\right)}{2^{\frac{n}{2}} \Gamma\left(\frac{n}{2}\right)} dy = 1-\alpha$$

and integrating by parts (8.4.16) will give

$$n(1-\alpha) = \int_{c_1}^{c_2} \frac{\left(e^{-\frac{y}{2}} y^{\frac{n}{2}-1}\right)}{2^{\frac{n}{2}} \Gamma\left(\frac{n}{2}\right)}$$

$$= \left[\frac{-2e^{-\frac{y}{2}} y^{\frac{n}{2}}}{2^{\frac{n}{2}} \Gamma\left(\frac{n}{2}\right)}\right]_{c_1}^{c_2} + \int_{c_1}^{c_2} \frac{\left(e^{-\frac{y}{2}} y^{\frac{n}{2}-1}\right)}{2^{\frac{n}{2}} \Gamma\left(\frac{n}{2}\right)} dy$$

$$= \frac{2e^{-\frac{c_1}{2}} c_1^{\frac{n}{2}} - 2e^{-\frac{c_2}{2}} c_2^{\frac{n}{2}}}{2^{\frac{n}{2}} \Gamma\left(\frac{n}{2}\right)} + n(1-\alpha)$$

$$\Rightarrow \qquad e^{-\frac{c_1}{2}} c_1^{\frac{n}{2}} = e^{-\frac{c_2}{2}} c_2^{\frac{n}{2}}$$

Example 8.6

Let $X_1, X_2, \ldots, X_n$ be a random sample from $b(1, p)$. Obtain the UMPU test for testing H: $p = p_0$ vs K: $p \neq p_0$.

As $T = \Sigma X_i$ is sufficient for p and its distribution is member of the exponential family of distributions, the UMPU test will have acceptance region $c_1 < X < c_2$, with

$$\sum_{x=c_1+1}^{c_2-1} \binom{n}{x} p_0^x (1-p_0)^{n-x} + \sum_{i=1}^{2} (1-r_i) \binom{n}{c_i} p_0^{c_i} (1-p_0)^{n-c_i} = 1-\alpha \quad \ldots(8.4.17)$$

and

$$\sum_{x=c_1+1}^{c_2-1} x \binom{n}{x} p_0^x (1-p_0)^{n-x} + \sum_{i=1}^{2} (1-r_i) c_i \binom{n}{c_i} p_0^{c_i} (1-p_0)^{n-c_i} = 1-\alpha \ldots(8.4.18)$$

By using individual probabilities and cumulative distribution function of X left hand side of (8.4.17) can be determined. Using

$$x \binom{n}{x} p_0^x (1-p_0)^{n-x} = np_0 \binom{n-1}{x-1} p_0^{x-1} (1-p_0)^{(n-1)-(x-1)}$$

(8.4.18) will be

$$np \sum_{x=c_1+1}^{c_2-1} \binom{n-1}{x-1} p_0^{x-1} (q_0)^{(n-1)-(x-1)} + \sum_{i=1}^{2} (1-r_i) \binom{n-1}{c_i-1} p_0^{c_i-1} (q_0)^{(n-1)-(c_i-1)}$$

$$= 1-\alpha$$

and left hand side of the latter equation can be computed from tables of Binomial distribution. If n is large this can be approximated by normal distribution.

8.5 LOCALLY MOST POWERFUL UNBIASED (LMPU) TEST

If the null hypothesis H: $\theta = \theta_0$ is locally tested against K: $\theta \neq \theta_0$ with size-α then one more restriction of unbiasedness is to be considered to get locally most powerful unbiased test. If $\beta_\phi(\theta)$ is the power function and if it admits continuous derivatives up to second order, that is,

$$\beta_\phi'(\theta) = \int \phi(x) \frac{\partial}{\partial \theta} p_\theta(x) dx$$

$$\beta_\phi''(\theta) = \int \phi(x) \frac{\partial^2}{\partial \theta^2} p_\theta(x) dx$$

then LMPU test of size α for testing H: $\theta = \theta_0$ vs K: $\theta \neq \theta_0$ is one which maximizes $\beta_\phi''(\theta)$ subject to $\beta_\phi(\theta_0) = \alpha$ and $\beta_\phi'(\theta_0) = 0$, which otherwise can be seen from the expansion of $\beta_\phi(\theta)$ about θ_0

$$\beta_\phi(\theta) = \alpha + (\theta - \theta_0) \frac{\partial}{\partial \theta_0} \beta_\phi(\theta) + \frac{(\theta - \theta_0)^2}{2!} \frac{\partial^2}{\partial \theta_0^2} \beta_\phi(\theta) + \ldots$$

taking $\beta_\phi(\theta_0) = \alpha$.

Using generalized Neyman-Pearson lemma the test $\phi_0(x)$ of the form

$$\phi_0(x) = \begin{cases} 1 & \text{if} \quad \dfrac{\partial^2}{\partial \theta_0^2} p_{\theta_0}(x) > k_1 p_{\theta_0}(x) + k_2 \dfrac{\partial}{\partial \theta_0} p_{\theta_0}(x) \\ \gamma & \text{if} \quad \dfrac{\partial^2}{\partial \theta_0^2} p_{\theta_0}(x) = k_1 p_{\theta_0}(x) + k_2 \dfrac{\partial}{\partial \theta_0} p_{\theta_0}(x) \\ 0 & \text{if} \quad \dfrac{\partial^2}{\partial \theta_0^2} p_{\theta_0}(x) < k_1 p_{\theta_0}(x) + k_2 \dfrac{\partial}{\partial \theta_0} p_{\theta_0}(x) \end{cases}$$

where k_1, k_2, γ are determined by solving $\beta_{\phi_0}(\theta_0) = \alpha$ and $\beta'_{\phi_0}(\theta_0) = 0$ for the LMPU test. If $p_\theta(x) > 0$, then

$$\phi_0(x)=\begin{cases}1 & \text{if } \dfrac{\partial^2}{\partial\theta_0^2}\log p_{\theta_0}(x)+\left[\dfrac{\partial}{\partial\theta_0}\log p_{\theta_0}(x)\right]^2 > k_1+k_2\dfrac{\partial}{\partial\theta_0}\log p_{\theta_0}(x)\\ \gamma & \text{if } \dfrac{\partial^2}{\partial\theta_0^2}\log p_{\theta_0}(x)+\left[\dfrac{\partial}{\partial\theta_0}\log p_{\theta_0}(x)\right]^2 = k_1+k_2\dfrac{\partial}{\partial\theta_0}\log p_{\theta_0}(x)\\ 0 & \text{if } \dfrac{\partial^2}{\partial\theta_0^2}\log p_{\theta_0}(x)+\left[\dfrac{\partial}{\partial\theta_0}\log p_{\theta_0}(x)\right]^2 < k_1+k_2\dfrac{\partial}{\partial\theta_0}\log p_{\theta_0}(x)\end{cases}$$

Example 8.7

Let $X_1, X_2, \ldots, X_n$ be a random sample from $N(\theta, 1)$. Then obtain LMPU test for testing H: $\theta = \theta_0$ vs K: $\theta \neq \theta_0$ with size α.

For n observations it is easy to verify that

$$\frac{\partial p_{\theta_0}(x)}{\partial\theta_0}\frac{1}{p_{\theta_0}(x)}=n(\bar{x}-\theta_0)$$

and

$$\frac{\partial^2 p_{\theta_0}(x)}{\partial\theta_0^2}\frac{1}{p_{\theta_0}(x)}=-n+n^2(\bar{x}-\theta_0)^2$$

so that LMPU test $\phi_0(x)$ will be

$$\phi_0(x)=\begin{cases}1 & \text{if } \ n(\bar{x}-\theta_0)\geq k_1(\bar{x}-\theta_0)+k_2\\ 0 & \text{otherwise}\end{cases}.$$

By symmetry of normal distribution the unbiasedness condition is satisfied when $k_1 = 0$, so that the LMPU test is

$$\phi_0(x)=\begin{cases}1 & \text{if } \ n(\bar{x}-\theta_0)^2\geq k_2\\ 0 & \text{otherwise}\end{cases}$$

As $n(\bar{x}-\theta_0)^2 \sim \chi^2_{(1)}$, one can get k_2.

PROBLEMS

8.1 For a random sample of size 2 from binomial distribution with mean θ, obtain UMP size α test for testing H: $\theta \le \frac{1}{3}$ or $\theta \ge \frac{1}{2}$ vs K: $\frac{1}{3} < \theta < \frac{1}{2}$.

8.2 Let

$$f(x) = \begin{cases} \frac{1}{4} & \text{if} \quad 0 \le x < \frac{1}{2} \\ \frac{7}{4} & \text{if} \quad \frac{1}{2} \le x < 1 \end{cases}$$

and

$$g(x) = \begin{cases} \theta x^{\theta-1} & \text{if} \quad 0 < x < 1, \theta > 0 \\ 0 & \text{otherwise} \end{cases}.$$

Then for testing H: $X \sim f(x)$ vs K: $X \sim g(x)$ a single observation rejects H when $x < \frac{1}{10}$ or $x > \frac{9}{10}$, examine whether the test is unbiased.

8.3 Let a single observation be available from the density

$$p_\theta(x) = \begin{cases} 2x & \text{if} \quad 0 < x < \theta \\ 1+\theta & \text{if} \quad \theta \le x < 1 \\ 0 & \text{otherwise} \end{cases}$$

Then reject H: $\theta \le \frac{1}{2}$ vs K: $\theta > \frac{1}{2}$ when, $x > c$ with size 0.1. Show that the test is unbiased.

8.4 Suppose an observation is taken from p.d.f. of X

$$p_\theta(x) = \begin{cases} \theta^{-1} e^{-\frac{x}{\theta}}, & \theta > 0, x > 0 \\ 0 & \text{otherwise} \end{cases}.$$

Examine whether a test which rejects $H: \theta \leq 1$ vs $K: \theta > 1$ when $x > c$ is unbiased for a given size a.

8.5 Obtain UMPU test at level 0.05 for testing $H: \theta = \frac{1}{2}$ vs $K: \theta \neq \frac{1}{2}$ based on $b(8, \theta)$.

8.6 Obtain the LMPU test for testing $H: \sigma^2 = \sigma_0^2$ vs $K: \sigma^2 \neq \sigma_0^2$ using a random sample from $N(0, \sigma^2)$ distribution.

9

CHAPTER

Similar Tests

9.1 INTRODUCTION

In many problems of testing of hypotheses we are interested in testing a single parameter when the distribution may depend on several parameters. The parameters other than parameter for which a hypothesis is tested hinder the testing by their presence and unknown nature. These parameters are called *nuisance parameters*. To resolve nuisance parameters and to get unbiased test a more simple way to get unbiased test is via a similar test. So theory of similar tests has been developed in this chapter and it is applied to multi-parameter exponential family of distributions. Especially, UMP unbiased tests have been developed for different hypotheses of $N(\mu, \sigma^2)$ distribution. Let us illustrate how the testing is not possible due to nuisance parameters.

Let $X_1, X_2, \ldots, X_n$ be a random sample from $N(\theta, \sigma^2)$ distribution. Then for testing the $H: \theta \le \theta_0$ vs $K: \theta > \theta_0$ due to distribution having MLR in $T = \bar{X}$, the UMP test will reject H, whenever $\bar{X} > c$ where, c can be determined by solving

$$\alpha = P_{\theta_0}\{\bar{X} > c\}$$

$$= P_{\theta_0}\left\{\frac{(\bar{X} - \theta_0)\sqrt{n}}{\sigma} > \frac{(c - \theta_0)\sqrt{n}}{\sigma}\right\}$$

$$= P_{\theta_0}\left\{Z > \frac{(c - \theta_0)\sqrt{n}}{\sigma}\right\}$$

where $$Z = \frac{(\bar{X} - \theta_0)\sqrt{n}}{\sigma} \sim N(0, 1).$$

If $P_{\theta_0}\{Z > Z_\alpha\} = \alpha$ or $\phi(Z_\alpha) = 1 - \alpha$ then

$$Z_\alpha = \frac{(c-\theta_0)\sqrt{n}}{\sigma} \Rightarrow c = \frac{Z_\alpha \sigma}{\sqrt{n}} + \theta_0$$

c cannot be determined as σ is unknown. For developing the theory to avoid nuisance parameters, we need following concepts.

Definition 9.1

The set (w) of points common to both Ω_H and Ω_K or limit points of both Ω_H and Ω_K is called boundary set.

Definition 9.2

A test ϕ is said to be α-similar on a boundary set w if $E_\theta \phi(X) = \alpha$ for all $\theta \in w$.

Note

It is clear that there exists at least one similar test on every ω, namely, $\phi(X) \equiv \alpha$.

Example 9.1

Given a random sample X_1 and X_2 from p.d.f.

$$p_\theta(x) = \frac{1}{\theta} e^{-\frac{x}{\theta}}, x > 0, \theta > 0$$

and given a test for testing $H: \theta \le 1$ vs $K: \theta > 1$, whenever $|x_1 - x_2| > \log \frac{1}{\alpha}$ is rejected with size α. Show that the test is α-similar.

Let $y = x_1$ and $z = x_1 - x_2$, i.e., $x_1 = y$ and $x_2 = y - z$. Then as Jacobian of the transformation being 1, the joint p.d.f. of X_1 and X_2 transforms to the joint p.d.f. of y and z as

$$f(y, z) = \frac{1}{\theta^2} e^{-\frac{z}{\theta}} \cdot e^{-\frac{2y}{\theta}}$$

which when integrated w.r.t. y, the marginal density of z will be

$$f(z)=\begin{cases}\frac{1}{2\theta}e^{\frac{z}{\theta}} & \text{when} \quad -\infty<z<0 \\ \frac{1}{2\theta}e^{-\frac{z}{\theta}} & \text{when} \quad 0<z<\infty\end{cases}.$$

Then the test function $\phi(z)$ will be

$$\phi(z)=\begin{cases}1 & \text{if} \quad |z|>\log\frac{1}{\alpha} \\ 0 & \text{otherwise}\end{cases}$$

i.e.,

$$\phi(z)=\begin{cases}1 & \text{if} \quad z<-\log\frac{1}{\alpha} \text{ or } z>\log\frac{1}{\alpha} \\ 0 & \text{otherwise}\end{cases}$$

The power function $\beta_\phi(\theta)$ is

$$\beta_\phi(\theta)=E_\theta\phi(z)$$

$$=\int_{-\infty}^{-\log\frac{1}{\alpha}}\frac{1}{2\theta}e^{\frac{z}{\theta}}dz+\int_{\log\frac{1}{\alpha}}^{\infty}\frac{1}{2\theta}e^{-\frac{z}{\theta}}dz$$

$$=\frac{1}{2}e^{\frac{1}{\theta}\log\alpha}+\frac{1}{2}e^{\frac{1}{\theta}\log\alpha}$$

$$=e^{\log\alpha^{\frac{1}{\theta}}}$$

$$=\alpha^{\frac{1}{\theta}}.$$

Here boundary set, $\omega = \Omega_H \cap \Omega_K$ or $\bar{\Omega}_H \cap \bar{\Omega}_K = \{1\}$. Thus, $\beta(\omega)=\beta_\phi\{1\}=\alpha$, which shows that the test is α-similar.

Example 9.2

For a random sample $X_1, X_2, \ldots, X_n$ of size n from $N(\theta, 1)$ distribution and for testing $H:\theta=0$ vs $K:\theta\neq 0$. Let the test of size α be

$$\phi(x) = \begin{cases} 1 & \text{if} \quad \bar{X} < -k \text{ or } \bar{X} > k \\ 0 & \text{if} \quad \text{otherwise} \end{cases}.$$

Show that the test is α-similar.
Here k is determined by

$$\begin{aligned} \alpha &= E_H \phi(X) \\ &= P_H(\bar{X} < -k) + P_H(\bar{X} > k) \\ &= 2P_H(\bar{X} > k), \text{ due to symmetry.} \end{aligned}$$

i.e., $$P_H(\bar{X} > k) = \frac{\alpha}{2}$$

i.e., $$P\{(\bar{X} - 0)\sqrt{n} > (k - 0)\sqrt{n}\} = \frac{\alpha}{2}$$

i.e., $$P\{Z > \sqrt{n}k\} = \frac{\alpha}{2}, \quad \text{where,} \quad z \sim N(0, 1)$$

$$\Rightarrow \sqrt{n}k = z_{\frac{\alpha}{2}}, \; z_{\frac{\alpha}{2}} \text{ is the upper } \frac{\alpha}{2}\text{th point of } N(0, 1)$$

$$\Rightarrow \quad k = \frac{z_{\frac{\alpha}{2}}}{\sqrt{n}}.$$

Thus, the test function is

$$\phi(x) = \begin{cases} 1 & \text{if} \quad \bar{X} < -\frac{z_{\frac{\alpha}{2}}}{\sqrt{n}} \text{ or } \bar{X} > \frac{z_{\frac{\alpha}{2}}}{\sqrt{n}} \\ 0 & \text{otherwise} \end{cases}$$

and the power function is

$$\begin{aligned} \beta_\phi(\theta) &= P_K\left(\bar{X} < -\frac{z_{\frac{\alpha}{2}}}{\sqrt{n}}\right) + P_K\left(\bar{X} > \frac{z_{\frac{\alpha}{2}}}{\sqrt{n}}\right) \\ &= 2P_K\left(\bar{X} > \frac{z_{\frac{\alpha}{2}}}{\sqrt{n}}\right) \end{aligned}$$

$$= 2P\left\{(\bar{X}-\theta)\sqrt{n} > \left(\frac{z_{\frac{\alpha}{2}}}{\sqrt{n}} - \theta\right)\sqrt{n}\right\}$$

$$= 2P\left\{z > z_{\frac{\alpha}{2}} - \sqrt{n}\theta\right\}.$$

As boundary set, $\omega = \{0\}$

$$\beta(\omega) = \beta_\phi(0)$$

$$= 2P\left\{z > z_{\frac{\alpha}{2}}\right\}$$

$$= 2\frac{\alpha}{2}$$

$$= \alpha$$

which shows that test is α-similar.

Section 9.2 deals with the technique of getting UMPU test from either UMP α-similar test or from α-similar Neyman structure test. Section 9.3 discusses the testing of one parameter of the multi-parameter exponential family of distributions whereas testing linear hypothesis of parameters of multi-parameter exponential family of distributions is dealt with in section 9.4. Various tests associated with parameters μ and σ in $N(\mu, \sigma^2)$ distribution have been given in section 9.5. Confidence bounds (confidence intervals) are discussed in section 9.6. So separate chapter on confidence interval estimation is not provided.

9.2 UMP α-SIMILAR AND NEYMAN STRUCTURE TESTS

In this section we search for UMPU test from either UMP α-similar test or from α-similar Neyman structure test. Several theorems are given to fulfill this purpose.

Theorem 9.1: Let power function, $\beta_\phi(\theta)$ be continuous in θ for any ϕ. If ϕ is unbiased test of size α for testing $H:\theta \in \Omega_H$ vs $K:\theta \in \Omega_K$, it is α-similar on the boundary.

Proof: Let ω be boundary and $\theta \in \omega$. Then there exists a sequence $\{\theta_n\}$, $\theta_n \in \Omega_H$ such that $\theta_n \to \theta$. Since $\beta_\phi(\theta)$ is continuous, $\beta_\phi(\theta_n) \to \beta_\phi(\theta)$ and as ϕ is unbiased and since $\beta_\phi(\theta_n) \leq \alpha$ for $\theta_n \in \Omega_H$, $\beta_\phi(\theta) \leq \alpha$ for

$\theta \in \Omega_H$. Similarly, there exists a sequence $\{\theta_n'\}$, $\theta_n' \in \Omega_K$ such that $\beta_\phi(\theta_n')$ is continuous $\beta_\phi(\theta_n') \to \beta_\phi(\theta)$ and hence $\beta_\phi(\theta) \geq \alpha$ for $\theta \in \Omega_K$. Thus, $\beta_\phi(\theta) = \alpha$ for $\theta \in \omega$, so ϕ is α-similar on ω.

Definition 9.3

Let $|\phi_\alpha|$ be a class of tests which are similar on ω and let $\phi_0 \in \{\phi_\alpha\}$. Then ϕ_0 is UMP α-similar if

$$E_\theta\phi_0(X) \geq E_\theta\phi(X) \ \forall\ \phi \in \{\phi_\alpha\},\ \forall\ \theta \in \Omega_K.$$

Note

It is easier to obtain UMP α-similar test than UMP unbiased test.

Theorem 9.2: Let $\beta_{\phi_0}(\theta)$ be the power function of ϕ_0 for testing $H{:}\theta \in \Omega_H$ vs $K{:}\theta \in \Omega_K$ and let it be continuous in θ and ϕ_0 be UMP unbiased α-similar level α test. Then ϕ_0 is UMP unbiased.

Proof: Since ϕ_0 is UMP α-similar level α test, $E_\theta\phi_0(X) \leq \alpha$ for $\theta \in \Omega_H$ and $E_\theta\phi_0(X) \geq E_\theta\theta(X)$ for $\theta \in \Omega_K$ or $\theta \in \omega$ and $\phi(X)$ is α-similar.

Taking $\phi(X) \equiv \alpha$, we get

$$E_{\theta_0}\phi_0(X) \leq \alpha \quad \text{for} \quad \theta \in \Omega_H$$

$$E_{\theta_0}\phi_0(X) \geq \alpha \quad \text{for} \quad \theta \in \Omega_K.$$

Thus, $\phi_0(X)$ is unbiased. By the continuity of $\beta_\phi(\theta)$ we see that the class of unbiased size α tests is a subclass of all α-similar tests. It follows that ϕ_0 is UMP unbiased size α test.

Definition 9.4

Let T be sufficient statistic for $\theta \in \omega$. Then a test ϕ is said to have Neyman structure if

$$E(\phi(X)/T = t) = \alpha \ \text{for a.e. } t.$$

Theorem 9.3: Any test having Neymann structure is α-similar on ω.

Proof: If $\phi(x)$ has Neyman structure then

$$E(\phi(X)/T = t) = \alpha \text{ for a.e. } t$$

$$\Rightarrow \quad EE(\phi(X)/T = t) = \alpha \text{ for a.e. } t$$

$$\Rightarrow \quad E_\theta \phi(X) = \alpha \text{ for all } \theta.$$

Note

Usually, it is easier to construct Neyman structure test and on constant surface $T = t$ we can obtain MP test separately. The resulting test is then MP among all similar tests provided every similar test has Neyman structure. The following theorem is useful in this connection.

Theorem 9.4: If T is boundedly complete sufficient statistic for $\theta \in \omega$ then every test similar on w has Neyman structure.

Proof: Sufficiency of T implies

$$E(\phi(X)/T = t) = \eta(t) \text{ independent of } \theta.$$

Let $\phi(x)$ be α-similar test. Then

$$\alpha = E_\theta \phi(X) \text{ for all } \theta \in \omega$$

$$\Rightarrow \quad = EE_\theta(\phi(X)/T = t)$$

$$\Rightarrow \quad = E_\theta \eta(t) \text{ for all } \theta \in \omega$$

$$\Rightarrow \quad E_\theta(\eta(t) - \alpha) = 0 \text{ for all } \theta \in w$$

$$\Rightarrow \quad \eta(t) - \alpha = 0, \text{ as } T \text{ is complete}$$

$$\Rightarrow \quad \eta(t) = E(\phi(X)/T = t) = \alpha \text{ for a.e. } t.$$

Note

If we are able to find complete sufficient statistic for θ then UMP α-similar test with Neyman structure can be obtained easily.

9.3 UMP UNBIASED TESTS FOR MULTI-PARAMETER EXPONENTIAL FAMILY

Let X have probability function

$$f_X(x;\theta) = c(\theta, \nu)\exp\left(\theta u(x) + \sum_{i=1}^{k} \nu_i t_i(x)\right)$$

where v is k component vector $(v_1, v_2, \ldots, v_k)$, and T is k component vector $(T_1, T_2, \ldots, T_k)$. We are interested in testing the hypotheses on θ when $v_1, v_2, \ldots, v_k$ are nuisance parameters such as

(i) $H_1 : \theta \leq \theta_0$ vs $K_1 : \theta > \theta_0$

(ii) $H_2 : \theta \leq \theta_1$ or $\theta \geq \theta_2$ vs $K_1 : \theta_1 < \theta < \theta_2$

(iii) $H_3 : \theta_1 < \theta < \theta_2$ vs $K_3 : \theta \leq \theta_1$ or $\theta \geq \theta_2$

(iv) $H_4 : \theta = \theta_0$ vs $K_4 : \theta \neq \theta_0$

For fixed θ, T is sufficient for v having probability function

$$f_X(x; v) = c(\theta, v)\left[\int e^{\theta u} du\right] e^{\Sigma v_i t_i}$$

and hence, conditional distribution of U given $T = t$ is

$$f_{U/T=t}(u/T = t) = e^{\theta u} \Big/ \int e^{\theta u} du = c(\theta) e^{\theta u}$$

which is a one-parameter exponential family of distributions independent of v, the nuisance parameters. Thus, we use conditional distribution in obtaining conditional tests for H_1 to H_4. From these conditional tests one can obtain unconditional tests.

Theorem 9.5: For testing $H_1 : \theta \leq \theta_0$ vs $K_1 : \theta > \theta_0$ in $(k + 1)$ parameter exponential family of distributions as $c(\theta, v) e^{\theta u + \Sigma v_i, t_i}$, a test $\phi_1(u, t)$ defined as

$$\phi_1(u, t) = \begin{cases} 1 & \text{if} \quad u > c_0(t) \\ \gamma_0(t) & \text{if} \quad u = c_0(t) \\ 0 & \text{if} \quad u < c_0(t) \end{cases}$$

such that $E_{\theta_0}(\phi_1(U, t)/T = t) = \alpha$ is UMP unbiased.

Proof: Since power function of exponential family is continuous, it is sufficient to show that ϕ_1 is UMP α-similar on the boundary $\omega = \{(\theta, v) : \theta = \theta_0\}$.

Let ϕ be any other α-similar test on w. Then because (U, T) is complete sufficient statistic for (θ, v), ϕ has Neyman structure, that is,

$$E_{\theta_0}(\phi(U, T)/T = t) = \alpha \text{ for a.e. } t.$$

Further, the conditional distribution of U given T being one-parameter exponential family and since the family possesses MLR, out of all tests ϕ for which

$$E_\theta(\phi(U,T)/T=t)=\alpha$$

ϕ_1 has maximum conditional power for $\theta > \theta_0$ and minimum conditional power for $\theta \le \theta_0$.

Thus, for $\theta > \theta_0$

$$E_\theta(\phi_1(U,T)/T=t) \ge E_\theta(\phi(U,T)/T=t)$$

$$\Rightarrow \quad EE_\theta(\phi_1(U,T)/T=t) \ge EE(\phi(U,T)/T=t)$$

$$\Rightarrow \quad E\phi_1(U,T) \ge E\phi(U,T)$$

which shows that $\phi_1(U,T)$ is UMP α-similar test. Further for $\theta \le \theta_0$,

$$E_\theta\phi_1(U,T) = E_\theta E(\phi_1(U,T)/T=t) \le E_\theta E(\phi(U,T)/T=t)$$
$$= \alpha$$

showing that ϕ_1 is of level α. Hence, ϕ_1 is UMP unbiased test.

Theorem 9.6: For testing $H_2:\theta \le \theta_1$ or $\theta \ge \theta_2$ vs $K_1:\theta_1 < \theta < \theta_2$ in $(k+1)$ parameter exponential family of distributions $c(\theta, v)e^{\theta u + \Sigma v_i, t_i}$, the test $\phi_2(u,t)$ defined as

$$\phi_2(u,t)=\begin{cases} 1 & \text{if} \quad c_1(t)<u<c_2(t) \\ \gamma_i(t) & \text{if} \quad u=c_i(t),\, i=1,2 \\ 0, & u<c_1(t) \text{ or } u>c_2(t) \end{cases}$$

satisfying

$$E_{\theta_1}(\phi_2(U,T)/T=t)=E_{\theta_2}(\phi_2(U,T)/T=t)=\alpha$$

is UMP unbiased.

Proof: We have to show equivalently that ϕ_2 is UMP α-similar on the boundary $\omega=\{(\theta,v):\theta=\theta_1 \text{ or } \theta=\theta_2\}$.

Let $\phi(u, t)$ be a similar test on w. Because T is complete sufficient for v, ϕ has Neyman structure on w and hence,

$$E_{\theta_1}(\phi(U,T)/T=t)=E_{\theta_2}(\phi(U,T)/T=t)=\alpha. \qquad \ldots(9.3.1)$$

Since the conditional distribution of U given T is independent of v and since it is a one-parameter exponential family of distributions out of all tests ϕ satisfying (9.3.1), the test ϕ_2 has maximum conditional power for $\theta \in \Omega_K$ and hence for $\theta_1 < \theta < \theta_2$,

$$E(\phi_2(U,T)/T = t) \geq E(\phi(U,T)/T = t)$$

$$E_\theta E(\phi_2(U,T)/T = t) \geq E_\theta E(\phi(U,T)/T = t)$$

$$\Rightarrow \qquad E_\theta \phi_2(U,T) \geq E_\theta \phi(U,T)$$

Showing that ϕ_2 is UMP α-similar and for $\theta \leq \theta_1$ or $\theta \geq \theta_2$ has minimum conditional power, that is,

$$E(\phi_2(U,T)/T = t) \leq E(\phi(U,T)/T = t)$$

$$\Rightarrow \qquad E_\theta E(\phi_2(U,T)/T = t) \leq E_\theta E(\phi(U,T)/T = t)$$

$$\Rightarrow \qquad E_\theta(\phi_2(U,T)/T = t) \leq E_\theta \alpha = \alpha$$

showing that ϕ_2 is of level-α test. As power function of exponential family is continuous UMP α-similar test is UMP unbiased and hence ϕ_2 is UMP unbiased level-α test.

Theorem 9.7: For testing $H_3 : \theta_1 < \theta < \theta_2$ vs $K_3 : \theta \leq \theta_1$ or $\theta \geq \theta_2$ in $(k + 1)$ parameter exponential family of distributions $c(\theta, v)e^{\theta u + \Sigma v_i, t_i}$, a test $\phi_3(u, t)$ of the form

$$\phi_3(u,t) = \begin{cases} 1 & \text{if } \; u < c_1(t) \text{ or } u > c_2(t) \\ \gamma_i(t) & \text{if } \; u = c_i(t), i = 1, 2 \\ 0 & \text{if } \; c_1(t) < u < c_2(t) \end{cases}$$

satisfying $E_{\theta_i}(\phi_2(U,t)/T = t) = \alpha, i = 1, 2$, is UMP unbiased.

Proof: Same as in Theorem 9.7 with interchange of H_3 with K_3.

Theorem 9.8: For testing $H_4 : \theta = \theta_0$ vs $K_4 : \theta \neq \theta_0$ in $(k+1)$ parameter exponential family of distributions $c(\theta, v)e^{\theta u + \Sigma v_i, t_i}$ the test $\phi_4 = \phi_3$ satisfying

$$E_{\theta_0}(\phi_4(U,T)/T = t) = \alpha$$

$$E_{\theta_0}(U\phi_4(U,T)/T = t) = \alpha E_{\theta_0}(U/T = t)$$

is UMP unbiased.

Proof: Let ϕ be any unbiased test of size α. Then ϕ must be α-similar on the boundary $\omega = \{(\theta, v): \theta = \theta_0\}$. Further, as ϕ is unbiased size α test.

$$\Rightarrow \quad \begin{cases} E_{\theta, v}\phi(U, t) = \alpha \text{ on } \omega \\ E_{\theta, v}\{U\phi(U, T) - \alpha U\} = 0 \end{cases}$$

Since T is complete for v for fixed θ on ω and since ϕ is α-similar it has Neyman structure, that is

$$\begin{cases} E_{\theta_0}(\phi(U,T)/T = t) = \alpha \text{ on } \omega \\ E_{\theta_0}[\{U\phi(U, t) - \alpha U\}/T = t] = 0 \end{cases} \quad \ldots(9.3.2)$$

Since the conditional distribution of U given $T = t$ is one parameter exponential distribution not involving nuisance parameters, out of all tests ϕ satisfying (9.3.2) ϕ_4 has maximum conditional power for $\theta \in \Omega_K$, that is for $\theta \neq \theta_0$

$$E_{\theta, v}(\phi_4(U, T)/T = t) \geq E_{\theta, v}(\phi(U, T)/T = t)$$

$$\Rightarrow \quad EE(\phi_4(U, T)/T = t) \geq EE_{\theta, v}(\phi(U, T)/T = t)$$

$$\Rightarrow \quad E\phi_4(U, T) \geq E_{\theta, v}\phi(U, T)$$

Hence, ϕ_4 is UMP unbiased.

9.4 TESTING LINEAR PARAMETRIC HYPOTHESIS

Let X have $(k + 1)$ parameter $(\theta, v_1, v_2, \ldots, v_k)$ exponential family of distributions $c(\theta, v)e^{\theta u + \Sigma v_i, t_i}$. Suppose we are interested in testing a hypotheses concerning linear function such as

$$\theta^* = a_0\theta + \Sigma v_i a_i.$$

We reparametrize the family with θ^*, v^* as

$$\theta = \frac{\theta^* - \Sigma v_i a_i}{a_0}, \quad \theta u = \frac{\theta^* u}{a_0} - \frac{\Sigma v_i a_i u}{a_0}$$

$$\Rightarrow \qquad \theta u + \Sigma v_i t_i = \frac{\theta^* u}{a_0} + \sum\left(t_i - \frac{a_i u}{a_0}\right)v_i$$

$$= \theta^* u^* + \Sigma v_i^* t_i^*$$

where $U^* = \dfrac{u}{a_0}$, $v_i^* = v_i$, $T_i^* = t_i - \dfrac{a_i u}{a_0}$, $i = 1, 2, \ldots, k$ and thus we want to test hypothesis concerning θ^* in the family

$$c_0(\theta^* v^*)\exp(\theta^* u^* + \Sigma v_i^* t_i^*)$$

and for which we have to obtain conditional distribution of U^* given T^*.

Comparison of Two Poisson Distributions

Comparision of radio activity of two substances will lead to equality of Poisson distributions and the beneficial effect of a treatment over other (control) treatment will lead to one-sided test. When two treatments on usual footing, the hypothesis of equality of two treatments is tested against two-sided alternative.

Let X, Y be independent Poisson variates with parameters λ, μ, respectively. Then their joint p.f.

$$P(X = x, Y = y) = \frac{e^{-(\lambda+\mu)}}{x!y!} e^{x\log\lambda + y\log\mu}$$

can be rewritten as

$$P(X = x, Y = y) = \frac{e^{-(\lambda+\mu)}}{x!y!} e^{y(\log\mu - \log\lambda) + (x+y)\log\lambda}$$

$$= c(\theta, v)e^{\theta u + vt}$$

where $\theta = \log\mu - \log\lambda$, $U = y$, $v = \log\lambda$, $T = X + Y$

[Another way of expressing the same with $v = \log\lambda$, $\theta = \log\mu$,

$$\theta^* = \log\mu - \log\lambda = a_0\theta + a_1 v$$

$$\Rightarrow \qquad a_0 = 1,\ a_1 = -1,\ \text{with } U^* = \frac{Y}{a_0},\ T = X - \frac{a_1 Y}{a_0} = Y + X]$$

For testing $\mu \le \lambda \Leftrightarrow \theta^* = \log\mu - \log\lambda \le 0$ or $\mu = \lambda \Leftrightarrow \theta^* = 0$ or $\mu = 2\lambda \Leftrightarrow \theta^* = \log 2$. We require conditional distribution of $U = y$ given $T = X + Y$ to avoid nuisance parameter, ν. Let us obtain conditional distribution of Y given $X + Y = t$,

$$P(Y = y/X + Y = t) = P\{Y = y, X = t - y\}/P(X + Y = t)$$

$$= \frac{\dfrac{e^{-\mu}\mu^y}{y!}\,\dfrac{e^{-\lambda}\mu^{t-y}}{(t-y)!}}{\dfrac{e^{-(\lambda+\mu)}(\lambda+\mu)^t}{t!}}$$

$$= \binom{t}{y}\frac{\mu^y\lambda^{t-y}}{(\lambda+\mu)^{t-y+y}}$$

$$= \binom{t}{y}\left(\frac{\mu}{\lambda+\mu}\right)^y\left(\frac{\lambda}{\lambda+\mu}\right)^{t-y}$$

which is $b(t, p)$ with $p = \dfrac{\mu}{\lambda+\mu}$.

Now $\mu = \lambda \Rightarrow p = \dfrac{1}{2}$, $\mu \le \lambda \Rightarrow p \le 1/2 \mu = 2\lambda \Rightarrow p = \dfrac{2}{3}$, etc.

Comparision of Two Binomial Distributions

Cures due to two treatments will lead to testing equality of two binomial distributions. Beneficial effect of one treatment with control or another treatment will lead to one sided test.

Let $X \sim b(m, p_1)$ and $Y \sim b(n, p_2)$ be independent. Then

$$P(X = x, Y = y) = \binom{m}{x}p_1^x q_1^{m-x}\binom{n}{y}p_2^y q_2^{n-y}$$

$$= \binom{m}{x}\binom{n}{y}q_1^m q_2^n e^{x\log\frac{p_1}{q_1} + y\log\frac{p_2}{q_2}}$$

$$= \binom{m}{x}\binom{n}{y}q_1^m q_2^n e^{y\left[\log\frac{p_2}{q_2} - \log\frac{p_1}{q_1}\right] + (x+y)\log\frac{p_1}{q_1}}$$

The hypothesis $p_2 \le p_1$ or $p_2 = p_1$, etc., $\Leftrightarrow \log\left(\frac{p_2 q_1}{p_1 q_2}\right)$. The distribution being a two-parameter exponential family with $\theta = \log\frac{p_2 q_1}{p_1 q_2}$, $U = Y$, $v = \log\frac{p_1}{q_1}$, $T = X + Y$, we need to obtain conditional distribution of U given T for testing hypothesis concerning θ.

$$P(Y = y/X + Y = t) = P(Y = y)P(X = t - Y)/P(X + Y = t)$$

$$= \frac{\binom{n}{y}\binom{m}{t-y}\left(\frac{p_2 q_1}{p_1 q_2}\right)^y}{\sum_{y=0}^{t}\binom{n}{y}\binom{m}{t-y}\left(\frac{p_2 q_1}{p_1 q_2}\right)^y}$$

$$= c_t(\rho)\binom{n}{y}\binom{m}{t-y}\rho^y$$

where

$$\rho = \frac{p_2 q_1}{p_1 q_2},\ c_t(\rho) = \frac{1}{\sum_{y=0}^{t}\binom{n}{y}\binom{m}{t-y}\left(\frac{p_2 q_1}{p_1 q_2}\right)^y}.$$

In particular, $p_2 = p_1 \Leftrightarrow \rho = 1$, the conditional distribution turns out to be

$$P(Y = y/X + Y = t) = \frac{\binom{n}{y}\binom{m}{t-y}}{\binom{m+n}{t}}$$

which is hypergeometric distribution.

Sign Test

Out of n individuals if x prefers attribute A with probability p, and y prefers attribute B with probability $1 - p$ then for testing equality of preference $\left(\text{i.e., } p_1 = p_2 = \frac{1}{2}\right)$ we use sign test which demands total number of X's following $b(n, p)$ distribution. When X exceeds c it results into rejection of the hypothesis.

Sometimes same individuals do not prefer neither A nor B then the test may be revised. Out of n individuals let x prefer A with probability p_1, y prefer B with probability p_2, and z neither prefer A nor B with probability p_3. The joint distribution of x, y, z will be

$$P(X = x, Y = y, Z = z) = \frac{n!}{x!y!z!} p_1^x p_2^y p_3^z,$$

$$n = x + y + z,\ x \geq 0,\ y \geq 0,\ z \geq 0,\ p_1 + p_2 + p_3 = 1$$

$$P(X = x, Y = y, Z = z) = \frac{n!}{x!z!(n - x - z)!}$$

$$\left(\frac{p_1}{1 - p_1 - p_3}\right)^x \left(\frac{p_3}{1 - p_1 - p_3}\right)^z (1 - p_1 - p_3)^n$$

$$= c(\theta, \nu) h(x, z) e^{\theta u + \nu T}$$

where

$$\theta = \log\left(\frac{p_1}{1 - p_1 - p_3}\right), U = X, \nu = \log\left(\frac{p_3}{1 - p_1 - p_3}\right), T = z.$$

Now

$$H: p_1 = p_2 \Leftrightarrow \log\frac{p_1}{p_2} = 0 \Rightarrow \log\left(\frac{p_1}{1 - p_1 - p_3}\right) = \theta = 0.$$

For testing $H: \theta = 0$, we need conditional distribution of x given z. The marginal distribution of z will be

$$\frac{n! p_3^z}{z!(n - z)!} \sum_{x=0}^{n-z} \frac{(n - z)!}{(n - x - z)! x!} p_1^x (1 - p_1 - p_3)^{n - z - x}$$

$$=\binom{n}{z}p_3^z(1-p_1-p_3+p_1)^{n-z}$$

$$=\binom{n}{z}p_3^z(1-p_3)^{n-z}$$

so that the conditional distribution of X given z is

$$\frac{\dfrac{n!}{x!z!(n-x-z)!}\left(\dfrac{p_1}{1-p_1-p_3}\right)^x\left(\dfrac{p_3}{1-p_1-p_3}\right)^z(1-p_1-p_3)^n}{\dfrac{n!p_3^z}{z!(n-z)!}(1-p_3)^{n-z}}$$

$$=\binom{n-z}{x}\left(\frac{p_1}{1-p_3}\right)^x\left(\frac{1-p_1-p_3}{1-p_3}\right)^{n-z-x}$$

$$=\binom{n-z}{x}\left(\frac{p_1}{1-p_3}\right)^x\left(1-\frac{p_1}{1-p_3}\right)^{n-Z-x}$$

which is $b\left(n-Z,\dfrac{p_1}{1-p_3}\right)$ distribution

Testing for Independence in 2 × 2 Table

We are interested in testing independence of two characteristics A, B, that is,

$$p_1=(B/A),\; p_2=P(B/\bar{A}).$$

Let

	A	$\bar{A}$	
B	X	X′	m
$\bar{B}$	Y	Y′	n
	t	t'	S

Then the joint distribution of X, X', Y, Y' is

$$f(x,x',y,y')=\frac{S!}{x!x'!y!y'!}P_{AB}^{x}P_{\bar{B}\bar{A}}^{x'}P_{\bar{B}A}^{y}P_{\bar{A}\bar{B}}^{y'},$$

$$x + x' + y + y' = S,\ P_{\bar{A}\bar{B}} = 1 - P_{AB} - P_{A\bar{B}} - P_{\bar{A}B}$$

$$f(x, x', y, y') = \frac{S!}{x!x'!y!(S - x - x' - y)!} P_{\bar{A}\bar{B}}^{S} \left(\frac{P_{AB}}{P_{\bar{A}\bar{B}}}\right)^{x} \left(\frac{P_{A\bar{B}}}{P_{\bar{A}\bar{B}}}\right)^{x'} \left(\frac{P_{\bar{A}B}}{P_{\bar{A}\bar{B}}}\right)^{y}$$

$$= ce^{x \log \frac{P_{AB}}{P_{\bar{A}\bar{B}}} + x' \log \frac{P_{A\bar{B}}}{P_{\bar{A}\bar{B}}} + y \log \frac{P_{\bar{A}B}}{P_{\bar{A}\bar{B}}}}$$

Let

$$\theta^* = a_0 \log \frac{P_{AB}}{P_{\bar{A}\bar{B}}} + a_1 \log \frac{P_{\bar{A}B}}{P_{\bar{A}\bar{B}}} + a_2 \log \frac{P_{A\bar{B}}}{P_{\bar{A}\bar{B}}}$$

Taking $a_1 = a_2 = 1,\ a_0 = -1,\ \Delta = e^{\theta^*} = \dfrac{P_{\bar{A}B}\,P_{A\bar{B}}}{P_{\bar{A}\bar{B}}\,P_{AB}}$

$$P_A = P_{AB} + P_{A\bar{B}},\ P_B = P_{AB} + P_{\bar{A}B}.$$

$$P_{AB} = P_A P_B + \frac{1-\Delta}{\Delta} P_{\bar{A}B} P_{A\bar{B}},\ P_{\bar{A}B} = P_{\bar{A}B} - \frac{1-\Delta}{\Delta} P_{\bar{A}B} P_{A\bar{B}},$$

$$P_{A\bar{B}} = P_A P_B - \frac{1-\Delta}{\Delta} P_{\bar{A}B} P_{A\bar{B}},\ P_{\bar{A}\bar{B}} = P_{\bar{A}} P_{\bar{B}} + \frac{1-\Delta}{\Delta} P_{\bar{A}B} P_{A\bar{B}}.$$

Independence of A and $B \Rightarrow \Delta = 1$. To test the independence we take conditional distribution of X given $X + X' = m,\ Y + Y' = t$. Thus,

$$P(X = x, Y = y / X + X' = m) = \binom{m}{x}\binom{n}{y}\left(\frac{P_{AB}}{P_B}\right)^{x}\left(\frac{P_{\bar{A}B}}{P_B}\right)^{m-x}\left(\frac{P_{A\bar{B}}}{P_{\bar{B}}}\right)^{y}\left(\frac{P_{\bar{A}\bar{B}}}{P_{\bar{B}}}\right)^{n-y}$$

$\Rightarrow$ two independent binomial variates with probability

$$p_1 = \frac{P_{AB}}{P_B},\ p_2 = \frac{P_{A\bar{B}}}{P_{\bar{B}}}.$$

If additional restriction $X + Y = t$ is imposed, the conditional distribution of X given $X + X' = m,\ X + Y = t$ is that of X given $X + Y = t$ in two independent binomial, i.e.,

$$P(X = x / X + X' = m,\ X + Y = t) = c_t(\rho)\binom{m}{x}\binom{n}{t-x}\rho^{t-x}$$

where

$$\rho = \frac{p_2 q_1}{p_1 q_2}, c_t(\rho) = \frac{1}{\sum_{x'=0}^{t} \binom{m}{x'} \binom{n}{t-x'} \rho^{x'}}$$

when $\rho = 1 \equiv$ independence,

$$P(X = x'/X + X' = m, X + Y = t) = \frac{\binom{m}{x}\binom{n}{t-x}}{\binom{m+n}{t}}$$

Example 9.3

Let $X = (X_1, X_2, \ldots, X_n)$ be a random sample from $u(\theta_1, \theta_2)$ distribution. Then obtain UMP test $H:\theta_1 \leq 0$ vs $K:\theta_1 \geq 0$.

Here

$$f(x_i) = \frac{1}{\theta_2 - \theta_1}, \theta_1 < x < \theta_2, i = 1, 2, \ldots, n$$

and

$$F(x) = \frac{x - \theta_1}{\theta_2 - \theta_1}.$$

The joint p.d.f. of $T_1 = \min X_i$, $T_2 = \max X_i$ is

$$f_{T_1, T_2}(t_1, t_2) = \frac{n(n-1)(t_2 - t_1)^{n-2}}{(\theta_2 - \theta_1)^n}, \theta_1 < t_1 < t_2 < \theta_2$$

and marginal p.d.f. of T_2 will be

$$f_{T_2}(t_2) = \frac{n(t_2 - \theta_1)^{n-1}}{(\theta_2 - \theta_1)^n}.$$

Hence, the conditional distribution of T_1 given T_2 is

$$f_{T_1/T_2}(t_1/t_2) = \frac{(n-1)(t_2-t_1)^{n-2}}{(t_2-\theta_1)^{n-1}}, \theta_1 < t_1 < t_2$$

which is p.d.f. of minimum of $(n-1)$ sample values taken from $u(\theta_1, t_2)$. The UMP unbiased test for testing $\theta_1 \le 0$ is to reject it when $T_1 \ge c(t_2)$ where $c(t_2)$ is such that

$$P_{\theta_1=0}\{T_1 \ge c(t_2)/T_2 = t_2\} = \alpha \Rightarrow \int_{c(t_2)}^{\infty} f_{T_1/T_2}(t_1/t_2)dt_1 = \alpha.$$

Instead of using conditional distribution, we use ancillary statistic

$$V = \frac{T_2 - T_1}{T_2 - \theta_1}, dV = \frac{dt_1}{t_2 - \theta_1},$$

$$f_V(v) = (n-1)v^{n-2}, 0 < v < 1$$

which being independent of t_2, the UMP unbiased test will reject the hypothesis if $V \ge c$, where c is such that

$$\int_c^1 (n-1)v^{n-1}dv = \alpha \Rightarrow c = (1-\alpha)^{\frac{1}{n-1}}.$$

Example 9.4

Let X, Y be independent variates with joint geometric distribution as $f_{X,Y}(x,y) = (1-\theta_1)(1-\theta_2)\theta_1^x\theta_2^y$, $x = 0, 1, \ldots,$ $y = 0, 1, \ldots, 0 < \theta_1, \theta_2 < 1$.

Find UMP unbiased test of size $\alpha = 0.2$ for testing

(a) $H_1 : \theta_1 \le \theta_2$ vs $K_1 : \theta_1 > \theta_2$

(b) $H_2 : \theta_1 = \theta_2$ vs $K_2 : \theta_1 \ne \theta_2$.

The joint p.f. of X and Y is a two-parameter exponential family such as

$$f_{X,Y}(x,y) = (1-\theta_1)(1-\theta_2)e^{x\log\theta_1 + y\log\theta_2} = c(\theta_1,\theta_2)e^{x\log\frac{\theta_1}{\theta_2} + (x+y)\log\theta_2}$$

$$H : \theta_1 \le \theta_2 \Rightarrow \theta^* = \log\frac{\theta_1}{\theta_2} \le 0, U = X, T = X + Y.$$

We have to obtain conditional distribution of X given $T = X + Y = t$

$$f_{X,T}(x,t) = (1-\theta_1)(1-\theta_2)\theta_1^x\theta_2^{t-x} = (1-\theta_1)(1-\theta_2)\left(\frac{\theta_1}{\theta_2}\right)^x (\theta_2)^t$$

Therefore,

$$f_T(t) = (1-\theta_1)(1-\theta_2)(\theta_2)^t \sum_{x=0}^{t} \rho^x, \text{ where } \rho = \frac{\theta_1}{\theta_2}$$

$$\Rightarrow \qquad f_{X/T}(x/t) = \frac{\rho^x}{\sum_{x=0}^{t} \rho^x} = \frac{1}{1+t}, \text{ if } \rho < 1.$$

We have UMP unbiased test for testing H_1 as

$$\phi_1(x,t) = \begin{cases} 1 & \text{if } \; x > c_0(t) \\ \gamma_0 & \text{if } \; x = c_0(t) \\ 0 & \text{if } \; x < c_0(t) \end{cases}$$

with $P_{\rho=1}\{X > c_0(t)/t\} + \gamma_0(t)P_{\rho=1}\{x = c_0(t)/t\} = 0.2$

i.e.,

$$\sum_{x=c_0}^{t} \frac{1}{1+t} + \gamma_0(t)\frac{1}{1+t} = 0.2$$

i.e.,

$$\frac{t-c_0}{1+t} + \frac{\gamma_0}{1+t} = 0.2.$$

For testing $\rho = 1$, the test is

$$\phi_2(x,t) = \begin{cases} 1 & \text{if } \; x > c_2(t) \text{ or } x < c_1(t) \\ \gamma_i & \text{if } \; x = c_i(t), i = 1, 2 \\ 0 & \text{if } \; c_1(t) < x < c_2(t) \end{cases}$$

where $c_1, c_2, \gamma_1, \gamma_2$ are such that

$$\sum_{c_1}^{c_2}\frac{1}{1+t}+\sum_{i=1}^{2}(1-\gamma_i)\frac{1}{1+t}=0.8$$

$$\sum_{x=c_1}^{c_2}\frac{x}{1+t}+\sum_{i=1}^{2}(1-\gamma_i)\frac{1}{1+t}=(0.8)E(X/t)$$

$$=(0.8)\frac{t+2}{2}$$

Example 9.5

Comparison of two exponential distributions

Lifetimes of electric bulbs produced by two different companies have to be tested for their equality of lifetimes which leads to testing equality of parameters of two independent exponential distributions.

Let X and Y be independently distributed as exponential distributions with p.d.f. as

$$f_X(x;\theta_1)=\frac{1}{\theta_1}e^{-\frac{x}{\theta_1}},\ x>0$$

$$f_Y(y;\theta_2)=\frac{1}{\theta_2}e^{-\frac{y}{\theta_2}},\ y>0$$

By making the transformation $X = x$ and $T = X + Y$ the joint p.d.f. of X, Y will be

$$f_{X,Y}(x;y)=\frac{1}{\theta_1\theta_2}e^{-\frac{x}{\theta_1}-\frac{y}{\theta_2}},\ 0<x<\infty,\ 0<y<\infty \qquad \text{...(9.4.1)}$$

transforms to the joint p.d.f. of X and T as

$$f_{X,T}(x;t)=\frac{1}{\theta_1\theta_2}e^{-\frac{x}{\theta_1}-\frac{t-x}{\theta_2}},\ t>x,\ x>0 \qquad \text{...(9.4.2)}$$

Integrating out (9.4.2) with respect to x, the marginal p.d.f. of T will be

$$f_T(t)=\frac{e^{-\frac{t}{\theta_1}-\frac{t}{\theta_2}}}{\theta_1-\theta_2},\ t>0,\ \theta_1\neq\theta_2 \qquad \text{...(9.4.3)}$$

Dividing (9.4.2) by (9.4.3) we get the conditional distribution of X given $T = t$ with p.d.f.

$$f_{X|T}(x \mid t) = \frac{\nu e^{-\nu x}}{1 - e^{-\nu t}},\ x > t \qquad \ldots(9.4.4)$$

where $$\nu = \frac{1}{\theta_1} - \frac{1}{\theta_2}$$

since the p.d.f. (9.4.1) can be written as

$$f_{X,Y}(x; y) = \frac{1}{\theta_1\theta_2} e^{-\left(\frac{1}{\theta_1} - \frac{1}{\theta_2}\right)x - \frac{1}{\theta_2}(x+y)}$$

Which is a exponential family of distributions with parameters $\frac{1}{\theta_1} - \frac{1}{\theta_2}$ and $\frac{1}{\theta_2}$ with corresponding complete sufficient statistics as X and $(X + Y) = T$ respectively, for testing $\theta_2 = \theta_1$ thus, we need conditional distribution of X given $T = t$ which we derived above in (9.4.4).

As (9.4.4) is p.d.f of X given $T = t$ which is truncated exponential distribution with monotone likelihood ratio in $(-X)$ the UMPU test level α for testing

$$H_1 : \theta_2 > \theta_1 \Leftrightarrow \nu > 0$$

is to reject H_1 whenever $X > c$ given $T = t$

where c is given by $c = \frac{1}{\nu} \log\left[\frac{1}{\alpha(1 - e^{-\nu t})}\right]$ for fixed t and given ν

Example 9.6

Let X, Y have p.d.f.

$$f_{X,Y}(x, y) = \lambda\mu e^{-\lambda x - \mu y},\ x > 0,\ y > 0$$

Obtain UMPU tests of size $\alpha = 0.2$ for testing

(a) $H_1 : \lambda \le \mu + 1$

(b) $H_2 : \lambda = \mu$

(c) $H_3 : \lambda \ge 2\mu$

For H_1,

$$f(x, y) = e^{-(\lambda-\mu)x-\mu(x+y)}$$

So we have to obtain conditional distribution of X given $X + Y$. Let $Z = X + Y$. Then $Y = Z - X$, $X = x$ and

$$f_{Z,X}(z, x) = \lambda\mu e^{-\lambda x-\mu(Z-X)}$$

$$= \lambda\mu e^{-\mu Z-(\lambda-\mu)x}.$$

The marginal distribution of Z will be

$$\lambda\mu e^{-\mu Z}\int_0^Z e^{-(\lambda-\mu)x}dx = \frac{\lambda\mu e^{-\mu Z}}{\lambda-\mu}[-e^{-(\lambda-\mu)x}]_0^Z$$

$$= \frac{\lambda\mu e^{-\mu Z}[1-e^{-(\lambda-\mu)Z}]}{\lambda-\mu}$$

And hence, the conditional distribution of X given $X + Y = Z$ is

$$f_{X/Z}(x/z) = \frac{\lambda\mu e^{-\mu Z}e^{-(\lambda-\mu)x}(\lambda-\mu)}{\lambda\mu e^{-\mu Z}[1-e^{-(\lambda-\mu)Z}]}$$

$$= \frac{(\lambda-\mu)e^{-(\lambda-\mu)x}}{[1-e^{-(\lambda-\mu)Z}]}, x < Z$$

The UMP unbiased test for (a) then will be $[\lambda - \mu = \theta \le 1 \text{ vs } \theta > 1]$ to reject H if $X > c(t)$ where c is such that

$$\alpha = P_{\theta=1}(X > c/Z = z) = \int_c^Z \frac{e^{-x}}{1-e^{-Z}}dx = \frac{e^{-c}-e^{-Z}}{1-e^{-Z}}$$

And for (b) $H:\theta = 0$, we accept H_2 when $c_1 < x < c_2$ where c_1, c_2 are such that

$$\int_{c_1}^{c_2}\frac{e^{-x}}{1-e^{-Z}}dx = 1-\alpha, \frac{1}{1-e^{-Z}}\int_{c_1}^{c_2} xe^{-x}dx = (1-\alpha)E(X/Z).$$

9.5 TESTS FOR THE PARAMETERS OF NORMAL POPULATION

Instead of finding conditional distribution of sufficient statistic of the parameter under testing given the sufficient statistics of the nuisance parameters and then getting conditional tests, we would like to construct an ancillary statistic $V = h(U, T)$, a function of (U, T) in $(k+1)$ parameter $(\theta, \nu_1, \ldots, \nu_k)$ exponential family with sufficient statistics (U, T) such that it is independent of T. The test will become unconditional.

Theorem 9.9: For testing $H_1 : \theta \le \theta_0$ vs $K_1 : \theta > \theta_1$ in $(k+1)$ parameter exponential family of distributions $c(\theta, \nu)e^{\theta u + \nu t}$, if $V = h(U, T)$ is monotonic function of U for fixed T and is independent of T when $\theta = \theta_0$ then a test

$$\phi_1(v) = \begin{cases} 1 & \text{if} \quad v > c_0 \\ \gamma_0 & \text{if} \quad v = c_0 \\ 0 & \text{if} \quad v < c_0 \end{cases}$$

with $E_{\theta_0}\phi_1(v) = \alpha$ is UMP unbiased. For testing $H_4 : \theta = \theta_0$, moreover if V is linear function of U for fixed T at $\theta = \theta_0$, then a test $\phi_4(V)$,

$$\phi_4(v) = \begin{cases} 1 & \text{if} \quad v < c_1 \text{ or } v > c_2 \\ \gamma_i & \text{if} \quad v = c_i, i = 1, 2 \\ 0 & \text{if} \quad c_1 < v < c_2 \end{cases}$$

with

$$\begin{cases} E_{\theta_0}\phi_4(V) = \alpha \\ E_{\theta_0}\{V\phi_4(V)\} = \alpha E_{\theta_0}V \end{cases}$$

is UMP unbiased.

Proof: For H_1 the constants c_0 and γ_0 are determined by

$$P_{\theta_0}\{U > c_0'(t)/T = t\} + \gamma_0 P_{\theta_0}\{U = c_0'(t)/T = t\} = \alpha.$$

Since V is monotonic function of U and it being independent of T, the above equation turns out be

$$P_{\theta_0}\{V > c_0\} + \gamma P_{\theta_0}\{V = c_0\} = \alpha.$$

For $H_2: \theta < \theta_1$ or $\theta > \theta_2$ and $H_3: \theta_1 < \theta < \theta_2$, similar proofs will work.

Now for H_4 the equations determining the constants are

$$E_{\theta_0}\{\phi_4(U,T)/T = t\} = \alpha$$

$$E_{\theta_0}\{U\phi_4(U,T)/T = t\} = \alpha E_{\theta_0}(U/T = t)$$

with the condition that V is a monotonic function of U for fixed T independent of T and V being linear function $V = a(t)U + b(t)$, U for fixed T but $\theta = \theta_0$, the equation determining the constants will turn out as

$$E_{\theta_0}\phi_4(V) = \alpha, U = \frac{V - b(t)}{a(t)}$$

and

$$E_{\theta_0}\left\{\frac{V - b(t)}{a(t)}\phi_4(V,t)/T = t\right\}$$

$$= \alpha E_{\theta_0}\left\{\frac{V - b(t)}{a(t)} \Big/ T = t\right\}$$

$\Rightarrow$ $$E_{\theta_0}\{V\phi_4(V,t)/t\} = \alpha E_{\theta_0}(V/t)$$

$\Rightarrow$ $$E_{\theta_0}\{V\phi_4(V)\} = \alpha E_{\theta_0}V$$

For showing independence of V and T, the following theorem of Basu is useful.

Theorem 9.10: If T is complete sufficient statistic for v in (θ, v) then $V = h(U, T)$ is independent of T iff the distribution of V does not involve v.

Proof: Let ϕ be a critical function. Then by assumption $E\phi(V)$ is independent of v, that is, $E_\theta\phi(V) = \alpha$, constant that is ϕ is α-similar test and T being complete sufficient it has Neyman structure, that is,

$$E_\theta\{\phi(V)/T = t\} = \alpha \text{ for a.e. } t$$

$\phi(V)$ is independent of T.

Remark

If the family of distributions is exponential one then T is always complete sufficient for v and hence we have to select V in such a way that its distribution is independent of v.

Example 9.7

Let $X_1, X_2, \ldots, X_n$ be independent, $N(\xi, \sigma^2)$. Then $T = \bar{X}$ is complete sufficient for ξ and $V = \Sigma(X_i - \xi) \sim N(0, \sigma^2)$ it does not involve ξ and hence, V is independent of T. This will hold for all σ.

Example 9.8

Let $X_1, X_2, \ldots, X_n$ be independent $N(\xi_0, \sigma^2)$. Then

$$V = \frac{\bar{X} - \xi_0}{\sqrt{\sum (X_i - \bar{X})^2}} \text{ or } V = \frac{\bar{X} - \xi_0}{\sqrt{\sum (X_i - \xi_i)^2}}$$

being invariant under the transformation $X_i - \xi_i \rightarrow \dfrac{X_i - \xi_0}{\sigma} \sim N(0, 1)$ does not involve σ and hence, it is independent of sufficient statistic $T = \sum (X_i - \xi_i)^2$ for v. This will not hold for all ξ.

Example 9.9

Let

$$\frac{U_1}{\sigma_1^2} \sim \chi^2_{f1}, \frac{U_2}{\sigma_2^2} \sim \chi^2_{f2}$$

are independent and let $\dfrac{\sigma_2^2}{\sigma_1^2} = a$, the joint density of U_1, U_2 is

$$c u_1^{\frac{f_1}{2}-1} u_2^{\frac{f_2}{2}-1} e^{-\frac{1}{2\sigma_2^2}(a u_1 + u_2)},$$

where $$T = a u_1 + u_2, v = -\frac{1}{2\sigma_2^2}.$$

The distribution of

$$V = \frac{U_2}{U_1} = a \frac{\frac{U_2}{\sigma_2^2}}{\frac{U_1}{\sigma_1^2}}$$

does not involve σ_2 and hence, V is independent of $aU_1 + U_2$. If $\sigma_1 = \sigma_2$, $\frac{U_1}{U_2}$ is independent of $U_1 + U_2$.

One Population Problem of $N(\xi, \sigma^2)$ Distribution

Let $X_1, X_2, \ldots, X_n$ be a random sample from $N(\xi, \sigma^2)$ distribution. We wish to test the following hypotheses.

(i) $H_1 : \sigma^2 \le \sigma_0^2$

(ii) $H_2 : \sigma^2 \le \sigma_1^2$ or $\sigma^2 \ge \sigma_2^2$

(iii) $H_3 : \sigma_1^2 < \sigma^2 < \sigma_2^2$

(iv) $H_4 : \sigma^2 = \sigma_0^2$

(v) $H_5 : \xi_1 \le \xi_0$

(vi) $H_6 : \xi \le \xi_1$ or $\xi \ge \xi_2$

(vii) $H_7 : \xi_1 < \xi < \xi_2$

(viii) $H_8 : \xi = \xi_0$

For testing H_1 to H_4 the joint density of $X_1, X_2, \ldots, X_n$ is exponential family

$$c(\xi, \sigma^2)e^{\theta u + vt}, \text{ with } \theta = -\frac{1}{2\sigma^2}, \; v = \frac{u\xi}{\sigma^2}, \; U = \Sigma X_i^2, \; T = \bar{X}.$$

Now

$$H_1 : \sigma^2 \le \sigma_0^2 \Rightarrow -\frac{1}{2\sigma^2} \le -\frac{1}{2\sigma_0^2} \Rightarrow \theta \le \theta_0, \theta_0 = -\frac{1}{2\sigma_0^2}.$$

Instead of finding the conditional distribution of U given T, we propose ancillary statistic

$$V = \sum (X_i - \bar{X})^2 = \sum X_i^2 - n\bar{X}^2 = U - nT^2,$$

which is a monotonic function of U for fixed T and is independent of T because distribution of V does not involve ξ as $X_i \to \frac{X_i - \xi_0}{\sigma} \sim N(0, 1)$,

V is invariant. So UMP unbiased test for testing H_1 is to reject H_1 whenever $V > c'$, that is $\frac{\sum(X_i - \bar{X})^2}{\sigma_0^2} > c_0$ with c_0 as

$$P_{\sigma_0^2}\left\{\frac{\sum(X_i - \bar{X})^2}{\sigma_0^2} > c_0\right\} = \alpha \Rightarrow P_{\sigma_0^2}\{\chi^2_{n-1} > \chi^2_{n-1;\,\alpha}\} = \alpha$$

where $c_0 = \chi^2_{n-1;\,\alpha}$. For H_2, H_3 determination of c_1, c_2 can be done in the same manner. Further V being linear function of U for fixed T at $\sigma^2 = \sigma_0^2$ the acceptance region for H_4 is

$$c_1 < \frac{\sum(X_i - \bar{X})^2}{\sigma_0^2} < c_2$$

with

$$\int_{c_1}^{c_2} f_{n-1}(y)dy = \frac{1}{n-1}\int_{c_1}^{c_2} y f_{n-1}(y)dy = 1 - \alpha$$

where $f_{n-1}(y)$ is p.d.f. of χ^2 on $n-1$ degree of freedom. For testing H_5 to H_8 the joint p.d.f. of $X_1, X_2, \ldots, X_n$ is exponential family with

$$\theta = \frac{n\xi}{\sigma^2}, U = \bar{X}, v = \frac{1}{2\sigma^2}, T = \sum X_i^2.$$

Now $H_5: \xi_1 \leq \xi_0 \Rightarrow \theta \leq \theta_0$. Let

$$V = \frac{\bar{X} - \xi_0}{\sqrt{\sum(X_i - \bar{X})^2}} = \frac{U^*}{\sqrt{T^* - nU^{*2}}}, U^* = \bar{X} - \xi_0, T^* = \sum(X_i - \xi_0)^2.$$

Then V being independent of T^* because the distribution of V does not involve σ^2 as V is invariant under transformation $X_i \to \frac{(X_i - \xi_0)}{\sigma} \sim N(0,1)$ under H_5 and further V being monotonic in U^* for fixed T^*, the UMP unbiased test for testing H_5 is to reject it whenever

$$V \geq c_0' \Rightarrow t(x) \geq c_0, \text{ with } t(x) = \frac{\sqrt{n(n-1)}(\bar{X} - \xi_0)}{\sqrt{\sum(X_i - \bar{X})^2}}.$$

For H_6, H_7; c_1, c_2 can be determined in the same manner. For H_8 as V is not a linear function of U^* for fixed T^*, we propose another ancillary statistic,

$$W = \frac{U^*}{\sqrt{T^*}} = \frac{\bar{X} - \xi_0}{\sqrt{\sum(X_i - \xi_0)^2}}.$$

Since W is a monotonic linear function of U^* for fixed T^* and due to its invariance property under the transformation

$$X_i \rightarrow \frac{(X_i - \xi_0)}{\sigma} \sim N(0, 1),$$

W being symmetric about ξ_0, the UMP unbiased test for H_8 will reject it whenever

$$|W| > c' \Rightarrow |t| > c$$

where

$$t = \frac{\sqrt{n(n-1)}(\bar{X} - \xi_0)}{\sqrt{\sum(X_i - \bar{X})^2}} = \frac{\sqrt{n(n-1)}W}{\sqrt{1 - nW^2}},$$

an increasing function of W

with

$$\int_{c_0}^{\infty} t_{n-1}(y)dy = \alpha, \int_{c}^{\infty} t_{n-1}(y)dy = \alpha/2,$$

where t_{n-1} is the p.d.f. of student's t-distribution on $(n - 1)$ degree of freedom.

Comparison of Two Normal Populations

Let $X_1, X_2, \ldots, X_m$ be a random sample from $N(\mu_1, \sigma_1^2)$ distribution and $Y_1, Y_2, \ldots, Y_n$ be an independent random sample from $N(\mu_2, \sigma_2^2)$. We are interested in testing

(i) $H_1: \dfrac{\sigma_2^2}{\sigma_1^2} \le \Delta_0$

(ii) $H_2: \dfrac{\sigma_2^2}{\sigma_1^2} = \Delta_0$, Δ_0 is specified.

The joint density of $X_1, X_2, \ldots, X_m$ and $Y_1, Y_2, \ldots, Y_n$ is

$$c(\mu_1, \mu_2, \sigma_1^2, \sigma_2^2) e^{-\frac{1}{2\sigma_1^2}\sum_{i=1}^{m} x_i^2 - \frac{1}{2\sigma_2^2}\sum_{j=1}^{n} y_j^2 + \frac{m\mu_1}{\sigma_1^2}\bar{x} + \frac{n\mu_2}{\sigma_2^2}\bar{y}}$$

which is a four-parameter exponential family of distributions with

$$\theta = -\frac{1}{2\sigma_2^2}, \nu_1 = -\frac{1}{2\sigma_1^2}, \nu_2 = \frac{m\mu_1}{\sigma_1^2}, \nu_3 = \frac{n\mu_2}{\sigma_2^2},$$

and $$U = \sum_{j=1}^{n} y_j^2, T_1 = \sum_{i=1}^{m} x_i^2, T_2 = \bar{x}, T_3 = \bar{y}.$$

Now

$$\frac{\sigma_2^2}{\sigma_1^2} \le \Delta_0 \Rightarrow -\frac{1}{2\sigma_2^2} \le -\frac{1}{2\sigma_1^2 \Delta_0} \Rightarrow \theta \le \frac{\nu_1}{\Delta_0}$$

$$\Rightarrow \qquad \theta^* = \theta - \frac{\nu_1}{\Delta_0} \le 0 \Rightarrow a_0\theta + a_1\nu_1 + a_2\nu_2 + a_3\nu_3 \le 0$$

where $a_0 = 1, a_1 = -\dfrac{1}{\Delta_0}, a_2 = a_3 = 0$, with this new parametrization, complete sufficient statistics are

$$U^* = \sum_{j=1}^{n} y_j^2, T_1^* = T_1 + \frac{1}{\Delta_0} U, T_2^* = \bar{X}, T_3^* = \bar{Y}.$$

Let

$$V = \frac{\dfrac{\sum (y_j - \bar{Y})^2}{\Delta_0}}{\sum (X_i - \bar{X})^2} = \frac{\dfrac{(U^* - nT_3^{*2})}{\Delta_0}}{T_1^* - \dfrac{1}{\Delta_0} U^* - mT_2^{*2}}.$$

Then V is monotonic function of U^* for fixed T_1^*, T_2^*, T_3^* and it is independent of T_1^*, T_2^*, T_3^* because V being invariant under the transformation $X_i \to \frac{X_i - \mu_1}{\sigma_1} \sim N(0,1),\ Y_j \to \frac{Y_j - \mu_2}{\sigma_2} \sim N(0,1)$ distribution of V does not involve $\mu_1, \mu_2, \sigma_1, \sigma_2$. Thus, the UMP unbiased test for testing H_1 rejects it if $V \geq c_0' \Rightarrow \frac{m-1}{n-1}V \geq c$, where, c is such that

$$\int_c^\infty F_{n-1,\,m-1}(y)dy = \alpha,$$

$\frac{m-1}{n-1}V = F_{n-1,\,m-1}$ has F distribution on $n-1$, $m-1$ degree of freedom and $F_{n-1,\,m-1}(y)$ is p.d.f. of F. For testing H_2, we propose ancillary statistic

$$W = \frac{\frac{\sum(y_j - \bar{Y})^2}{\Delta_0}}{\sum(X_i - \bar{X})^2 + \frac{1}{\Delta_0}\sum(y_j - \bar{Y})^2} = \frac{\frac{(U^* - nT_3^{*2})}{\Delta_0}}{T_1^* - \frac{nT_3^{*2}}{\Delta_0} - \frac{mT_2^{*2}}{\Delta_0}},$$

which is a linear function of U^* independent of T_1^*, T_2^*, T_3^*. Then UMP unbiased test for testing H_2 will accept it if

$$c_1 < W < c_2 \Rightarrow c_1 < \frac{\frac{\sum(y_j - \bar{Y})^2}{\sigma_2^2}}{\frac{\sum(X_i - \bar{X})^2}{\sigma_1^2} + \frac{\sum(y_j - \bar{Y})^2}{\sigma_2^2}} < c_2$$

where $\Delta_0 = \frac{\sigma_2^2}{\sigma_1^2}$

$$\Rightarrow \qquad c_1 < \frac{W_1}{W_1 + W_2} < c_2,$$

where

$$W_1 = \frac{\sum (y_j - \bar{Y})^2}{\sigma_2^2} \sim \chi^2_{m-1}, W_2 = \frac{\sum (x_j - \bar{X})^2}{\sigma_1^2} \sim \chi^2_{n-1}.$$

Thus, W has $\beta_{\frac{n-1}{2}, \frac{m-1}{2}}$ distribution with

$$EW = \frac{n-1}{m+n-2} \text{ and } [E(W)]\beta_{\frac{n-1}{2}, \frac{m-1}{2}}(w) = \frac{n-1}{m+n-2}\beta_{\frac{n+1}{2}, \frac{m-1}{2}}(w).$$

So c_1, c_2 are such that

$$\int_{c_1}^{c_2} \beta_{\frac{n-1}{2}, \frac{m-1}{2}}(w)dw = \int_{c_1}^{c_2} w\beta_{\frac{n-1}{2}, \frac{m-1}{2}}(w)dw = 1-\alpha.$$

$$\Rightarrow \qquad \int_{c_1}^{c_2} \beta_{\frac{n-1}{2}, \frac{m-1}{2}}(w)dw = \int_{c_1}^{c_2} \beta_{\frac{n+1}{2}, \frac{m-1}{2}}(w)dw = 1-\alpha.$$

Let $X_1, X_2, \ldots, X_m$ be a random sample from $N(\mu_1, \sigma^2)$ and let $Y_1, Y_2, \ldots, Y_n$ be an independent random sample from $N(\mu_2, \sigma^2)$ distribution. Then we wish to test

(i) $H_1 : \mu_2 \leq \mu_1$

(ii) $H_1 : \mu_2 = \mu_1$.

The joint density of $X_1, X_2, \ldots, X_m$ and $Y_1, Y_2, \ldots, Y_n$ is a three-parameter exponential family such as

$$c(\mu_1, \mu_2, \sigma)e^{\theta u + v_1 t_1 + v_2 t_2}$$

with

$$\theta = \frac{n\mu_2}{\sigma^2}, v_1 = \frac{m\mu_1}{\sigma^2}, v_2 = -\frac{1}{2\sigma^2}, U = \bar{Y}, T_1 = \bar{X}, T_2 = \Sigma X_i^2 + \Sigma Y_i^2.$$

Let us reparametrize it by expressing

$$m\mu_1\bar{x} + n\mu_2\bar{y} = k_1(\bar{y} - \bar{x}) + k_2(m\bar{x} + n\bar{y})$$

$$\Rightarrow \qquad \left.\begin{aligned} k_1 + nk_2 &= n\mu_2 \\ mk_2 - k_1 &= m\mu_1 \end{aligned}\right\}$$

$$\Rightarrow \qquad \begin{cases} k_2 = \dfrac{(m\mu_1 + n\mu_2)}{m+n} \\[2ex] k_1 = \dfrac{(\mu_2 - \mu_1)}{\left(\dfrac{1}{m} + \dfrac{1}{n}\right)} \end{cases}$$

Therefore,

$$m\mu_1\bar{x} + n\mu_2\bar{y} = \frac{(\bar{y} - \bar{x})(\mu_2 - \mu_1)}{\left(\dfrac{1}{m} + \dfrac{1}{n}\right)} + \frac{(m\bar{x} + n\bar{y})(m\mu_1 + n\mu_2)}{m+n}$$

and the exponential family is rewritten as

$$c(\theta^*, v_1^*, v_2^*)e^{\theta^* u^* + v_1^* t_1^* + v_2^* t_2^*}$$

with

$$\theta^* = \frac{(\mu_2 - \mu_1)}{\left(\dfrac{1}{m} + \dfrac{1}{n}\right)\sigma^2}, U^* = (\bar{Y} - \bar{X}), v_1^* = \frac{(m\mu_1 + n\mu_2)}{(m+n)\sigma^2}, v_2^* = -\frac{1}{2\sigma^2},$$

$$T_1^* = (m\bar{x} + n\bar{y}), T_2^* = \Sigma X_i^2 + \Sigma Y_i^2.$$

Let

$$V = \frac{(\bar{Y} - \bar{X})}{\sqrt{\sum(x_i - \bar{x})^2 + \sum(y_i - \bar{y})^2}} = \frac{U^*}{\sqrt{T_2^* - \dfrac{T_1^{*2}}{m+n} - \dfrac{mn}{m+n}U^{*2}}},$$

as

$$\Sigma(x_i - \bar{x})^2 + \Sigma(y_i - \bar{y})^2 = \Sigma X_i^2 + \Sigma Y_i^2 - (m\bar{x}^2 + n\bar{y}^2),$$

$$T_1^{*2} = (m\bar{x} + n\bar{y})^2 = m^2\bar{x}^2 + n^2\bar{y}^2 - 2mn\overline{xy},$$

$$mnU^{*2} = mn(\bar{Y} - \bar{X})^2 = mn\bar{Y}^2 + mn\bar{X}^2$$

$$-2mn\bar{Y}\bar{X},$$

$$\Rightarrow \qquad T_1^{*2} + mnU^{*2} = n(m+n)\bar{y}^2 + m(m+n)\bar{x}^2$$

$$\Rightarrow \qquad \frac{(T_1^{*2} + mnU^{*2})}{(m+n)} = m\bar{x}^2 + n\bar{y}^2.$$

Then V being invariant under the transformation

$$X_i \to \frac{X_i - \mu_1}{\sigma} \sim N(0,1),\ Y_j \to \frac{Y_j - \mu_2}{\sigma} \sim N(0,1),$$

which does not involve μ_1, μ_2, σ^2 and hence it does not involve v_1^*, v_2^*. So V is independent of T_1^*, T_2^*. Further, it being monotonic in U^* for fixed T_1^*, T_2^*, the UMP unbiased test will reject it when $V \geq c_0' \Rightarrow t \geq c_0$, where, t is

$$t = \frac{\dfrac{(\bar{Y} - \bar{X})}{\sqrt{\left(\dfrac{1}{m} + \dfrac{1}{n}\right)}}}{\sqrt{\dfrac{\sum(x_i - \bar{x})^2 + \sum(y_i - \bar{y})^2}{(m+n-2)}}},$$

and it has student's t distribution on $(m + n - 2)$ degree of freedom and c_0 is determined by

$$\int_{c_0}^{\infty} t_{m+n-2}(y)dy = \alpha,$$

$t_{m+n-2}(y)$ is density of student's t distributions on $(m+n-2)$ degree of freedom.

For testing $H_2 : \mu_2 = \mu_1$ as V being not linear function of V^*, we propose another statistic

$$W = \frac{(\bar{Y} - \bar{X})}{\sqrt{\sum X_i^2 + \sum Y_i^2 - \dfrac{(m\bar{x} + n\bar{y})^2}{m+n}}} = \frac{U^*}{\sqrt{T_2^* - \dfrac{T_1^{*2}}{m+n}}},$$

W is invariant under the transformation

$$X_i \rightarrow \frac{X_i - \mu_1}{\sigma} \sim N(0,1),\ Y_j \rightarrow \frac{Y_j - \mu_2}{\sigma} \sim N(0,1),$$

and hence its distribution does not involve μ_1, μ_2, σ, it is independent of T_1^*, T_2^*. Further, W being increasing linear function of U^* for fixed T_1^*, T_2^* at $\mu_2 = \mu_1$ the UMP unbiased test will reject H_2 if

$$|W| > c \Rightarrow |V| > c_1 \text{ as } |V| = \frac{|W|}{\sqrt{1 - \frac{mn}{m+n} W^2}},$$

a monotonic increasing function of $|W|$ and V is symmetric about $\mu_2 - \mu_1 = 0$, we reject H_2 if $|t| > c$ where c is such that

$$\int_c^{\infty} f_{m+n-2}(y)dy = \alpha/2.$$

Testing Independence in Bivariate Normal Population

Let $(X_1, Y_1), (X_2, Y_2), \ldots, (X_n, Y_n)$ be a random sample from bivariate normal distribution with parameters $\theta_1, \theta_2, \sigma_1, \sigma_2, \rho$. To test $\rho \leq 0$ and $\rho = 0$.

The joint p.d.f. of $(X_1, Y_1), (X_2, Y_2), \ldots, (X_n, Y_n)$ is

$$(2\pi\sigma_1\sigma_2\sqrt{1-\rho^2})^{-\frac{n}{2}}$$

$$\exp\left[-\frac{1}{2(1-\rho^2)}\left\{\frac{\sum(x_i-\theta_1)^2}{\sigma_1^2} + \frac{\sum(y_j-\theta_2)^2}{\sigma_2^2} - \frac{2\rho}{\sigma_1\sigma_2}(x_i-\theta_1)(y_j-\theta_2)\right\}\right]$$

which is a five-parameter exponential family with

$$v_1 = \frac{\rho}{\sigma_1\sigma_2(1-\rho^2)},\ v_2 = -\frac{1}{2\sigma_1^2(1-\rho^2)},\ v_3 = -\frac{1}{2\sigma_2^2(1-\rho^2)},$$

$$v_4 = \frac{1}{(1-\rho^2)}\left(\frac{\theta_1}{\sigma_1^2} - \frac{\rho\theta_2}{\sigma_1\sigma_2}\right),$$

$$\nu_5 = \frac{1}{(1-\rho^2)}\left(\frac{\theta_2}{\sigma_2^2} - \frac{\rho\theta_1}{\sigma_1\sigma_2}\right), T_1 = \Sigma X_i Y_i, T_2 = \Sigma X_i^2, T_3$$

$$= \Sigma Y_i^2, T_4 = \Sigma X_i, T_5 = \Sigma Y_i.$$

For $\rho \le 0 \Leftrightarrow \nu_i \le 0$, we suggest the statistic R,

$$R = \frac{\sum(X_i - \bar{X})(Y_i - \bar{Y})}{\sqrt{\sum(X_i - \bar{X})^2 \sum(Y_i - \bar{Y})^2}}$$

which is invariant under transformation

$$X_i \to \frac{X_i - \theta_1}{\sigma_1} \sim N(0,1), Y_j \to \frac{Y_j - \theta_2}{\sigma_2} \sim N(0,1),$$

R is independent of $\theta_1, \theta_2, \sigma_1, \sigma_2$ and hence by Basu's theorem R is independent of T_2, T_3, T_4, T_5 when $\rho = 0$. Then UMP unbiased test rejects $\rho \le 0$,

whenever $R \ge c_0 \Rightarrow \frac{R\sqrt{n-2}}{\sqrt{1-R^2}} \ge k_0$. Since R is a linear function of T_1 for

$\rho = 0$, it rejects $\rho = 0$ whenever

$$\frac{|R|\sqrt{n-2}}{\sqrt{1-R^2}} > k_1$$

where k_0, k_1 are such that

$$\int_{k_0}^{\infty} t_{n-2}(y)dy = \alpha, \int_{k_1}^{\infty} t_{n-2}(y)dy = \alpha/2.$$

9.6 CONFIDENCE BOUNDS

The theory of testing of hypotheses will be helpful in setting confidence bounds. In the case of breaking of strength of a new alloy one needs lower bound and for the toxicity of a drug one needs upper bound.

If $\underline{\theta}(x)$ is a function of X and if $P_\theta\{\underline{\theta}(x) \le \theta\} \ge 1 - \alpha \; \forall \; \theta$, then $\underline{\theta}(x)$ is called the lower bound of θ at confidence level $1 - \alpha$.

If $\{S_l\}$ is the class of lower bounds of θ at confidence level 1 – α and if $\underline{\theta}^*(x) \in \{S_l\}$ such that $P_\theta\{\underline{\theta}^*(x) \le \theta\} \le P\{\underline{\theta}(x) \le \theta\}\ \forall\ \underline{\theta}(x) \in \{S_l\}$ then $\underline{\theta}^*(x)$ is called uniformly most accurate(UMA) lower confidence bound.

More generally, a family of subsets $S(x)$ of parameter space Ω is said to constitute a family of confidence sets, at level 1 – α if

$$P\{\theta \in S(x)\} \ge 1 - \alpha\ \forall\ \theta \in \Omega.$$

If $\{S\}$ is the class of confidence sets at level (1 – α) and if $S^*(x) \in \{S\}$ then $S^*(x)$ is the UMA confidence bound for θ if

$$P\{\theta \in S^*(x)\} \le P\{\theta \in S(x)\}\ \forall\ S(x) \in \{S\}.$$

If $S(x) = \{\theta : \theta(X) \le \theta\}$ then we get lower bound. Similarly, for upper bound.

	H is True	**K is True**
Accept H	$1 - \alpha$	B
Reject H	α	$1 - \beta$

If A, R are acceptance and rejection regions, respectively, for testing H vs K then the Neyman's testing of hypothesis can be formulized as: the acceptance region A_0 of MP test satisfies $P_k\{X \in A_0\} \le P_k\{X \in A\}$ where, A is acceptance region satisfying $P_H\{X \in A\} \ge 1 - \alpha$.

Theorem 9.11:

(i) For each $\theta_0 \in \Omega$ let $A(\theta_0)$ be the acceptance region of a level α test for testing $H(\theta_0) : \theta \ge \theta_0$ and for each sample point x let $S(x)$ be

$$S(x) = \{\theta : \theta \in A(\theta_0), \theta \in \Omega\}$$

Then $S(x)$ is a family of confidence sets of θ at confidence level (1 – α).

(ii) If for θ_0, $A(\theta_0)$ is acceptance region of UMP test the probability

$$P_\theta\{\theta_0 \in S(x)\}\ \forall\ \theta \in K(\theta_0),$$

among all (1 – α) level for testing $H(\theta_0)$ at level α against the alternatives $K(\theta_0)$ then for each θ_0 in Ω, $S(X)$ minimizes the probability $P_\theta(\theta_0 \in S(x)\}\ \forall\ \theta \in K(\theta_0)$ among all (1 – α) level families of confidence sets for θ.

Proof:

(i) By definition $\theta \in S(x) \Leftrightarrow x \in A(\theta)$ and hence

$$P_\theta\{\theta \in S(x)\} = P_\theta\{x \in A(\theta)\} \geq (1-\alpha).$$

(ii) If $S^*(x)$ is any family of confidence sets at level $(1 - \alpha)$ and if

$$A^*(\theta) = \{x : \theta \in S^*(x)\} \geq (1-\alpha), \text{ then}$$

$$P_\theta\{x \in A^*(\theta)\} = P_\theta\{\theta \in S^*(x)\} \geq (1-\alpha)$$

so that $A^*(\theta_0)$ is the acceptance region of a level α test for testing $H(\theta_0)$. Since $A(\theta_0)$ is the acceptance region of UMP test of $H(\theta_0)$ then for any $\theta \in K(\theta_0)$,

$$P_\theta\{\theta_0 \in S^*(x)\} = P_\theta\{X \in A^*(\theta_0)\} \geq P_\theta\{X \in A(\theta_0)\}$$
$$= P_\theta\{\theta_0 \in S(x)\}$$

Definition 9.5

A family $\{S(x)\}$ of confidence sets at confidence level $(1 - \alpha)$ is said to be a uniformly most accurate (UMA) family of confidence sets at level $(1 - \alpha)$ if for any other family $S^*(x)$ of confidence sets at level $(1 - \alpha)$ and for each θ, θ'

$$P_{\theta'}\{\theta \in S(x)\} \leq P_{\theta'}\{\theta \in S^*(x)\}.$$

Unbiased Confidence Sets

Definition 9.6

Let $\{P_{\theta,\nu}(\theta,\nu) \in \Omega\}$ be a family of distributions. $\{S(x)\}$ is said to be an unbiased family of confidence sets for θ at confidence level $(1 - \alpha)$ if

$$P_{\theta,\nu}\{\theta \in S(x)\} \geq (1-\alpha) \text{ for all } \theta, \forall\, \theta \in H(\theta', \nu) \text{ and}$$

$$P_{\theta,\nu}\{\theta' \in S(x)\} \leq (1-\alpha) \text{ for all } \theta \in K(\theta') \text{ and } \forall\, \theta, \nu.$$

Theorem 9.12: If $A(\theta_0)$ is the acceptance region of a UMP unbiased test of size α for testing $H(\theta_0): \theta = \theta_0$ vs $K(\theta_0): \theta \neq \theta_0$ then $S(x) = \{\theta : x \in A(\theta)\}$ is UMA unbiased family of confidence sets, that is, $S(x)$ is unbiased and $P_{\theta',\nu}\{\theta \in S(x)\}$ is minimum for all θ, θ', ν and for all unbiased family of confidence sets at level $(1 - \alpha)$.

Proof: $S(x)$ is unbiased because

$$P_{\theta,v}\{\theta \in S(x)\} = P_{\theta,v}\{X \in A(\theta)\} \leq (1-\alpha)\ \forall\ \theta,\ v.$$

Let $S^*(x)$ be any other unbiased family of confidence sets at level $(1-\alpha)$ and let $A^*(\theta) = \{x : \theta \in S^*(x)\}$. Then $P_{\theta',v}\{X \in A^*(\theta)\} = P_{\theta',v}\{\theta \in S^*(x)\} \leq (1-\alpha)$, as $A^*(\theta)$ is the acceptance region of the unbiased test of size α. Hence,

$$P_{\theta',v}\{\theta \in S^*(x)\} = P_{\theta',v}\{X \in A^*(\theta)\} \geq P_{\theta',v}\{X \in A(\theta)\}$$
$$= P_{\theta',v}\{\theta \in S(x)\}$$

because $A(\theta)$ is the acceptance region of a UMP unbiased test.

PROBLEMS

9.1 For the following distribution obtain UMP/UMPU test at level α for testing $H:\theta = 0$ vs $K:\theta \neq 0$, based on a random sample $X = (X_1, X_2, \ldots, X_n)$ and also obtain UMA confidence bounds with confidence coefficient $(1-\alpha)\%$ for θ.

(a) $N(\xi, \sigma^2)$;

(i) $\theta = \dfrac{\xi}{\sigma}$,

(ii) $\theta = \dfrac{|\xi|}{\sigma}$

(b) $f(x;\theta) = \theta e^{-\theta x}, 0 < x < \infty$

(c) $f(x;\theta) = \dfrac{1}{\theta} e^{-\frac{x}{\theta}}, 0 < x < \infty$

(d) $u(0,\theta), 0 < x < \theta$

(e) $u(\theta, \theta+1), 0 < x < \theta$

(f) $f(x;\theta) = e^{-(x-\theta)}, \theta < x < \infty$.

9.2 Let $Z_1, \ldots, Z_n$ be independent normally distributed with common variance σ^2 and mean $E(Z_i) = \theta_i (i = 1, 2, \ldots, s)$, $E(Z_i) = 0 (i = s+1, \ldots, n)$. Obtain UMPU tests for testing $\theta_1 \leq \theta_1^0$ and $\theta_1 = \theta_1^0$ and also obtain UMA confidence bounds for θ_1 with $(1-\alpha)$ confidence coefficient.

9.3 Let $X_1, X_2, \ldots, X_m$ and $Y_1, Y_2, \ldots, Y_n$ be independent samples from $N(\xi, \sigma^2)$ and $N(\eta, \tau^2)$, respectively. Obtain UMPU test for testing

$$H{:}\frac{\tau}{\sigma} = \Delta(\text{specified}) \text{ vs } K{:}\frac{\tau}{\sigma} > \Delta.$$

Also obtain (1 – α)% UMA confidence bounds for $\frac{\tau}{\sigma}$.

9.4 Let $X_1, X_2, \ldots, X_n$ be a random sample from the distribution

$$f_X(x; a, b) = \frac{1}{a} e^{-\frac{(x-b)}{a}}, x \geq b.$$

Obtain UMPU test at level α for testing

(i) $H{:}a = 1$ vs $K{:}a \neq 1$

(ii) $H{:}b = 0$ vs $K{:}b \neq 0$.

Also obtain UMA confidence bounds for a and b with confidence coefficient (1 – α).

9.5 Suppose $X = (X_1, X_2, \ldots, X_n)$ is a random sample from p.d.f

$$f(x; \alpha) = \alpha e^{-\alpha x}, 0 < x < \alpha,$$

and $Y = (Y_1, Y_2, \ldots, Y_n)$ is an independent random sample from p.d.f.

$$f(x; \beta) = \beta e^{-\beta x}, 0 < x < \beta.$$

Obtain UMPU tests for testing $H{:}\alpha = \beta$ vs $K{:}\alpha \neq \beta$, at level α. Also obtain UMA confidence bounds for $\frac{\alpha}{\beta}$ with confidence coefficient (1 – α).

10

CHAPTER

Sequential Probability Ratio Test (SPRT) and Likelihood Ratio Test (LRT)

10.1 INTRODUCTION

Under a particular type of sampling scheme, sequential probability ratio test (SPRT) has been developed for testing simple null hypothesis against simple alternative hypothesis. Boundary points are approximated and OC, ASN functions of SPRT have been derived. Instead of using fixed sample size one may take observations one after another and after every observation decision is taken for acceptance or rejection of H or continue sampling. Such type of sampling is called *sequential sampling* and the analysis is called *sequential analysis* which was developed by Wald in 1947.

Wald Procedure of SPRT

If $X_1, X_2, X_3, \ldots$ are i.i.d. random variables having p.d.f. $f(x;\theta)$ and if we want to test $H:\theta=\theta_0$ vs $K:\theta=\theta_1$, by using the observations $x_1, x_2, x_3, \ldots$ sequentially we construct the ratio

$$\lambda_n = \prod_{i=1}^{n} \frac{f(x_i, \theta_1)}{f(x_i, \theta_0)}, \; n = 1, 2, 3, \ldots$$

and choose two numbers B and A, $0 < B < 1 < A < \infty$ and adopt the following procedure

(a) Continue sampling (compute λ_{n+1}) if $B < \lambda_n < A$.

(b) If for $k = 1, 2, \ldots, n-1$, $B < \lambda_k < A$ and if $\lambda_n \le B$, accept H,

(c) If for $k = 1, 2, \ldots, n-1$, $B < \lambda_k < A$ and if $\lambda_n \ge A$, reject H,

Example 10.1

For the exponential family

$$f(x;\theta)=c(\theta)h(x)e^{q(\theta)t(x)}$$

continue sampling till

$$c_1+D_n<\Sigma T(x_i)<c_2+D_n$$

where c_1, c_2, D_n depend on θ_0, θ_1.

Notes

(1) Writing $Z_i=\log\frac{f(x_i,\theta_1)}{f(x_i,\theta_0)}$, the governing inequality of SPRT is $\log B<\Sigma Z_i=S_n<\log A$.

(2) If $f(x_n,\theta_1)=0, f(x_n,\theta_0)>0$, we define $\lambda_n=1$. If $f(x_n,\theta_1)>0$, $f(x_n,\theta_0)=0$, we reject H. If $f(x_n,\theta_1)=f(x_n,\theta_0)$, we accept H.

(3) In SPRT with strength (α, β), the error probabilities are given in advance and it is hoped that we may arrive at the decision earlier than the fixed sample size test. In fixed sample size test we fix n and α and minimize β. In SPRT instead of fixed sample size it varies stochastically, denoted by $N(\alpha, \beta)$, we obtain n for fixed (α, β), say $n(\alpha, \beta)$. Since in SPRT, N, the sample size being random variable let $E[N(\alpha, \beta)]$ be the sample size. By the following illustration we show that $E[N(\alpha, \beta)]<n(\alpha, \beta)$.

Taking n observations from $N(\theta,1)$ for testing $H:\theta=\theta_0$ vs $K:\theta=\theta_1$ the test rejects H if $\bar{X}>\frac{Z_\alpha}{\sqrt{n}}$, where, $\phi(Z_\alpha)=1-\alpha$ and the power is $\phi(\sqrt{n}(\theta_1-\theta_0)-Z_\alpha)$. If $n=9, \alpha=0.005, \theta_0=0, \theta_1=\frac{1}{2}$ then $Z_\alpha=2.57$ and power is $\phi(-1.076)=0.141$.

Alternatively, we toss an unbiased coin. If we get head then n_1 observations are taken and tail then n_2 observations are taken and reject H when $\bar{X}>\frac{Z_\alpha}{\sqrt{n_2}}+\theta_0$. The average sample size is $\frac{n_1+n_2}{2}$ and power is

$$\frac{1}{2}\phi(\sqrt{n_1}\,(\theta_1-\theta_0)-Z_\alpha)+\frac{1}{2}\phi(\sqrt{n_2}\,(\theta_1-\theta_0)-Z_\alpha).$$

For $\alpha = 0.005$, $n_1 = 2$, $n_2 = 16$, the power is

$$\frac{1}{2}\phi(-1.869)+\frac{1}{2}\phi(0.576)=0.51655$$

which is larger than the [power (0.141) of fixed size.

(4) Randomization is applicable if λ_n is discrete.

(5) SPRT is a random walk problem with initial point S_0 undergoing a jump Z_1 at time $t = 1$, a jump Z_2 at time $t = 2$ and so on,

$$S_n = S_0 + \sum_{i=1}^{n} Z_i = S_{n-1} + Z_n.$$

If $Z_1, Z_2, Z_3, \ldots$ have common p.f. $P(Z_i = 1) = p$, $P(Z_i = -1) = q$ and $P(Z_i = 0) = 1-p-q$, $0 < p < 1, 0 < q < 1$, $p + q \leq 1$. S_n performs simple random walk. a, $-b$ are absorbing barriers and $S_n \leq -b$ is attained with probability $1 - \alpha$ and $S_n \geq \alpha$ is attained with probability α.

Section 10.2 deals with the determination of boundaries, A and B. Some properties of SPRT have been obtained in section 10.3. Operating characteristic (OC) and average sample number (ASN) functions have been derived in section 10.4 whereas Wald's fundamental identity of SPRT is discussed in section 10.5. By using the principle of likelihood ratio method some tests have been derived in section 10.6. One-way and Two-way classification tests along with Bartlett's test are given in section 10.7

10.2 BOUNDARIES

Because of the difficulty of determining exact boundaries B, A for which α, β take preassigned values the following inequalities are useful.

Theorem 10.1: For Wald's SPRT (B, A) with strength (α, β),

$$A \leq \frac{1-\beta}{\alpha}, B \geq \frac{\beta}{1-\alpha}.$$

Proof: Let $X = (X_1, X_2, X_3, \ldots, X_k)$ denote k dimensional point in the k-dimensional space R_k.

Let

$$E_k = \{X : N = k \text{ and } \lambda_k \geq A\}$$

$$F_k = \{X : N = k \text{ and } \lambda_k \leq B\}$$

and E_k and F_k be mutually disjoint. Then

$$\alpha = P_{\theta_0}\ \{\text{reject } H\}$$

$$= P_{\theta_0}\left\{\bigcup_{k=1}^{\infty} E_k\right\}$$

$$= \sum_{k=1}^{\infty} P_{\theta_0}\{E_k\}$$

$$1 - \beta = P_{\theta_1}\ \{\text{reject } H\} = \sum_{k=1}^{\infty} P_{\theta_1}\{E_k\}.$$

Since $P_{\theta_1}(x) \geq A P_{\theta_0}(x)$ holds at every $x \in E_k$.

$$\alpha = \sum_{k=1}^{\infty} P_{\theta_0}\{E_k\}$$

$$= \sum_{k=1}^{\infty} \int_{E_k} P_{\theta_0}(x)dx \leq \frac{1}{A} \sum_{k=1}^{\infty} \int_{E_k} P_{\theta_1}(x)dx$$

$$= \frac{1}{A} \sum_{k=1}^{\infty} P_{\theta_1}(E_k)$$

$$= \frac{1}{A}(1 - \beta)$$

$$\Rightarrow \qquad A \leq \frac{1 - \beta}{\alpha}$$

Similarly, as $P_{\theta_1}(x) \le BP_{\theta_0}(x)$ for $x \in F_k$,

$$\beta = \sum_{k=1}^{\infty} P_{\theta_1}(F_k)$$

$$= \sum_{k=1}^{\infty} \int_{F_k} P_{\theta_1}(x)dx \le B\sum_{k=1}^{\infty} \int_{F_k} P_{\theta_0}(x)dx$$

$$= B\sum_{k=1}^{\infty} P_{\theta_0}(F_k)$$

$$= B(1-\alpha)$$

$$\Rightarrow \qquad B \ge \frac{\beta}{1-\alpha}.$$

By taking approximate values of A and B, that is, neglecting the excess part just as

$$B = \frac{\beta}{1-\alpha},\ A = \frac{1-\beta}{\alpha}$$

the changes in error probabilities are negligible which has been proved in the following theorem.

Theorem 10.2: Let α', β' be error probabilities corresponding to the boundaries, $B' = \dfrac{\beta}{1-\alpha},\ A' = \dfrac{1-\beta}{\alpha}$.

Then $$\alpha' \le \frac{\alpha}{1-\beta},\ \beta' \le \frac{\beta}{1-\alpha},\ \alpha' + \beta' \le \alpha + \beta.$$

Proof: By the above theorem

$$B' = \frac{\beta}{1-\alpha} \ge \frac{\beta'}{1-\alpha'},\ A' = \frac{1-\beta}{\alpha} \le \frac{1-\beta'}{\alpha'} \qquad \ldots(10.2.1)$$

i.e.,

$$\beta' \le (1-\alpha')\frac{\beta}{1-\alpha},\ \alpha' \le (1-\beta')\frac{\alpha}{1-\beta} \qquad \ldots(10.2.2)$$

$$\Rightarrow \qquad \beta' \leq \frac{\beta}{1-\alpha}, \alpha' \leq \frac{\alpha}{1-\beta}$$

From (10.2.2) we get

$$\beta'(1-\alpha) \leq (1-\alpha')\beta \qquad \ldots(10.2.3)$$

$$\alpha'(1-\beta) \leq (1-\beta')\alpha \qquad \ldots(10.2.4)$$

Adding (10.2.3) and (10.2.4), we get

$$\beta' + \alpha' \leq \alpha + \beta \qquad \ldots(10.2.5)$$

Note

The values of β', α' will be different from the values β, α but the differences are negligible because

$$\alpha' \leq \frac{\alpha}{1-\beta} = \alpha(1-\beta)^{-1} \cong \alpha(1+\beta) \cong \alpha$$

$$\beta' \leq \frac{\beta}{1-\alpha} = \beta(1-\alpha)^{-1} \cong \beta(1+\alpha) \cong \beta.$$

From (10.2.5) it is clear that at least one of the inequalities $\alpha' \leq \alpha, \beta' \leq \beta$ is true.

In fact, $\alpha' \leq \alpha, \beta' \leq \beta$ may happen with appreciable increase in critical region.

Further A', B' are functions of β, α and hence they are free from the distribution. In fixed sample size test the boundaries depend on the distribution. Both the theorems are true even if $X_1, X_2, X_3, \ldots$ are dependent.

10.3 SOME PROPERTIES OF SPRT

Lemma 10.1: For testing $H: \theta = \theta_0$ vs $K :: \theta = \theta_1$ by SPRT (B, A) of strength (α, β), let $X_1, X_2, X_3, \ldots$ be a sequence of i.i.d. random variables and let

$Z_i = \log \frac{f(x_i; \theta_1)}{f(x_i; \theta_0)}, i = 1, 2, 3, \ldots, S_i = \sum_{j=1}^{i} Z_j$, and let N be the number of

observations needed to arrive at the decision. If $P(|Z| > 0) > 0$ then

(i) $P(N < \infty) = 1$

(ii) $Ee^{tN} < \infty$.

Proof: As $X_1, X_2, X_3, \ldots$ form a sequence of i.i.d. random variables $Z_1, Z_2, Z_3, \ldots$ also form a sequence of i.i.d. random variables. Let $r = \left[\frac{m}{k}\right]$ = greatest integer in $\frac{m}{k}$, where m, k are fixed integers.

Let $T_1 = Z_1 + \ldots + Z_k$, $T_2 = Z_{k+1} + \ldots + Z_{2k}, \ldots,$

$T_i = Z_{(i-1)k+1} + \ldots + Z_{ik}$ and $S_{ik} = Z_1 + \ldots + Z_{ik} = T_1 + T_2 + \ldots + T_{(i-1)} + T_i$, $S_{(i-1)k} = T_1 + T_2 + \ldots + T_{(i-1)}$, therefore $T_i = S_{ik} - S_{(i-1)k}$, SPRT does not terminate so long as $m < N$ and $\log B < S_m < \log A$,

$$\Rightarrow \quad \log B < S_{ik} < \log A$$

$$\Rightarrow \quad \log B < S_{(i-1)k} < \log A$$

$$\text{i.e.} \quad -\log A < -S_{(i-1)k} < -\log B$$

$$\Rightarrow \quad \log B - \log A < S_{ik} - S_{(i-1)k} < \log A - \log B$$

$$\Rightarrow \quad -c < T_i < c, \text{ where } c = \log A - \log B$$

Therefore,

$$P\{N > m\} \le P\{|T_i| < c\}, i = 1, 2, \ldots, r$$

$$= [P\{|T_1| < c\}]^r, \text{ since } T_i \text{ are i.i.d.}$$

Thus,

$$\lim_{m\to\infty} P\{N > m\} \le \lim_{r\to\infty} [P\{|T_1| < c\}]^r$$

$$= 0, \text{ as } P\{|T_1| < c\} < 1.$$

Now

$$P\{|Z| > 0\} > 0, ET_1 = kEZ_1 > 0$$

$$V(T_1) = k\sigma^2, \text{letting } P\{|T_i| < c\} = \delta$$

$$\lim_{m\to\infty} P\{N > m\} \le \lim_{r\to\infty} \delta^r.$$

Select k such that $k > \frac{c^2}{\sigma^2}$ then $ET_1^2 \geq k\sigma^2 > c^2$

$$\Rightarrow \qquad E(T_1^2 < c^2) = 0 \Rightarrow P\{|T_1| < c\} < 1$$

$$\Rightarrow \qquad \delta < 1$$

For showing (ii), we note that $P\{N > m\} < \delta^2$

$$E(e^{tN}) = \sum_{m=1}^{\infty} e^{tm} P(N = m) \leq \sum_{m=1}^{\infty} e^{tm} P(N > m-1)$$

$$\leq \sum e^{tm} \delta^r, \; r = \left[\frac{m-1}{k}\right], \delta = P\{|T_1| < c\}$$

$$= \sum e^{tm} \delta^{\frac{m}{k}} \delta^{r-\frac{m}{k}} \leq \delta^{-\frac{1}{k}} \sum e^{tm} \delta^{\frac{m}{k}}$$

$$= \delta^{-\frac{1}{k}} \sum \left(e^t \delta^{\frac{1}{k}}\right)^m, \; r < \frac{m}{k} - \frac{1}{k} < \infty$$

If $e^t \delta^{\frac{1}{k}} < 1$, i.e., if $e^t < \delta^{-\frac{1}{k}}$, i.e., if $t < -\frac{1}{k} \log \delta = t_0$.

10.4 OPERATING CHARACTERISTIC (OC) FUNCTION AND AVERAGE SAMPLE NUMBER (ASN) FUNCTION OF SPRT

For testing $H:\theta = \theta_0$ vs $K::\theta = \theta_1$ if $L(\theta)$ denotes the probability of accepting H when $X_1, X_2, X_3, \ldots$ have p.f. $f(x;\, \theta)$ then $L(\theta)$ is called OC function and $1 - L(\theta)$ is the power function of the test.

For obtaining OC function we shall prove the following lemma.

Lemma 10.2: Let Z be a random variable satisfying

(i) $P\{|Z| > 0\} > 0$

(ii) $M(t) = Ee^{tZ} < \infty$

(iii) $EZ \neq 0$.

Then there exists $t^* \neq 0$ such that $M(t^*) = 1$.

Proof: Due to assumption (i) there exists $\in > 0$ such that $P\{Z > \in\} = \delta > 0$. If $t > 0$ then

$$M(t) = \int e^{tZ} f(Z)dz \geq \int_{Z > \in} e^{tZ} f(Z)dz > e^{t\in} P\{Z > \in\}.$$

Similarly, if $t < 0$, $M(t) > e^{-t\in} P\{Z < -\in\}$. Thus, $M(t) \to \infty$ as $t \to \pm\infty$.

Now condition (ii) implies that moments of all order of Z exists and may be obtained by differentiating $M(t)$ under the expectation sign.

Thus, $M''(t) = E(e^{tZ} Z^2) > 0 \Rightarrow M(t)$ is convex. Since $M(t)$ is convex and $\to \infty$ as $t \to \pm\infty$, it must have minimum at $t = t_0$, $M'(t_0) = 0$.

Further as

$$M'(t)\Big|_{t=0} = EZe^{tZ}\Big|_{t=0} = E(Z) \neq 0,$$

According to third assumption, $t_0 \neq 0$. Also $M(t_0) < M(0) = 1$. Thus, M is strictly decreasing over $(-\infty, t_0)$ and increasing over (t_0, ∞). Due to convexity of $M(t)$ and minimum occurring at $t_0 \neq 0$ and due to $M(0) = 1$, there exists another point t^* such that $M(t^*) = 1$. Further, if $t_0 > 0$, $M'(t_0) < 0$, $E(Z) < 0$ and if $t_0 < 0$, $M'(t) > 0$, $E(Z) = 0$.

Now let us obtain approximate value of OC function $[L(\theta)]$. According to the above lemma there exists $h \neq 0$ such that $Ee^{hZ} = 1$. Taking $Z = \log \dfrac{f(x;\theta_1)}{f(x;\theta_0)}$, we shall have non-zero h such that $E_\theta \left[\dfrac{f(x;\theta_1)}{f(x;\theta_0)}\right]^h = 1$.

Define $f^*(x;\theta) = \left[\dfrac{f(x;\theta_1)}{f(x;\theta_0)}\right]^h f(x;\theta)$ and consider auxillairy problem of testing $H^*: X_1, X_2, X_3, \ldots$ have p.f. f vs $K^*: X_1, X_2, X_3, \ldots$ have p.f. f^*, with boundaries (B^h, A^h) in SPRT with strength (α^*, β^*). Then the likelihood ratio λ_n^* at n^{th} stage is

$$\lambda_n^* = \prod_{i=1}^{n} \frac{f^*(x_i;\theta)}{f(x_i;\theta)} = \prod_{i=1}^{n} \left[\frac{f(x_i;\theta_1)}{f(x_i;\theta_0)}\right]^h = \lambda_n^h$$

where λ_n is the likelihood ratio for testing $H: \theta = \theta_0$ vs $K: \theta = \theta_1$, on SPRT (B, A) of strength (α, β). If $h > 0$ then $B^h < \lambda_n^* < A^h \Rightarrow B < \lambda_n < A$

and if $h<0$ then $B^h<\lambda_n^*<A^h \Rightarrow A<\lambda_n<B$. Thus, governing inequality for SPRT (B^h, A^h) is same as that of SPRT (B, A). So $L(\theta)$ = probability of accepting H = probability of accepting H^* when θ is true. Now the approximate values of B^h and A^h in SPRT (B^h, A^h) being

$$B^h \cong \frac{\beta^*}{1-\alpha^*},\ A^h \cong \frac{1-\beta^*}{\alpha^*},$$

$$\alpha^* \cong \frac{1-B^h}{A^h-B^h} \Rightarrow L(\theta)=1-\alpha^* \cong \frac{A^h-1}{A^h-B^h}, \text{ if } h\neq 0.$$

Further if $h = 0$,

$$L(\theta)=\frac{A^h-1}{A^h-B^h}=\frac{\log A}{\log A-\log B}.$$

As

$$L(\theta_0)=1-\alpha,\ L(\theta_1)=\beta,\ \lim_{n\to\infty} L(\theta)=1,\ \lim_{n\to-\infty} L(\theta)=0,$$

we have five points of OC function

h	$-\infty$	-1	0	1	∞
θ		θ_1	θ'	θ_0	
$L(\theta)$	0	B	$\frac{\log A}{\log A-\log B}$	$1-\alpha$	1

Average Sample Number (ASN) Function of SPRT

Since N is a random variable we need to obtain its average value, $E_\theta N$, which is called ASN function. To obtain this ASN we need the following Wald's equations.

Lemma 10.3: Let $Z_1, Z_2, Z_3, \ldots$ be a sequence of i.i.d. random variables. For any sequencial stopping rule

(i) $E(Z_1+Z_2+\ldots+Z_N)=E_\theta NEZ$, if $EZ\neq 0$

(ii) $E(Z_1+Z_2+\ldots+Z_N)^2=E_\theta NEZ^2$, if $EZ=0$, $EZ^2<\infty$.

Proof: Let $N < n$, where n is sufficiently large given integer. Then

$$E(Z_1 + Z_2 + \ldots + Z_n) = E(Z_1 + Z_2 + \ldots + Z_N) + E(Z_{N+1} + Z_{N+2} + \ldots + Z_n)$$

$$\Rightarrow \quad nEZ = E(Z_1 + Z_2 + \ldots + Z_N) + E(n - N)EZ$$

$$\Rightarrow \quad E(Z_1 + \ldots + Z_N) = E_\theta NEZ.$$

Further [for proving (ii)],

$$E(Z_1 + Z_2 + \ldots + Z_n)^2 = E\sum_{i=1}^{n} Z_i^2 + E\sum_{i \neq j} Z_i Z_j$$

$$= \sum_{i=1}^{n} EZ_i^2 + \sum_{i \neq j} EZ_i EZ_j$$

$$= nE_\theta Z_i^2$$

Now

$$E(Z_1 + \ldots + Z_n)^2 = E(Z_1 + \ldots + Z_N)^2 + E(Z_{N+1} + \ldots + Z_N)^2 + 2\sum_{i \neq j} E(Z_i Z_j)$$

$$\Rightarrow \quad nE_\theta Z^2 = E(Z_1 + \ldots + Z_N)^2 + E(n - N)EZ^2$$

$$\Rightarrow \quad E(Z_1 + Z_2 + \ldots + Z_N)^2 = E_\theta NEZ^2.$$

Let us obtain ASN of SPRT (B, A) with strength (α, β). Let $Z_i = \log \frac{f(x_i; \theta_1)}{f(x_i; \theta_0)}$, $i = 1, 2, \ldots$ for testing $H: \theta = \theta_0$ vs $K: \theta = \theta_1$ for $f(x; \theta)$

and let us approximate $S_N = \sum_{i=1}^{n} Z_i$ by $\log B$ when H is true and by $\log A$ when K is true. Then

$$E_\theta S_N \cong \log BP(S_N = \log B) + \log AP(S_N = \log A)$$

$$\cong \log BL(\theta) + \log A(1 - L(\theta)).$$

Using Wald's equation (i)

$$E_\theta N = \frac{E_\theta S_N}{E_\theta Z} \cong \frac{\log BL(\theta) + \log A(1 - L(\theta))}{E_\theta Z}, \; E_\theta Z \neq 0.$$

If $E_\theta Z = 0$, approximating S_N^2 by $(\log B)^2$ when H is true and by $(\log A)^2$ when K is true, we have

$$ES_N^2 \cong (\log B)^2 P_\theta(\text{accept } H) + (\log A)^2 P_\theta(\text{accept } K)$$

$$\cong (\log B)^2 \frac{\log A}{\log A - \log B} + (\log A)^2 \frac{\log B}{\log A - \log B}$$

$$\cong -\log A \log B$$

$$\Rightarrow \qquad E_{\theta'} N \cong \frac{-\log A \log B}{E_{\theta'} Z^2}$$

But using Wald's second equation where, θ' is such that $E_{\theta'} Z = 0$.

10.5 WALD'S FUNDAMENTAL IDENTITY OF SPRT

Theorem 10.3 Let $X_1, X_2, X_3, \ldots$ be i.i.d. random variables having common p.f. $f(x; \theta_0)$ under H and $f(x; \theta_1)$ under K, and let N be stopping rule, $Z_i = \log \frac{f(x_i; \theta_1)}{f(x_i; \theta_0)}$, $S_N = \sum_{i=1}^{N} Z_i$. Assuming that $M(t) = Ee^{tZ_i} < \infty$, $i = 1, 2, \ldots$

$$E\{e^{S_N t}[M(t)]^{-N}\} = 1.$$

Proof: Consider the identity

$$E\{e^{S_N t + (S_n - S_N)t}\} = Ee^{S_n t} = [M(t)]^n \qquad \ldots(10.5.1)$$

where n is a fixed positive integer. Let us denote

$$p_n = P(N \le n),\ E_n U = E(U/N \le n),\ E_n^*(U) = E(U/N > n).$$

Then (10.5.1) will be written as

$$p_n E_n e^{S_N t + (S_n - S_N)t} + (1 - p_n) E_n^* e^{S_n t} = [M(t)]^n \qquad \ldots(10.5.2)$$

In the subpopulation $N \le n$, $S_n - S_N$ is independent of S_N and hence

$$E_n e^{S_N t + (S_n - S_N)t} = E_n e^{S_N t} E_n e^{(S_n - S_N)t}$$
$$= E_n e^{S_N t}[M(t)]^{n-N},$$

$$p_n E_n\{e^{S_N t}[M(t)]^{n-N}\} + (1-p_n)E_n^* e^{S_n t} = [M(t)]^n$$

$$\Rightarrow \qquad p_n E_n\{e^{S_N t}[M(t)]^{-N}\} + (1-p_n)E_n^* e^{S_n t}[M(t)]^{-n} = 1. \qquad \ldots(10.5.3)$$

Since $P(N < \infty) = 1$, $\lim_{n\to\infty}\{1 - P(N < n)\} = 0$ and for $n < N$,

$$\log B < S_n < \log A \;\Rightarrow\; e^{\log Bt} < e^{S_n t} < e^{\log At} \Rightarrow E_n e^{S_n t}$$

is bounded for n and so $\lim_{n\to\infty} p_n E_n\{e^{S_N t}[M(t)]^{-n}\} = 0$. Since

$$\lim_{n\to\infty} p_n E_n\{e^{S_N t}[M(t)]^{-N}\} = Ee^{S_N t}[M(t)]^{-N},$$

will reduce to $E\{e^{S_N t}[M(t)]^{-N}\} = 1$.

Note

Wald has shown that this identity can be differentiated any number of times w.r.t. t under expectation such that $|M(t)| \le 1$.

Application of the Identity

(i) To obtain ASN function

Differentiating the identity $E\{e^{S_N t}[M(t)]^{-N}\} = 1$ w.r.t. t, we get

$$E\left\{\left[S_N - \frac{NM'(t)}{M(t)}\right]e^{S_N t}[M(t)]^{-N}\right\} = 0 \qquad \ldots(10.5.4)$$

Again differentiating (10.5.4) w.r.t. t,

$$E\left\{\left[S_N - \frac{NM'(t)}{M(t)}\right]^2 - \frac{NM''(t)M(t) - N[M'(t)]^2}{[M(t)]^2}\right\}e^{S_N t}[M(t)]^{-N} = 0$$

$$\ldots(10.5.5)$$

As $M(0) = 1$, $M'(t) = EZe^{tZ}$, $M''(t) = EZ^2e^{tZ}$, $M'(0) = EZ$, $M''(0) = EZ^2$, (10.5.4), (10.5.5) at $t = 0$, will give $ES_N = ENEZ$, if $EZ \neq 0$, $ES_N^2 = ENEZ^2$, if $EZ = 0$.

(ii) To get OC function
If $h \neq 0$ such that $M(h) = Ee^{hZ} = 1$, so that the fundamental identity will reduce to

$$1 = Ee^{S_N h}$$

$$= P(S_N \leq \log B)E(e^{S_N h}/S_N \leq \log B) + P(S_N \geq \log A)E(e^{S_N h}/S_N \geq \log A)$$

$$\cong L(\theta)e^{h \log B} + [1 - L(\theta)]e^{h \log A}$$

$$\cong L(\theta)B^h + [1 - L(\theta)]A^h$$

$$\Rightarrow \qquad L(\theta) \cong \frac{A^h - 1}{A^h - B^h}, \text{ if } h \neq 0.$$

(iii) To obtain (B, A)
For $h \neq 0$, $M(h) = 1$ and the identity reduces to

$$1 = Ee^{S_N h} = E_\theta\left[\frac{f(x;\theta_1)}{f(x;\theta_0)}\right]^h$$

$$\Rightarrow \qquad h(\theta_0) = 1 \quad \text{and} \quad h(\theta_1) = -1, \text{ so that}$$

$$\left.\begin{aligned} L(\theta_0) &= 1 - \alpha \cong \frac{A-1}{A-B} \\ L(\theta_1) &= \beta \cong \frac{\frac{1}{A} - 1}{\frac{1}{A} - \frac{1}{B}} \end{aligned}\right\} \Rightarrow \left\{\begin{aligned} B &\cong \frac{\beta}{1-\alpha} \\ A &\cong \frac{1-\beta}{\alpha} \end{aligned}\right. .$$

Example 10.2

Let $X_1, X_2, X_3, \ldots$ have common p.m.f.

$$f(x;\theta) = \theta^x (1-\theta)^{1-x}, x = 0, 1, \ldots$$

Then obtain SPRT (α, β) for testing $H: \theta = \theta_0$ vs $K: \theta = \theta_1 > \theta_0$, and obtain OC, ASN functions.

Here,

$$\frac{f(x;\theta_1)}{f(x;\theta_0)} = \left[\frac{\theta_1(1-\theta_0)}{\theta_0(1-\theta_1)}\right]^x \frac{1-\theta_1}{1-\theta_0} = D^x c,$$

where

$$D = \frac{\theta_1(1-\theta_0)}{\theta_0(1-\theta_1)}, c = \frac{1-\theta_1}{1-\theta_0} > 1,$$

so that

$$Z = \log\frac{f(x;\theta_1)}{f(x;\theta_0)} = x\log D - \log c$$

and governing inequality of SPRT is

$$\log B < \Sigma Z_i < \log A$$

$$\Rightarrow \quad \log B < \log D \Sigma x_i - n\log c < \log A$$

$$\Rightarrow \quad \frac{\log B + n\log c}{\log D} < \Sigma x_i < \frac{\log A + n\log c}{\log D}$$

$$\Rightarrow \quad h_1 n + h_2 < \Sigma x_i < h_1 n + h_3$$

where

$$h_1 = \frac{\log c}{\log D} > 0, h_2 = -\frac{\log B}{\log D} > 0, h_3 = \frac{\log A}{\log D} > 0.$$

For OC function we should obtain $h \neq 0$, such that

$$E_\theta\left[\frac{f(x;\theta_1)}{f(x;\theta_0)}\right]^h = 1 \text{ i.e. } \sum_{x=0}^{n}\left[\frac{f(x;\theta_1)}{f(x;\theta_0)}\right]^h f(x;\theta) = 1$$

i.e., $$\frac{1-\theta}{c^h} + \frac{\theta D^h}{c^h} = 1 \Rightarrow \theta = \frac{c^h - 1}{D^h - 1}.$$

Hence, OC function $\cong \dfrac{A^h - 1}{A^h - B^h}$ can be plotted for h values

$$\lim_{h \to 0} \frac{c^h - 1}{D^h - 1} = \frac{\log C}{\log D} = \theta', \text{ say, } L(\theta') = \frac{\log A}{\log A - \log B}.$$

For ASN function we require $EZ = \theta \log D - \log c \neq 0$, then

$$E_\theta N \cong \frac{L(\theta)\log B + [1 - L(\theta)]\log A}{\theta \log D - \log c}, \text{ if } \theta \neq \frac{\log c}{\log D},$$

$$E_{\theta'} N \cong \frac{-\log A \log B}{E_\theta Z^2}, E_\theta Z^2 = \text{var}(Z) = (\log D)^2, \theta' = \frac{\log c}{\log D}.$$

Illustration for using SPRT Instead of Fixed Sample Size

Let $\alpha = \beta = 0.05$, $H: \underline{\theta} = 0.05$, $K: \theta = 0.1$, $E_{\theta_0} N = 159$, $E_{\theta_1} N = 128$, average $N = 144$. For non-sequential test of size 144, $\alpha = 0.05$ we solve

$$E_{\theta_0}(\Sigma X_i > k) = 0.05 \Rightarrow \sum_{j=K+1}^{144} \binom{144}{j} (0.05)^j (0.95)^{144-j} \leq 0.05 \Rightarrow k = 12.$$

Then

$$\beta = 1 - \sum_{j=K+1}^{144} \binom{144}{j} (0.1)^j (0.9)^{144-j} = 0.32097.$$

If we use normal approximation $\Sigma X_i \sim N_{\theta_0}(0.05N, 0.0475N)$, $\Sigma X_i \sim N_{\theta_1}(0.1N, 0.09N)$. The critical region is $\Sigma X_i \geq nc$, then n is such that

$$\frac{(c - 0.05)\sqrt{n}}{\sqrt{0.0475}} = 1.645, \frac{(c - 0.1)\sqrt{n}}{\sqrt{0.09}} = -1.645 \Rightarrow n \cong 272.$$

Example 10.3

Let $X_1, X_2, \ldots$ be i.i.d. random variables distributed as $N(\theta, 1)$. Then test $H: \theta = \theta_0$ vs $K: \theta = \theta_1$ by SPRT. Since

$$\frac{f(x;\theta_1)}{f(x;\theta_0)} = e^{(\theta_1-\theta_0)x + \frac{1}{2}(\theta_0^2 - \theta_1^2)}, Z = (\theta_1 - \theta_0)\left\{x - \frac{(\theta_0 + \theta_1)}{2}\right\},$$

and

$$\Sigma Z_i = (\theta_1 - \theta_0)\left\{\Sigma x_i - \frac{n}{2}(\theta_0 + \theta_1)\right\}.$$

So sampling continues so long as

$$b < \Sigma Z_i < a \Rightarrow \frac{b}{(\theta_1 - \theta_0)} + \frac{n}{2}(\theta_0 + \theta_1) < \Sigma x_i < \frac{a}{(\theta_1 - \theta_0)} + \frac{n}{2}(\theta_0 + \theta_1)$$

is satisfied. For OC function we have to solve

$$e^{\frac{1}{2}(\theta_0^2 - \theta_1^2)h} \int_{-\infty}^{\infty} \frac{1}{\sqrt{2\pi}} e^{(\theta_1-\theta_0)hx} e^{-\frac{1}{2}(x-\theta)^2} dx = 1 \qquad \ldots(10.5.6)$$

The integral being m.g.f. of $N(0, 1)$ with argument $(\theta_1 - \theta_0)$, (10.5.6) gives

$$e^{\frac{1}{2}(\theta_0^2 - \theta_1^2)h} e^{(\theta_1-\theta_0)h\theta + \frac{(\theta_0+\theta_1)^2 h^2}{2}} = 1$$

$$\Rightarrow \qquad \frac{h}{2}(\theta_1 - \theta_0) + \theta - \frac{(\theta_1 + \theta_0)}{2} = 0$$

$$\Rightarrow \qquad h = \frac{2\left(\frac{(\theta_1 + \theta_0)}{2} - \theta\right)}{(\theta_1 - \theta_0)} = \frac{(\theta_1 + \theta_0) - 2\theta}{(\theta_1 - \theta_0)} \neq 0$$

provided $\qquad \theta \neq \dfrac{(\theta_1 + \theta_0)}{2}.$

For ASN function

$$E_\theta Z = (\theta_1 + \theta_0)\theta + \frac{1}{2}(\theta_1 - \theta_0)(\theta_1 + \theta_0)$$

$$= (\theta_1 - \theta_0)\left\{\theta - \frac{(\theta_1 + \theta_0)}{2}\right\} \neq 0 \text{ if } \theta \neq \frac{(\theta_1 + \theta_0)}{2}$$

$$E_\theta N = \frac{\log B L(\theta) + [1 - L(\theta)]\log A}{\frac{(\theta_1 - \theta_0)}{2}(2\theta - \theta_1 - \theta_0)} \text{ if } \theta \neq \frac{(\theta_1 + \theta_0)}{2}$$

and

$$L(\theta) = \frac{A^h - 1}{A^h - B^h}.$$

Particularly,

$$E_{\theta_0} N = \frac{(1-\alpha)\log B + \alpha \log A}{\left[-\frac{(\theta_1 - \theta_0)}{2}\right]^2}$$

$$E_{\theta_1} N = \frac{\beta \log B + (1-\beta)\log A}{\left[\frac{(\theta_1 + \theta_0)}{2}\right]^2}.$$

If $E_\theta Z = 0$, $\theta' = \frac{(\theta_1 + \theta_0)}{2}$, or $h = 0$,

$$L(\theta') = \frac{\log A}{\log A - \log B}, \quad E_{\theta'} N = -\frac{\log A \log B}{\text{var}_{\theta'} Z} = -\frac{\log A \log B}{(\theta_1 - \theta_0)^2}.$$

Saving of Observations by Using SPRT

For $N(\theta, 1)$ distribution testing $H{:}\theta = \theta_0$ vs $K{:}\theta = \theta_1$ sequentially yield $E_{\theta_0} N$, $E_{\theta_1} N$ as above. Now let us obtain sample size in fixed sample method given α, β. As we reject H when $\bar{X} \geq \frac{Z_\alpha}{\sqrt{n}} + \theta_0$ the power of the test is

$$1 - \beta = \phi((\theta_1 - \theta_0)\sqrt{n} - Z_\alpha)$$

$$\Rightarrow \qquad (\theta_1 - \theta_0)\sqrt{n} - Z_\alpha = Z_\beta$$

$$\Rightarrow \qquad n = \frac{(Z_\alpha + Z_\beta)^2}{(\theta_1 - \theta_0)^2} = n(\alpha, \beta).$$

Therefore,

$$\frac{E_{\theta_1} N}{n(\alpha, \beta)} = \frac{2[\beta \log B + (1-\beta)\log A]}{(Z_\alpha + Z_\beta)^2}.$$

For $\alpha = \beta = 0.025$, $A = \frac{1}{\beta} = 39$, $Z_\alpha = Z_\beta = 1.96$, then

$$\frac{E_{\theta_1} N}{n(\alpha, \beta)} = \frac{2 \log 39(0.96)}{(3.92)^2} = 0.453.$$

Hence, 55% saving of observations.

10.6 LIKELIHOOD RATIO TEST (LRT)

In many problems of testing of hypotheses, no UMP or UMPU tests are available and in such cases likelihood ratio test is useful. Further in dealing with non-parametric tests, the method of likelihood ratio test method is also useful. Although this method does not guarantee that the test is MP or unbiased, it is the method which gives a test function quickly.

Let $L(x/\theta)$ be likelihood of random sample $X = (X_1, X_2, \ldots, X_n)$ with k-parameter vector $\theta = (\theta_1, \theta_2, \ldots, \theta_r, \theta_{r+1}, \ldots, \theta_{r+s})$ and let $r + s = k$. To test $H: \theta_i = \theta_{i_0}$ vs $K: \theta_i \neq \theta_{i_0}$, $i = 1, 2, \ldots, r$. Find m.l.e. $\hat{\theta}_i$, $i = 1, 2, \ldots, r+s$ in the unconditional space $\Omega = \{\theta_j, j = 1, 2, \ldots, r+s\}$ and $L(\hat{\Omega}) = \sup_{\Omega} L(x/\theta_j) = L(x/\hat{\theta}_j)$, $j = 1, 2, \ldots, r+s$.

Also find conditional m.l.e. $\hat{\theta}_l^*$ of $\theta_l, l = r+1, r+2, \ldots, r+s$ in $\omega = \{\theta_l\}$ space giving $L(\omega) = \sup_{\omega} L(x/\theta_l) = L(x/\hat{\theta}_l), l = r+1, r+2, \ldots, r+s$.

Then the likelihood ratio criterion λ is

$$\lambda = \frac{L(x/\hat{\theta}_l)}{L(x/\hat{\theta}_j)}, 0 \le \lambda \le 1, \text{ as } L(\hat{\omega}) \le L(\hat{\Omega}).$$

The large value of λ signifies that H is acceptable so the critical region of the test is $\lambda \le c$, where, c is such that

$$\int_0^c g(\lambda)d\lambda = \alpha,$$

$g(\lambda)$ is the p.d.f. of λ under H.

Asymptotic Distribution of λ

Consider the family of distributions $\{P_\theta, \theta \in \Omega\}$ of $X = (X_1, \ldots, X_n)$. If T is m.l.e. of θ and if T is MVB estimator of θ and if L is likelihood of $x_1, \ldots, x_n$ then due to necessary and sufficient condition that T to be MVB estimator of θ is

$$\frac{\partial \log L}{\partial \theta} = -E\left(\frac{\partial^2 \log L}{\partial^2 \theta}\right)(T - \theta)$$

i.e.,
$$L \propto e^{-\frac{1}{2}E\left(\frac{\partial^2 \log L}{\partial^2 \theta}\right)(t-\theta)^2}$$

$\Rightarrow t$ has normal distribution.

For k component vector $\underline{\theta}$, the analogue is

$$\frac{\partial \log L}{\partial \theta} = (T-\theta)'V^{-1}(T-\theta), \text{ where, } V^{-1} = (V_{ij})^{-1}, V_{ij} = -E\left(\frac{\partial^2 \log L}{\partial \theta_i \partial \theta_j}\right)$$

Therefore,

$$L \propto e^{-\frac{1}{2}(\underline{t}-\underline{\theta})'V^{-1}(\underline{t}-\underline{\theta})} \quad \ldots(10.6.1)$$

which implies that T has k variate normal distribution. When K is true.

$$L(\hat{\Omega}) = L(x/\theta_r, \theta_s) \propto 1, \qquad \text{...(10.6.2)}$$

$(\underline{t} - \underline{\theta}) = \underline{0}$, when H is true, the s components of $(\underline{t} - \underline{\theta})$ corresponding to θ_s will still be zero and in ω space the m.l.e. of θ_s being $\hat{\theta}_s^* = t$, say, (10.6.1) reduces to

$$L(\hat{\omega}) = L(x/\theta_{r_0}, \theta_s^*) \propto e^{-\frac{1}{2}(\underline{t}_r - \underline{\theta}_{r_0})' V_r^{-1} (\underline{t}_r - \underline{\theta}_{r_0})} \qquad \text{...(10.6.3)}$$

where V_r denotes $r \times r$ matrix of V by taking first r rows and first r columns. Thus, from (10.6.3) and (10.6.2) λ criterion is

$$\lambda = \frac{L(x/\theta_{r_0}, \hat{\theta}_s^*)}{L(x/\hat{\theta}_r, \hat{\theta}_s)} = e^{-\frac{1}{2}(\underline{t}_r - \underline{\theta}_{r_0})' V_r^{-1} (\underline{t}_r - \underline{\theta}_{r_0})}.$$

Therefore, $-2\log\lambda = (\underline{t}_r - \underline{\theta}_{r_0})' V_r^{-1} (\underline{t}_r - \underline{\theta}_{r_0})$.

Now as $\underline{t}$ has asymptotically k variate normal distribution, $\underline{t}_r$ has r variate normal distribution asymptotically with mean vector $\underline{\theta}_{r_0}$ and covariance matrix V_r. Thus, $-2\log\lambda$ has asymptotically non-central χ^2 distribution on r degree of freedom and non-centrality parameter $(\theta_r - \theta_{r_0})' V_r^{-1} (\theta_r - \theta_{r_0})$. Under H, $\theta_r = \theta_{r_0}$ and the non-centrality parameter is zero, so that

$$-2\log\lambda \xrightarrow{d} \chi_r^2.$$

Example 10.4

Let $X = (X_1, X_2, \ldots, X_n)$ be a random sample from $N(\mu, 1)$ distribution.

Test $H: \mu \le \mu_0$ vs $K: \mu > \mu_0$ by LRT method.

Here

$$L(x, \mu) = \frac{1}{(2\pi)^{\frac{n}{2}}} e^{-\frac{1}{2}\sum(x_i - \bar{x})^2 - \frac{n}{2}(\bar{x} - \mu)^2}, \; \hat{\mu}_\Omega = \bar{X}.$$

Therefore,

$$L(\hat{\Omega}) = \frac{1}{(2\pi)^{\frac{n}{2}}} e^{-\frac{1}{2}\sum(x_i - \bar{x})^2} \qquad \text{...(10.6.4)}$$

Under ω space, $\mu \leq \mu_0$, so if $\bar{x} \leq \mu_0$, $\hat{\mu} = \bar{x}$ and if $\bar{x} > \mu_0$ then $\hat{\mu} = \mu_0$, $(\bar{x} - \mu)^2$ is minimum. Thus,

$$\hat{\mu} = \begin{cases} \bar{x} & \text{if} \quad \bar{x} \leq \mu_0 \\ \mu_0 & \text{if} \quad \bar{x} > \mu_0 \end{cases},$$

and

$$L(\hat{\omega}) = \begin{cases} \dfrac{1}{(2\pi)^{\frac{n}{2}}} e^{-\frac{1}{2}\Sigma(x_i - \bar{x})^2} & \text{if} \quad \bar{x} \leq \mu_0 \\ \dfrac{1}{(2\pi)^{\frac{n}{2}}} e^{-\frac{1}{2}\Sigma(x_i - \bar{x})^2 - \frac{n}{2}(\bar{x} - \mu_0)^2} & \text{if} \quad \bar{x} > \mu_0 \end{cases}$$

Therefore,

$$\lambda(x) = \begin{cases} 1 & \text{if} \quad \bar{x} \leq \mu_0 \\ e^{-\frac{n}{2}(\bar{x} - \mu_0)^2} & \text{if} \quad \bar{x} > \mu_0 \end{cases},$$

$$\lambda(x) \leq c \Leftrightarrow \bar{x} - \mu_0 \geq k \text{ when } c < 1, \bar{x} \geq k + \mu_0.$$

Power function is

$$\begin{aligned} \beta(\mu) &= P_K(\bar{x} \geq k + \mu_0) \\ &= P\{\sqrt{n}(\bar{x} - \mu) \geq \sqrt{n}(\mu_0 - \mu + k)\} \\ &= 1 - \phi\{\sqrt{n}(\mu_0 - \mu + k)\}, \end{aligned}$$

where ϕ is distribution function of $N(0, 1)$ and

$$\alpha = \beta(\mu_0) = 1 - \phi(k\sqrt{n}).$$

If $\qquad \phi(Z_\alpha) = 1 - \alpha$ then $Z_\alpha = k\sqrt{n}, k = \dfrac{Z_\alpha}{\sqrt{n}}$

Reject H whenever $\bar{x} \geq \mu_0 + \dfrac{Z_\alpha}{\sqrt{n}}$.

Example 10.5

Let $X = (X_1, X_2, \ldots, X_n)$ be a random sample from $N(\mu, 1)$ distribution.

Test $H: \mu = \mu_0$ vs $K: \mu \neq \mu_0$ by LRT method.

Here

$$L(\widehat{\Omega}) = \frac{1}{(2\pi)^{\frac{n}{2}}} e^{-\frac{1}{2}\sum(x_i - \bar{x})^2},$$

$$L(\widehat{\omega}) = \frac{1}{(2\pi)^{\frac{n}{2}}} e^{-\frac{1}{2}\sum(x_i - \bar{x})^2 - \frac{n}{2}(\bar{x} - \mu_0)^2},$$

$$\lambda(x) = e^{-\frac{n}{2}(\bar{x} - \mu_0)^2} \leq c \Leftrightarrow |\bar{X} - \mu_0| \geq k,\ P_{\mu_0}\{|\bar{X} - \mu_0| \geq k\} = \alpha,$$

$$\Rightarrow \quad \phi(k\sqrt{n}) - \phi(-k\sqrt{n}) = 1 - \alpha$$

$$\Rightarrow \quad 2\phi(k\sqrt{n}) - 1 = 1 - \alpha$$

$$\Rightarrow \quad \phi(k\sqrt{n}) = 1 - \frac{\alpha}{2} \Rightarrow k = \frac{Z_{\frac{\alpha}{2}}}{\sqrt{n}}.$$

Test is to reject H when $|\bar{X} - \mu_0| \geq \frac{1}{\sqrt{n}} Z_{\frac{\alpha}{2}}$ and power function is

$$\beta(\mu) = P_\mu\left(|\bar{X} - \mu_0| \geq \frac{1}{\sqrt{n}} Z_{\frac{\alpha}{2}}\right)$$

$$= P_\mu\left(\bar{X} \geq \mu_0 + \frac{1}{\sqrt{n}} Z_{\frac{\alpha}{2}}\right) + P_\mu\left(\bar{X} \leq \mu_0 - \frac{1}{\sqrt{n}} Z_{\frac{\alpha}{2}}\right)$$

$$= 1 - \phi\left[\sqrt{n}(\mu_0 - \mu) + Z_{\frac{\alpha}{2}}\right] + \phi\left[\sqrt{n}(\mu_0 - \mu) - Z_{\frac{\alpha}{2}}\right].$$

As

$$\frac{d}{d\mu}\beta(\mu) = \sqrt{n}\varphi\left[\sqrt{n}(\mu_0 - \mu) + Z_{\frac{\alpha}{2}}\right] - \sqrt{n}\varphi\left[\sqrt{n}(\mu_0 - \mu) - Z_{\frac{\alpha}{2}}\right],$$

where $\varphi(Z)$ is p.d.f. of $N(0, 1)$ which is symmetric and is decreasing function of $|Z|$, so

$$\frac{d}{d\mu}\beta(\mu) = \gtreqless 0, \text{ when } \mu \gtreqless \mu_0.$$

Example 10.6

Let $X = (X_1, X_2, \ldots, X_n)$ be a random sample from $N(\mu, \sigma^2)$ distribution.

Test $H: \mu = \mu_0$ vs $K: \mu \neq \mu_0$ by LRT method.

Let

$$\hat{\mu}_\Omega = \bar{x}, \hat{\sigma}_\Omega^2 = \frac{\sum (x_i - \bar{x})^2}{n}$$

$$L(\hat{\Omega}) = (\hat{\sigma}_\Omega \sqrt{2\pi})^{-n} e^{-\frac{n}{2}}, \hat{\sigma}_\omega^2 = \frac{1}{n}(x_i - \mu_0)^2, L(\hat{\omega}) = (\hat{\sigma}_\omega \sqrt{2\pi})^{-n} e^{-\frac{n}{2}},$$

$$[\lambda(x)]^{\frac{2}{n}} = \frac{\hat{\sigma}_\Omega^2}{\hat{\sigma}_\omega^2} = \frac{\sum (x_i - \bar{x})^2}{\sum (x_i - \bar{x})^2 + n(\bar{x} - \mu_0)^2} = \left[1 + \frac{n(\bar{x} - \mu_0)}{\sum (x_i - \bar{x})^2}\right],$$

$$\lambda(x) \leq c \Rightarrow \frac{n(\bar{x} - \mu_0)^2}{\sum (x_i - \bar{x})^2} \geq c \Rightarrow |T| \geq c,$$

where T has student's t distribution on $(n - 1)$ degree of freedom.

Example 10.7

Let $X = (X_1, X_2, \ldots, X_m)$ be a random sample from $N(\mu_1, \sigma^2)$ distribution and $Y = (Y_1, Y_2, \ldots, Y_n)$ be another independent random sample from $N(\mu_2, \sigma^2)$.

Test $H: \mu_1 = \mu_2$ vs $K: \mu_1 \neq \mu_2$ by LRT method.

Let

$$\hat{\mu}_{1_\Omega} = \bar{X},$$

$$\hat{\mu}_{2_\Omega} = \bar{Y},$$

$$\hat{\sigma}_\Omega^2 = \frac{1}{m+n}[\Sigma(x_i - \bar{x})^2 + \Sigma(y_j - \bar{y})^2],$$

$$L(\hat{\Omega}) = \frac{e^{-\frac{m+n}{2}}}{(\sqrt{2\pi}\hat{\sigma}_\Omega)^{\frac{m+n}{2}}} \qquad \ldots(10.6.5)$$

$$\hat{\mu}_1 = \hat{\mu}_2 = \hat{\mu}_\omega = \frac{\sum x_i + \sum y_j}{m+n}, \hat{\sigma}_\omega^2 = \frac{1}{m+n}[\Sigma(x_i - \hat{\mu}_1)^2 + \Sigma(y_j - \hat{\mu}_2)^2]$$

$$L(\hat{\omega}) = \frac{e^{-\frac{m+n}{2}}}{(\sqrt{2\pi}\hat{\sigma}_\omega)^{\frac{m+n}{2}}} \qquad \ldots(10.6.6)$$

$$[\lambda(x)]^{\frac{2}{m+n}} = \frac{\hat{\sigma}_\Omega^2}{\hat{\sigma}_\omega^2},$$

$$\Sigma(x_i - \hat{\mu}_\omega)^2 + \Sigma(y_j - \hat{\mu}_\omega)^2 = (m+n)\hat{\sigma}_\Omega^2 + \frac{mn}{m+n}(\bar{x} - \bar{y})^2.$$

Therefore,

$$\lambda^{\frac{2}{m+n}} = \left[1 + \frac{mn(\bar{x} - \bar{y})^2}{(m+n)^2\hat{\sigma}_\Omega^2}\right]^{-1},$$

H is rejected when $|T| > c$

where $$T = \frac{\bar{x} - \bar{y}}{\sqrt{\left(\frac{1}{m} + \frac{1}{n}\right)\frac{\Sigma(x_i - \bar{x})^2 + \Sigma(y_j - \bar{y})^2}{m+n-2}}} \sim t_{m+n-2}$$

10.7 ONE-WAY CLASSIFICATION

Let $X_{ij}(i = 1, 2, \ldots, k, j = 1, 2, \ldots, n_i)$ be a random sample form $N(\mu_i, \sigma^2)$ distribution. Test $H: \mu_1 = \ldots = \mu_k$ vs K:any μ.

$$\sum_{i=1}^{k} n_i = n,$$

$$\bar{X}_i = \frac{\sum_{j=1}^{n_i} X_i}{n_i},$$

$$s_i^2 = \Sigma(X_{ij} - \bar{X}_i)^2,$$

$$\bar{X} = \frac{\sum_i \sum_j X_{ij}}{n},$$

$$s^2 = \sum_i \sum_j (X_{ij} - \bar{X})^2,$$

$$\hat{\mu}_{i_\Omega} = \bar{X}_i,$$

$$\hat{\sigma}_\Omega^2 = \frac{\sum s_i^2}{n}.$$

As $$\Omega = \{\mu_1, \mu_2, \ldots, \mu_k, \sigma^2\}$$

$$L(\hat{\Omega}) = \frac{1}{(2\pi)^{\frac{n}{2}}(\hat{\sigma}_\Omega)^n} e^{-\frac{n}{2}},$$

As $\omega = \{\mu, \sigma^2\}, \hat{\mu}_\omega = \bar{X}, \hat{\sigma}_\omega^2 = \frac{s^2}{n}$, where $\mu_1 = \mu_2 = \ldots = \mu_k = \mu$

$$L(\hat{\omega}) = \frac{1}{(2\pi)^{\frac{n}{2}}(\hat{\sigma}_\omega)^n} e^{-\frac{n}{2}}.$$

$$\lambda^{\frac{2}{n}} = \frac{L(\hat{\omega})}{L(\hat{\Omega})} = \frac{\sum s_i^2}{s^2}. \text{ As } s^2 = \sum_i \sum_j (X_{ij} - \bar{X}_i + \bar{X}_i - \bar{X})^2$$

$$= \sum s_i^2 + \sum n_i(\bar{X}_i - \bar{X})^2$$

$$= \left(1 + \frac{\sum n_i(\bar{X}_i - \bar{X})^2}{\sum s_i^2}\right)^{-1}.$$

As $\frac{\sum n_i(\bar{X}_i - \bar{X})^2}{\sum s_i^2}$ and Σs_i^2 are independent, under the null hypothesis

$$\sum_{i=1}^{k} n_i(\bar{X}_i - \bar{X})^2 \sim \sigma^2 \chi^2_{k-1} \text{ and } \sum s_i^2 \sim \sigma^2 \chi^2_{n-k},$$

and hence,

$$F = \frac{\dfrac{\sum_{i=1}^{k} n_i(\bar{X}_i - \bar{X})^2}{(k-1)}}{\dfrac{\sum_i \sum_j (X_{ij} - \bar{X})^2}{(n-k)}} \sim F_{k-1,\, n-k} \text{ dustribution.}$$

Reject H when $F \geq F_{k-1,\, n-k,\, \alpha}$ at level α.

10.8 TWO-WAY CLASSIFICATION

Let $X_{ij}(i = 1, 2, \ldots, k, j = 1, 2, \ldots, l)$ be a r.s. from $N(\mu + \alpha_i + \beta_j, \sigma^2)$ distribution.

Test $H_A : \alpha_1 = \ldots = \alpha_k = 0, \Sigma \alpha_i = \Sigma \beta_j = 0.\ \hat{\mu}_\Omega = \bar{X},$

$$\hat{\sigma}^2_\Omega = \frac{1}{kl} \sum_i \sum_j (X_{ij} - \bar{X}_{i.} - \bar{X}_{.j} + \bar{X})^2, \hat{\alpha}_{i_\Omega} = \bar{X}_{i.} - \bar{X}, \hat{\beta}_{j\Omega} = \bar{X}_{.j} - \bar{X},$$

$$\bar{X} = \frac{\sum_i \sum_j X_{ij}}{kl} = \frac{\sum_i \bar{X}_{i.}}{k} = \frac{\sum_j \bar{X}_{.j}}{l}$$

where

$$\bar{X}_{i.} = \frac{\sum_j^{l} \bar{X}_{ij}}{l}, \bar{X}_{.j} = \frac{\sum_i^{k} \bar{X}_{ij}}{k}$$

$$L(\hat{\Omega}) = \frac{e^{-\frac{kl}{2}}}{(2\pi)^{\frac{kl}{2}} (\hat{\sigma}_\Omega)^{kl}}.$$

Now

$$\omega = \{\mu, \beta_1, \beta_2, \ldots, \beta_l, \sigma^2\}, \hat{\mu}_\omega = \bar{X}, \hat{\beta}_{j_\omega} = \bar{X}_{.j} - \bar{X}, j = 1, 2, \ldots, l,$$

$$\hat{\sigma}_{\omega}^{2} = \frac{1}{kl}\sum_{i}\sum_{j}(X_{ij} - \bar{X}_{.j})^{2}.$$

$$L(\hat{\omega}) = \frac{e^{-\frac{kl}{2}}}{(2\pi)^{\frac{kl}{2}}(\hat{\sigma}_{\omega})^{kl}}, \lambda = \left(\frac{\hat{\sigma}_{\Omega}}{\hat{\sigma}_{\omega}}\right)^{kl},$$

Now

$$\sum_{i}\sum_{j}(X_{ij} - \bar{X}_{.j})^{2} = l\sum_{i}(\bar{X}_{i.} - \bar{X})^{2} + \sum_{i}\sum_{j}(X_{ij} - \bar{X}_{i.} - \bar{X}_{.j} + \bar{X})^{2},$$

$$Q_1 = Q_2 + Q_3, \text{ say.}$$

Since X_{ij} are independently normally distributed

$$Q_1 \sim \sigma^2\chi^2_{l(k-1)}, Q_2 \sim \sigma^2\chi^2_{(k-1)}, Q_3 \sim \sigma^2\chi^2_{(k-1)(l-1)}$$

They are independently distributed.

Then

$$\lambda^{\frac{2}{kl}} = \frac{\hat{\sigma}_{\Omega}}{\hat{\sigma}_{\omega}} = \frac{Q_3}{Q_1} = \frac{Q_3}{Q_2 + Q_3} = \frac{1}{\left(1 + \frac{Q_2}{Q_3}\right)}.$$

if $$F_A = \frac{\frac{Q_2}{(k-1)}}{\frac{Q_3}{(k-1)(l-1)}} \geq F_{(k-1), [(k-1)(l-1), \alpha]}; \text{ reject } H_A.$$

For testing $H_B: \beta_1 = \beta_2 = \ldots = \beta_l = 0$ if $Q_4 = k\Sigma(\bar{X}_{.j} - \bar{X})^2$ then reject H_B when

$$F_B = \frac{\frac{Q_4}{(l-1)}}{\frac{Q_3}{(k-1)(l-1)}} \geq F_{(l-1), [(k-1)(l-1), \alpha]}$$

Bartlett's Test

Let $x_{i1}, x_{i2}, \ldots, x_{in}$ be a random sample from $N(\mu_i, \sigma_i^2), i = 1, 2, \ldots, k$. To test $H: \sigma_1^2 = \ldots = \sigma_k^2 = \sigma^2$, say. Then in Ω space, $\hat{\mu}_i = \bar{x}_i, \hat{\sigma}_i = s_i^2 = \frac{1}{n}\Sigma(x_{ir} - \bar{x}_i)^2$.

$$L(\hat{\Omega}) = \frac{1}{(2\pi)^{\frac{n}{2}}} \frac{1}{\left(\prod_{i=1}^{k} s_i^2\right)^{\frac{n}{2}}} e^{-\frac{n}{2}}, n = \sum n_i,$$

and in ω space, $\hat{\mu}_i = \bar{x}_i, \hat{\sigma}^2 = \frac{\sum n_i s_i^2}{n} = s^2$, so that

$$L(\hat{\omega}) = \frac{1}{(2\pi)^{\frac{n}{2}}} \frac{1}{(s^2)^{\frac{n}{2}}} e^{-\frac{n}{2}},$$

Hence,

$$\lambda = \prod \left(\frac{s_i^2}{s^2}\right)^{\frac{n_i}{2}} \quad \ldots(10.8.1)$$

$$-2\log\lambda = n\log s^2 - \sum_{i=1}^{k} n_i \log s_i^2. \quad \ldots(10.8.2)$$

When H is true, each of $\frac{n_i s_i^2}{2\sigma^2} \sim G\left(\frac{1}{2}, \frac{n_i - 1}{2}\right)$ and hence $\frac{\sum n_i s_i^2}{2\sigma^2} \sim G\left(\frac{1}{2}, \frac{n-k}{2}\right)$ distribution. Now for a Gamma variate with parameter p

$$E(\log aX) = \frac{1}{\Gamma(p)} \int_0^{\infty} \log(ax) e^{-x} x^{p-1} dx$$

$$= \log a + \frac{d}{dp} \log\Gamma(p).$$

Using Sterling approximation

$$\frac{d}{dp}\log\Gamma(p) \cong \log p - \frac{1}{2p} - \frac{1}{12p^2} + 0\left(\frac{1}{p^3}\right)$$

$$E\log aX \cong \log a + \log p - \frac{1}{2p} - \frac{1}{12p^2} + 0\left(\frac{1}{p^3}\right) \qquad \ldots(10.8.3)$$

Using (10.8.3) in (10.8.2), we have

$$E(-2\log\lambda) \cong n\left\{\log\left(\frac{2\sigma^2}{n}\right) + \log\left(\frac{n-k}{2}\right) - \frac{1}{n-k} - \frac{1}{3(n-k)^2} + 0\left(\frac{1}{n^3}\right)\right\}$$

$$-\sum_{i=1}^{k} n_i\left\{\log\left(\frac{2\sigma^2}{n_i}\right) + \log\left(\frac{n_i-1}{2}\right) - \frac{1}{n_i-1} - \frac{1}{3(n_i-1)^2} + 0\left(\frac{1}{n_i^3}\right)\right\}$$

$$\cong n\left\{\log\left(1-\frac{k}{n}\right) - \frac{1}{n-k} - \frac{1}{3(n-k)^2} + 0\left(\frac{1}{n^3}\right)\right\}$$

$$-\sum_{i=1}^{k} n_i\left\{\log\left(1-\frac{1}{n_i}\right) - \frac{1}{n_i-1} - \frac{1}{3(n_i-1)^2} + 0\left(\frac{1}{n_i^3}\right)\right\}$$

$$\cong (k-1) + \left[\left(\sum_{i=1}^{k}\frac{1}{(n_i-1)} - \frac{k}{n-k}\right) + \frac{1}{2}\left(\sum_{i=1}^{k}\frac{1}{n_i} - \frac{k^2}{n}\right)\right.$$

$$\left. + \frac{1}{3}\left\{\sum\frac{n_i}{(n_i-1)^2} - \frac{n}{(n-k)^2}\right\}\right] + 0\left(\frac{1}{N^3}\right)$$

Bartlett's modified criterion is to replace $n_i - 1$ by v_i, so that

$$E(-2\log\lambda^*) \cong (k-1) + \frac{1}{3}\left(\sum\frac{1}{v_i} - \frac{1}{v}\right)$$

Therefore,

$$\frac{-2\log\lambda^*}{1+\frac{1}{3(k-1)}\left(\sum\frac{1}{v_i}-\frac{1}{v}\right)}\xrightarrow{d}\chi^2_{k-1}.$$

PROBLEMS

10.1 Let X have Poisson distribution with parameter θ. Obtain SPRT (B, A) test for testing $H:\theta=\theta_0$ vs $K:\theta=\theta_1>\theta_0$ with strength, (α, β). Also obtain OC and ASN functions.

10.2 For testing $H:\theta=\theta_0$ vs $K:\theta=\theta_1>\theta_0$ with SPRT (B, A) of strength (α, β) and when X has the following distribution, obtain approximate values of A and B. Also obtain OC and ASN functions of them,

(a) $f(x;\theta)=(1-\theta)\theta^x,\ x=0,1,2,\ldots$

(b) $f(x;\theta)=\theta e^{-\theta x}, 0<x<\infty.$

10.3 For $N(\theta, 1)$ distribution, SPRT (B, A) with strength (α, β), obtain approximate values of b of B and a of A for testing $H:\theta=\theta_0$ vs $K:\theta=\theta_1>\theta_0$ and show that the power function is $\beta(\theta)=\frac{1-e^{2(c-\theta)b}}{e^{2(c-\theta)a}-e^{2(c-\theta)b}}$, where $c=\frac{\theta_0+\theta_1}{2}$ and show that average sample number function is $E_\theta(N)\cong\frac{b+\beta(\theta)(b-a)}{\theta-c}$.

10.4 For the gamma distribution with p.d.f.

$$f(x;\theta)=\frac{\theta^p}{\Gamma(p)}e^{-\theta x}x^{p-1}, 0<x<\infty$$

show that function $h(\theta)$ satisfies

$$\theta\left[\left(\frac{\theta_1}{\theta_0}\right)^{h(\theta)}-1\right]=(\theta_1-\theta_0)^{h(\theta)},$$

when p is known.

10.5 For the negative binomial distribution with p.f.

$$P(X=x)=\binom{x+r-1}{x}(1-\theta)^r\theta^x, x=0,1,2,\ldots$$

show that function $h(\theta)$ satisfies

$$\theta=\frac{c^{h(\theta)}-1}{c^{h(\theta)}-D^{h(\theta)}},$$

where, $c=\dfrac{1-\theta_1}{1-\theta_0}$, $D=\dfrac{\theta_1}{\theta_0}$, when r is known.

10.6 Given a random sample $X=(x_1,\ldots,x_n)$ from the following distributions, obtain tests for testing

(i) $H:\theta=\theta_0$ vs $K:\theta>\theta_0$

(ii) $H:\theta=\theta_0$ vs $K:\theta\neq\theta_0$

by using LRT principle. Also examine whether the tests are UMPU.

(a) $N(\mu,\theta)$ when μ is known

(b) $N(\mu,\theta)$ when μ is unknown

(c) $f(x;\theta)=\theta e^{-\theta x}, x>0, \theta>0$

(d) $f(x;\theta,\sigma)=\sigma e^{-\sigma(x-\theta)}, \theta<x<\infty$

10.7 Let $X_1, X_2,\ldots, X_m$ and $Y_1, Y_2,\ldots, Y_n$ be two independent random samples from $N(\mu_1,\sigma^2)$ and $N(\mu_2,\sigma^2)$, respectively. Obtain tests for testing $\mu_1\leq\mu_2$ vs $\mu_1>\mu_2$

(i) when σ in known

(ii) when σ in unknown, by using LRT principle.

10.8 Suppose $X=(X_1, X_2,\ldots, X_m)$ is a random sample from p.d.f.

$$f(x;\alpha)=\alpha e^{-\alpha x}, 0<x<\infty,$$

and $Y=(Y_1, Y_2,\ldots, Y_n)$ is an independent random sample from p.d.f.

$$f(x;\beta)=\beta e^{-\beta x}, 0<x<\infty.$$

By using LRT principle, test the hypothesis $H:\alpha=\beta$ vs $K:\alpha\neq\beta$.

10.9 Let $Y_{ij} \sim N(\alpha + \beta(x_i - \bar{x}), \sigma^2)$, $i = 1, 2, \ldots, k$, $j = 1, 2, \ldots, n_i$ be independent observations. By LRT method obtain the tests for testing

(i) $H: \beta = 0$ vs $K: \beta \neq 0$

(ii) $H: \beta = \beta_0$ vs $K: \beta \neq \beta_0$.

10.10 Suppose $Y_{ij} (i = 1, 2, \ldots, k, j = 1, 2, \ldots, n_i)$ are independent random variables having $N(\mu_i, \sigma^2)$ distribution. Then test the hypothesis $H: \mu_i = \alpha + \beta t_i^2$ by LRT principle, where, α, β are parameters and t_i are known constants.

10.11 Describe the method of testing the goodness of fit of distribution for the given data and by the use of LRT principle show that the test statistic is asymptotically χ^2 variate.

10.12 With the use of LRT method show that the test statistic for testing independence of two attributes follows χ^2 distribution for large n. Obtain exact test for 2×2 contingency table.

10.13 If $N_1, N_2, \ldots, N_k$ follows multinomial distribution with parameters $n, p_1, \ldots, p_k$ then for testing $H: p_i = p_{i0}, i = 1, 2, \ldots, k$ vs K any p_i by LRT principle show that the test statistic is

$$\lambda = n^n \prod_{i=1}^{k} \left(\frac{p_{i0}}{N_i} \right)^{N_i}$$

10.14 Let $\underline{X}_1, \underline{X}_2, \ldots, \underline{X}_n$ be a random sample from $N_p(\underline{\mu}, \Sigma)$. Then obtain LRT statistic for testing $\underline{\mu} = \underline{\mu}_0$,

(i) When Σ is known

(ii) When Σ is unknown.

11

CHAPTER

Non-Parametric Methods

11.1 INTRODUCTION

So far we have seen the methods of testing parameters of specified distribution, that is, the form of the distribution is known. If it is not known, then the problem is related to non-parametric methods. Whenever distribution is not specified the test is regarding the form of the distribution or the median of the distribution. In parametric methods the comparision of test is done by power functions whereas in non-parametric methods power function is not available and hence comparison between two non-parametric tests is not possible. This is the greatest drawback of the non-parametric methods; only large sample comparision is possible. In the parametric methods the test heavily depends on the parameter, it may not be robust whereas in non-parametric methods all the tests are robust, which is the greatest advantage of non-parametric methods.

In this chapter the following non-parametric methods have been discussed in sections (11.2) to (11.7) as under:

1. Sign test (section 11.2)
2. Median test (section 11.3)
3. Run test (section 11.4)
4. Mann-Whitney *U*-Test (Wilcoxon *T*-Test) (section 11.5)
5. Wilcoxon signed Rank test (section 11.6)
6. The empirical distribution function and Kolmogorov-Smirnov test. (section 11.7)

11.2 SIGN TEST

Let $X = (X_1, X_2, \ldots, X_n)$ be a random sample from a population F_X, with unknown median M. We wish to test the position of median, that is, of M when the distribution is symmetric. Thus, we wish to test

$H: M = M_0$ vs $K: M > M_0$ at level α. We have not assumed any form of the distribution of X and hence the problem becomes non-parametric.

If the hypothesis is true, on the average, half of the values will lie above M_0 and half of them below it. Let T be the number of values of X_i which are greater than M_0. Since T is actually the number of plus signs among n differences $(X_i - M_0)$, $i = 1, 2, \ldots, n$. When H is true, $P(X_i > M_0) = \frac{1}{2}$ and T has $b\left(n, \frac{1}{2}\right)$ distribution and when K is true, $P(X_i > M_0) > \frac{1}{2}$ and hence large value of T gives evidence that H is not true. So we reject H when $T \geq c$, where, c is such that

$$\sum_{k=c}^{n} \binom{n}{k} 2^{-n} \leq \alpha,$$

and the power function is

$$\beta(\theta) = \sum_{k=c}^{n} \binom{n}{k} \theta^k (1-\theta)^{n-k}, \text{ where, } \theta = P(X > M_0)$$

which can be calculated.

If n is small, c can be obtained from the table of binomial distribution, but if n is large one may use normal approximation.

$$\frac{T - \frac{n}{2}}{\sqrt{\frac{n}{4}}} \cong N(0,1) \Rightarrow \frac{c - \frac{n}{2}}{\frac{1}{2}\sqrt{n}} \cong Z_\alpha$$

$$\Rightarrow \qquad c \cong \frac{\sqrt{n} Z_\alpha}{2} + \frac{n}{2}.$$

where Z_α is upper α% point of $N(0, 1)$ distribution.

If the alternative is two-sided such as $K: M \neq M_0$ then we reject H when T is too small or too large, i.e., when $T \leq r$ or $T \geq (n - r)$, where r is obtained by the use of binomial distribution.

The confidence interval for M can be set. Let $X_{(1)} < X_{(2)} < \ldots < X_{(n)}$ be the corresponding order statistics. When $r < T < (n - r)$, we accept H, i.e., accept H when $X_{(r+1)} \leq M_0 \leq X_{(n-r)}$ and hence $(X_{(r+1)}, X_{(n-r)})$

is confidence set with confidence coefficient $\sum_{k=r+1}^{n-r-1}\binom{n}{k}2^{-n}$. For $n = 5$, $(X_{(1)}, X_{(5)})$ is confidence interval with confidence coefficient

$$\sum_{k=1}^{4}\binom{5}{k}2^{-5} = \frac{1}{32}(5+10+10+5) = \frac{30}{32} = \frac{15}{16}$$

and $(X_{(2)}, X_{(4)})$ with confidence coefficient

$$\sum_{k=2}^{3}\binom{5}{k}2^{-5} = \frac{20}{32} = \frac{5}{8}.$$

Example 11.1

For the following observations 7, 4.4, 4.9, 1.3, 2.6, 6.2, 3.5, 3.4, 2.1, 2.7 determine the sign test.

Let M be median. To test $H{:}M = 2$ vs $K{:}M > 2$. The number of observations, T larger than 2 is 9.

As $P(T \geq 9) = \dfrac{11}{2^{10}} = 0.0107$, we reject H at 0.1% level of significance.

Paired Sign Test

Sometimes paired observations are natural, e.g., twin children, the effect of fertilizers by taking adjacent plots. Let $(X_1, Y_1) \ldots (X_n, Y_n)$ be a random sample of n pairs and let $P(X_i, Y_i) = \theta$ for every i, but we don't know the distribution of X and Y. Thus, this question is non-parametric. We wish to test $H{:}\theta = \dfrac{1}{2}$ vs $K; \theta > \dfrac{1}{2}$. If T denotes the total number of $D_i = (Y_i - X_i) > 0$ then T follows binomial distribution with parameters $n, \dfrac{1}{2}$ under H. Reject H when $T \geq c$.

Example 11.2

For the following observations

$$X{:}6.2, 1.0, 3.4, 3.7, 3.4, 3.9, 1.3, 3.3, 2.7, 1.5$$

and

$$Y:7.0, 1.7, 4.4, 3.3, 4.9, 4.4, 2.2, 4.6, 2.6, 1.7,$$

obtain the paired sign test.

Here $T = 8$ and $P(T \geq 8) = \frac{56}{2^{10}} = 0.0547$ we reject H.

11.3 MEDIAN TEST

The Kolmogorov-Smirnov test defined later compares the proportion of observations from each sample which do not exceed some number x. The test criteria is the maximum difference (absolute) between the two distributions, instead of using all possible differences, if we choose only some specific number, say, median and compare only the proportion of observations which are less than median, we get a better test called median test.

Median test is constructed for testing the difference between the location of two distributions. Let $X = (X_1, \ldots, X_n)$ be a random sample from random variable X having p.d.f. f and let $Y = (Y_1, \ldots, Y_n)$ be a random sample from p.d.f. g; f and g may have different locations even though they are same in shape. We assume that $g(x) = f(x-\Delta)$ for some constant Δ.

We wish to test $H:\Delta = 0$ vs $K:\Delta \neq 0$. The procedure for median test is as follows:

If $\tilde{Z}$ is the median of the combined sample $(X_1, \ldots, X_m)$ and $(Y_1, \ldots, Y_n)$, then under H, $\frac{m}{2}$ observations of X must be larger than $\tilde{Z}$. If T is the total number of observations of X which are greater than $\tilde{Z}$ then we reject H when T is too large or too small.

We require the distribution of T under H.

If X, X, Y, Y, Y, X, Y, X, X. ...(11.3.1)

is the ordered combined sample of $X_1, \ldots, X_m, Y_1, \ldots, Y_n$ then there are $\binom{m+n}{m}$ ways of these arrangements. So each arrangement has probability $\frac{1}{\binom{m+n}{m}}$.

Case I Let $(m + n)$ be even and let $a = \dfrac{(m+n)}{2}$.

Then for the arrangement in (11.3.1), if $(m - t)$ are the numbers of X's in first 'a' observations and t are the numbers of X's in last 'a' observations, then

$$P_H(T=t) = \frac{\binom{a}{t}\binom{a}{m-t}}{\binom{m+n}{m}} = \frac{\binom{m}{t}\binom{n}{a-t}}{\binom{m+n}{a}}.$$

Case II Let $(m + n)$ be odd, that is, let $a = \dfrac{(m+n-1)}{2}$.

For the terms given in (11.3.1), let there be $(m - 1)$ observations of X's be from first $(a + 1)$ terms and let t observations of X's be in the last a terms. Then

$$P_H(T=t) = \frac{\binom{a+1}{m-t}\binom{a}{t}}{\binom{m+n}{m}} = \frac{\binom{m}{t}\binom{n}{a-t}}{\binom{m+n}{a}}.$$

For large values, following 2×2 contingency table is useful:

	X observation	***Y* observation**	**Total**
≤ Median	$m - t$	$n - a + t$	$m + n - a$
> Median	T	$a - t$	a
Total	M	n	$m + n$

If H: $\Delta > 0$, then reject H when T is too small.

Example 11.3

For the following observations

$$X: 60, 62, 63, 67, 68, 70, 73, 78$$

and

$$Y: 64, 71, 72, 75, 80, 81, 87.$$

Test $H: f(x) = g(y)$ by median test.

Here the median of these 15 observations is $Z = 71$. There are $T = 2$ observations of X which are larger than 71.

$$P_H(T \leq 2) = \frac{1 + 56 + 588}{6435} = \frac{645}{6435} = 0.1$$

Therefore, H is accepted at 10% level of significance.

11.4 RUN TEST

Persons purchasing ration are in a queue such as *MFMFMFMFMF*. Is this random? No, because there is intentional mixing. Similarly, *MMMMMFFFFF* is also not random. Randomness can be tested by the theory of runs.

Given an ordered sequence of two or more types of symbols, a run is defined as a succession of one or more identical symbols which are followed or preceded by a different symbol or no symbol at all, e.g., If the arrangement is *MMFFMMMF*, there are four runs.

Few runs, too many runs, a run of excessive length or too many runs of excessive length can be used as statistical criteria for rejection of the null hypothesis of randomness. This run test is also useful to test equality of two populations.

Let $X = (X_1, X_2, \ldots, X_m)$ be a random sample from p.d.f. f and let $Y = (Y_1, Y_2, \ldots, Y_n)$ be an independent random sample from p.d.f. g. We wish to test $H: f = g$ vs $K: f \neq g$. Let both samples be mixed and arranged in the ascending order of magnitude such as: *XYXXYYYXXYYYY* ... and let T be the number of runs. We argue as follows.

If H is true, then the observations are well mixed and T will be expected to be large, and if $f \neq g$, then T will be small due to ranges which are wide and different and if the location is different then the large run of Y at the end of the queue gives small number of runs. If the total number of runs is small then reject H.

To use this test, we need the distribution of T under H. Under H, each arrangement of X's in $(m + n)$ observations has probability $\frac{1}{\binom{m+n}{m}}$. For getting $P_H(T = t)$, we have to find the arrangements of t runs.

Case I Let t be even, say $t = 2k + 1$.

Then we shall have k runs of X and k runs of Y. Let $X_{(1)}, \ldots, X_{(m)}$ be ordered sample and to get k runs, we need to separate mX's , that is, $(m - 1)$ spaces between X's will be filled in by $(k - 1)$ separators in $\binom{m-1}{k-1}$ ways. Similarly, k runs of Y's will be done in $\binom{n-1}{k-1}$ ways.

The total number of runs starting with X is $\binom{m-1}{k-1}\binom{n-1}{k-1}$. Similarly, starting with a run of Y, the total number of runs is $\binom{m-1}{k-1}\binom{n-1}{k-1}$.

Thus, the total number of arrangement is $2\binom{m-1}{k-1}\binom{n-1}{k-1}$ and hence

$$P_H(T = 2k) = \frac{2\binom{m-1}{k-1}\binom{n-1}{k-1}}{\binom{m+n}{m}}.$$

Case II Let t be odd, that is, $t = 2k + 1$.

Then there will be k runs of X and $(k + 1)$ runs of Y, or $(k + 1)$ runs of X and k runs of Y. So the number of arrangements will be

$$\binom{m-1}{k-1}\binom{n-1}{k} + \binom{m-1}{k}\binom{n-1}{k-1}$$

and hence,

$$P_H(T = 2k+1) = \frac{\binom{m-1}{k-1}\binom{n-1}{k} + \binom{m-1}{k}\binom{n-1}{k-1}}{\binom{m+n}{m}}.$$

Moments of the distribution of T

Let $T = I_1 + I_2 + I_3 + \ldots + I_{m+n}$,

where

$$I_k = \begin{cases} 1 & \text{if } k^{\text{th}} \text{ element} \neq (k-1)^{st} \text{ element} \\ 0, & \text{otherwise} \end{cases}$$

It is Bernoulli random variable with parameter

$$p = \frac{mn}{\binom{m+n}{2}},$$

so that, $$E(I_k) = E(I_k^2) = \frac{2mn}{(m+n)(m+n-1)}.$$

Since T is a linear combination of I_k, we have

$$E(T) = 1 + \sum_{k=2}^{m+n} E(I_k) = 1 + \sum_{k=2}^{m+n} \frac{2mn}{(m+n)(m+n-1)} = 1 + \frac{2mn}{m+n},$$

$$\text{var}(T) = \text{var}\left(\sum_{k=2}^{m+n} I_k\right) = (m+n-1)\,\text{var}(I_k) + \sum_{2 \le j \neq k \le m+n} \sum \text{cov}(I_k, I_j)$$

$$= (m+n-1)E(I_k^2) - (m+n-1)[E(I_k)]^2 + \sum_{j \neq k}\sum E(I_k I_j).$$

To evaluate the $(m + n - 1)(m + n - 2)$ joint moments of the type $E(I_k I_j)$ for $j \neq k$, the subscript choices are

(1) for the $2(m+n-2)$ selections, where $j = k-1$ or $j = k+1$

$$E(I_k I_j) = \frac{mn(m-1) + mn(n-1)}{(m+n)(m+n-1)(m+n-2)} = \frac{mn}{(m+n)(m+n-1)}.$$

(2) for the remaining $(m+n-1)(m+n-2) - 2(m+n-2)$ selections of $j \neq k$ that is for $(m+n-2)(m+n-3)$ selections

$$E(I_k I_j) = \frac{mn(m-1)(n-1) + \ldots + 4 \text{ times}}{(m+n)(m+n-1)(m+n-2)(m+n-3)}$$

$$= \frac{4mn(m-1)(n-1)}{(m+n)(m+n-1)(m+n-2)(m+n-3)}.$$

As $(m + n) \to \infty$ such that,

$$\frac{m}{m+n} = \lambda \text{ and } \frac{n}{m+n} = 1 - \lambda$$

remain constant, then

$$\lim_{(m+n)\to\infty} E\left(\frac{T}{m+n}\right) = 2\lambda(1-\lambda)$$

$$\lim_{(m+n)\to\infty} \text{var}\left(\frac{T}{\sqrt{m+n}}\right) = 4\lambda^2(1-\lambda)^2.$$

so that,

$$Z = \frac{T - 2(m+n)\lambda(1-\lambda)}{2\sqrt{m+n}\,\lambda(1-\lambda)} \xrightarrow{d} N(0, 1)$$

We may use median test but it is sensitive only to location and it will fail for scale but the run test detects the differences in location and scale both.

Example 11.4

For the following observations

$$X: 60, 62, 63, 67, 68, 70, 73, 78$$

and

$$Y: 64, 71, 72, 75, 80, 81, 87$$

the ordered set will be *XXXYXXXYYXYXYYY*. Test randomness of this queue by run test at 5% level.

Here the number of runs, $T = 8$.

$$P(T \le 8): m = 8 \text{ and } \lambda = \frac{8}{15}$$

$$Z = \frac{T - 2(m+n)\lambda(1-\lambda)}{2\sqrt{m+n}\ \lambda(1-\lambda)}$$

$$= \frac{8 - 2(15)\frac{8}{15}\frac{7}{15}}{2\sqrt{15}\frac{8}{15}\frac{7}{15}}$$

$$= \frac{8 - \frac{112}{15}}{\frac{112}{15\sqrt{15}}} = \frac{\sqrt{15}}{14} = 0.2766 < 1.96$$

We accept H.

11.5 MANN-WHITNEY *U* TEST (WILCOXON *T* – TEST)

For testing the difference between the locations of two distributions, Mann-Whitney test is used. If $X = (X_1, X_2, \ldots, X_m)$ and $Y = (Y_1, Y_2, \ldots, Y_n)$ are two independent random samples from the distributions F and G with same shape but different locations in their p.d.f. such as $g(x) = f(x - \Delta)$ for some constant Δ. To test $H: \Delta = 0$ vs $K: \Delta \neq 0$.

Combine $(m + n)$ observations and arrange them in ascending order of magnitude assign rank to X observations such as in a combined sample $YXXXYYX$, the ranks S_i of X_i observations are $S_1 = 2, S_2 = 3, S_3 = 4,\ S_4 = 7$. If H is not true the probability that X observations are larger than Y observations will be high and hence ranks of X observations will be large.

If $T_X = \sum_{i=1}^{m} S_i$ (called Wilcoxon-T statistic) is large, we would like to reject H.

If U_X denotes the total number of (ij) pairs in which $Y_j < X_i$, $i = 1, 2, \ldots, m$ and $j = 1, 2, \ldots, n$ then U_X is called the U statistic of Mann-Whitney. If X_i has rank S_i then there are $S_i - 1$ observations left to X_i, out of which there are $(i - 1)$ observations of X and there are $(S_i - 1)$ observations of Y, so that

$$U_X = \sum_{i=1}^{m}(S_i - i) = \sum_{i=1}^{m} S_i - \frac{m(m+1)}{2} = T_X - \frac{m(m+1)}{2}.$$

Thus,

$$T_X = U_X + \frac{m(m+1)}{2}. \text{ Similarly, } T_Y = U_Y + \frac{n(n+1)}{2}, U_X + U_Y = mn.$$

Large value of U_X discredits H.

Consistency of *U*-Statistic and its Asymptotic Distribution

Let us introduce indicator variables

$$D_{ij} = \begin{cases} 1 & \text{if} \quad Y_j < X_i, i = 1, 2, \ldots, m, j = 1, 2, \ldots, n \\ 0 & \text{otherwise} \end{cases}.$$

Then Mann-Whitney U-statistic will be $U = \sum_{i=1}^{m}\sum_{j=1}^{n} D_{ij}$. Let

$$\pi = P(Y < X)$$

$$= \int_{-\infty}^{\infty}\int_{-\infty}^{x} f_Y(y) f_X(x) dy dx$$

$$= \int_{-\infty}^{\infty} F_Y(x) f_X(x) dx$$

$$= \int_{-\infty}^{\infty} F_X(x) f_X(x) dx$$

when $H: F_X(x) = F_Y(x) \; \forall \; x$ is true

$$= \frac{F^2(\infty) - F^2(-\infty)}{2} = \frac{1}{2}.$$

Also random variables D_{ij} have Bernoulli distribution with probability of success (π) and hence,

$$E(D_{ij}) = E(D_{ij}^2) = \pi, \text{ var}(D_{ij}) = \pi(1-\pi)$$

$$\text{cov}(D_{ij}, D_{hk}) = 0, \text{ for } i \neq h, j \neq k$$

$$\text{cov}(D_{ij}, D_{ik}) = \pi_1 - \pi^2, \text{ for } j \neq k$$

$$\text{cov}(D_{ij}, D_{hj}) = \pi_2 - \pi^2, \text{ for } i \neq h$$

where

$$\pi_1 = P[(Y_j < Y_k < X_i) \cup (Y_k < Y_j < X_i)]$$

$$= P(Y_j < Y_k < X_i) + P(Y_k < Y_j < X_i)$$

$$= \int_{-\infty}^{\infty} \int_{-\infty}^{x_i} \int_{-\infty}^{x_k} f_Y(y_j) f_Y(y_k) f_X(x_i) dy_j \, dy_k \, dx_i$$

$$+ \int_{-\infty}^{\infty} \int_{-\infty}^{x_i} \int_{-\infty}^{y_j} f_Y(y_k) f_Y(y_j) f_X(x_j) dy_k \, dy_j \, dx_i$$

$$= 2\int_{-\infty}^{\infty} \left[\int_{-\infty}^{x_i} F_Y(y_k) f_Y(y_k) dy_k \right] f_X(x_i) dx_i$$

$$= 2\int_{-\infty}^{\infty} \frac{[F_Y(x_i)]^2}{2} f_X(x_i) dx_i$$

$$= \int_{-\infty}^{\infty} [F_Y(x)]^2 f_X(x) dx.$$

Similarly,

$$\pi_2 = P[(X_i > Y_j) \cap (X_h > Y_j)] = \int_{-\infty}^{\infty} [1 - F_X(y)]^2 f_Y(y) dy.$$

Since U is defined as a linear combination of mn random variables, D_{ij}

$$EU = \sum_i \sum_j E(D_{ij}) = mn\pi.$$

$$\text{var}(U) = \sum_i \sum_j \text{var}(D_{ij}) + \sum_{i=1}^{m} \sum_{j \neq k} \sum \text{cov}(D_{ij}, D_{ik}) + \sum_{i=1}^{m} \sum_{j \neq h} \sum \text{cov}(D_{ij}, D_{hj})$$

$$+ \sum_{i \neq h} \sum \sum_{j \neq k} \sum \text{cov}(D_{ij}, D_{hk})$$

$$= mn\pi(1-\pi) + mn(n-1)(\pi_1 - \pi^2) + mn(m-1)(\pi_2 - \pi^2),$$

by using above results.

Thus, $$E\left(\frac{U}{mn}\right) = \pi$$

and

$$\text{var}\left(\frac{U}{mn}\right) = \frac{\pi(1-\pi)}{mn} + \frac{(n-1)}{mn}(\pi_1 - \pi^2) + \frac{(m-1)}{mn}(\pi_2 - \pi^2)$$

$$\to 0 \text{ as } m \to \infty, n \to \infty.$$

which proves that $\frac{U}{mn}$ is consistent for π. When H is true $F_Y(x) = F_X(x)$ and hence,

$$\pi_1 = \int_{-\infty}^{\infty} [F_X(x)]^2 f_X(x)dx = \left[\frac{[F_X(x)]^3}{3}\right]_{-\infty}^{\infty} = \frac{1}{3}$$

and

$$\pi_2 = \int_{-\infty}^{\infty} [1 - F_X(x)]^2 f_X(x)dx = \left[-\frac{[1 - F_X(x)]^3}{3}\right]_{-\infty}^{\infty} = \frac{1}{3}, \pi = \frac{1}{2}.$$

Using these results, under H,

$$EU = \frac{mn}{2}, \text{var}(U) = \frac{mn}{4} + \frac{mn(n-1)}{12} + \frac{mn(n-1)}{12} = \frac{mn(m+n-1)}{12}.$$

We shall use asymptotic distribution of U when m and n are large such as

$$Z = \frac{\left|U - \frac{mn}{2}\right|}{\sqrt{\frac{mn(m+n-1)}{12}}} \xrightarrow{d} N(0, 1).$$

If $Z > 1.96$, reject H at 5% level of significance.

Example 11.5

For the following observations

$$X: 52, 54, 55, 60, 61, 62 \quad \text{and} \quad Y: 56, 63, 64, 67, 71.$$

Obtain Mann-Whitney U statistic and test $H: f(x) = g(y)$ at 5% level.

The combined sample is (the underlined are Y values)

$$52, 54, 55, \underline{56}, 60, 61, 62, \underline{63}, \underline{64}, \underline{67}, \underline{71}$$

$$U_X = \text{Total number of } Y \text{ Values less than } X \text{ values} = 3$$

$$Z = \frac{|U_X - 15|}{\sqrt{30 \times \frac{12}{12}}} = \frac{|3 - 15|}{\sqrt{30}} = 2.01 > 1.96 \text{ m, Reject } H$$

or

Y ranks are $4, 8, 9, 10, 11$ so that $T_y = 42$. Also

$T_y = 42$ will have arrangements $\{5, 7, 9, 10, 11\}, \{6, 7, 8, 10, 11\}$

$T_y = 43$ have arrangements $\{5, 8, 9, 10, 11\}, \{6, 7, 9, 10, 11\}$,

$T_y = 44$ have arrangements $\{6, 8, 9, 10, 11\}$,

$T_y = 45$ have arrangements $\{7, 8, 9, 10, 11\}$.

Thus, in all we have 7 possible arrangements for $T_y \geq 42$ and hence

$$P_H\{T_y \geq 42\} = \frac{7}{\binom{11}{5}} = 0.015 < 0.020. \text{ Reject } H \text{ at } \alpha = 0.02.$$

Example 11.6

For the following observations

$$X{:}60, 62, 63, 67, 68, 70, 73, 78 \quad \text{and} \quad Y{:}64, 71, 72, 75, 80, 81, 87.$$

Obtain Mann-Whitney U statistic and test $H: f(x) = g(y)$ at 5% level.

The ordered combined sample is

$$60, 62, 63, \underline{64}, 67, 68, 70, \underline{71}, \underline{72}, 73, \underline{75}, 78, \underline{80}, \underline{81}, \underline{87}.$$

$$U_X = 10 \text{ and } Z = \frac{\left|U - \frac{mn}{2}\right|}{\sqrt{\frac{mn(m+n-1)}{12}}} = \frac{|10-28|}{\sqrt{\frac{56*16}{12}}} = 1.2 < 1.96$$

and hence reject K and accept H at $\alpha = 0.05$ level.

11.6 WILCOXON SIGNED RANK TEST

The ordinary sign test utilizes only signs but not magnitudes. Wilcoxon signed rank test utilizes both. If $(X_i, Y_i), i = 1, 2, \ldots, n$ are n pairs of observations from two distributions F and G and we wish to test $H{:}F = G$.

Let $Z_i = X_i - Y_i$ and $|Z_1| < |Z_2| < \ldots < |Z_n|$ be ordered sample for the absolute values of $Z_1, Z_2, \ldots, Z_n$. Retain the signs of Z_i in $|Z_i|$. Let T^+ be total number of positive Z_i, e.g., if $6, -3, -5, 1, -4$ are the values of Z_i, the ranks of positive Z_i are 1 and 5, so that $T^+ = 6$. If H is not true then the probability that Z_i is positive is larger than it is negative. So reject H for large value of T^+. Similarly, define T^-. Note that

$$T^+ + T^- = \sum_{i=1}^{n} i = \frac{n(n+1)}{2}$$

and $$T^+ = \sum_{i=1}^{n} D_i r\,|Z_i|,\; T^- = \sum_{i=1}^{n} (i - D_i) r\,|Z_i|.$$

Therefore,

$$T^{+} - T^{-} = 2\sum_{i=1}^{n} D_i r|Z_i| - \frac{n(n+1)}{2},$$

where

$$D_i = \begin{cases} 1 & \text{if} \quad Z_i > 0 \\ 0 & \text{if} \quad Z_i < 0 \end{cases}, \; r|Z_i| = \text{rank of } |Z_i|.$$

Let $Z_{(1)} < Z_{(2)} < \ldots < Z_{(n)}$ be ordered sample for the differences $Z_i = X_i - Y_i$, $i = 1, 2, \ldots, n$, where,

$$Z_{(i)} = \begin{cases} 1 & \text{if} \quad Z_i \text{ is positive and has rank } i \\ 0 & \text{otherwise} \end{cases}.$$

Then

$$T^{+} = \sum_{i=1}^{n} iZ_{(i)},\; E(Z_{(i)}) = \theta_i,\; \text{var}(Z_{(i)}) = \theta_i(1-\theta_i)\sum_{i=1}^{n} i^2$$

where

$$\theta_i = P(Z_{(i)} = 1) = P\{i^{\text{th}} \text{ ordered statistic } Z_{(i)} \text{ has positive value}\}$$

$$= \int_0^{\infty} \frac{n!}{(i-1)!(n-i)!} [F_{|Z_{(i)}|}(z)]^{i-1} [1 - F_{|Z_{(i)}|}(z)]^{n-i} f_{|Z_{(i)}|}(z) dz$$

$$= n\binom{n-1}{i-1} \int_0^{\infty} [F_Z(z) - F_Z(-z)]^{i-1} [1 - F_Z(z) + F_Z(-z)]^{n-i} f_Z(z) dz$$

$$= n\binom{n-1}{i-1} \int_0^{\infty} [2F_Z(z) - 1]^{i-1} [2 - 2F_Z(z)]^{n-i} f_Z(z) dz,$$

due to symmetry

$$= \frac{n}{2}\binom{n-1}{i-1} \int_0^{1} V^{i-1}(1-V)^{n-i} dv$$

$$= \frac{n}{2}\binom{n-1}{i-1} B(i, n-i+1) = \frac{1}{2}.$$

So

$$E(T^+) = \sum_{i=1}^{n} i\theta_i = \frac{n(n+1)}{4}$$

and

$$\text{var}(T^+) = \sum_{i=1}^{n} \theta_i(1-\theta_i)i^2 = \frac{n(n+1)(2n+1)}{24}$$

$$W = \frac{\left(T^+ - \frac{n(n+1)}{4}\right)}{\sqrt{\frac{n(n+1)(2n+1)}{24}}} \xrightarrow{d} N(0, 1).$$

Example 11.7

For the following observations

$$X: 7.0, 1.7, 4.4, 3.3, 4.9, 4.4, 2.2, 4.6, 2.6, 1.7$$
$$Y: 6.2, 1.0, 3.4, 3.7, 3.4, 3.9, 1.3, 3.3, 2.7, 1.5$$

Use Wilcoxon sign rank test to test $H: f(x) = g(y)$.

Let the difference be $Z = X - Y$

$$Z = 0.8, 0.7, 1.0, -0.4, 1.5, 0.5, 0.9, 1.3, -0.1, 0.2$$

and Z_i in ascending order of magnitude ignoring the sign, i.e., $|Z_i|$ the sample is

$$-0.1, 0.2, -0.4, 0.5, 0.7, 0.8, 0.9, 1.0, 1.3, 1.5.$$

Thus,

$$T^+ = 2+4+5+6+7+8+9+10 = 51$$
$$T^- = 1+3 = 4.$$

Therefore,

$$W = \frac{\left(T^+ - \frac{n(n+1)}{4}\right)}{\sqrt{\frac{n(n+1)(2n+1)}{24}}} = \frac{2[51-27.5]}{\sqrt{55\times 7}} = 2.4 > 1.96$$

So we reject H at 5% level.

11.7 EMPIRICAL DISTRIBUTION FUNCTION

In χ^2 goodness of fit test only $k \leq n$ comparisons are made. Instead of it, we wish to have comparison for all categories, comparison between observed and expected cumulative relative frequencies is done. The sample cumulative distribution function is called the empirical distribution function and it is an estimate of population c.d.f. The tests are based on functions of deviations (S.S. of deviations, absolute deviation, maximum deviation).

The empirical distribution of a random sample of size n denoted by $S_n(x)$ on real x is the proportion of sample values which do not exceed the number x. Thus, $S_n(x)$ is a step function with a jump $\frac{1}{n}$. Let $X_{(1)}, X_{(2)}, \ldots, X_{(n)}$ be order statistics of $X_1, X_2, \ldots, X_n$. Then

$$S_n(x) = \begin{cases} 0 & \text{if} \quad x < X_1 \\ \frac{k}{n} & \text{if} \quad X_{(k)} < x < X_{(k+1)}; k = 1, 2, \ldots, n-1. \\ 1 & \text{if} \quad x \geq X_{(n)} \end{cases}$$

Theorem 11.1 If $S_n(x)$ is empirical distribution of $X_1, X_2, \ldots, X_n$. Then

$$P\left[S_n(x) = \frac{j}{n}\right] = \binom{n}{j} [F_X(x)]^j [1 - F_X(x)]^{n-j}, \ j = 0, 1, 2, \ldots, n.$$

Proof: Define

$$\delta_i(t) = \begin{cases} 1 & \text{if} \quad X_i \leq t \\ 0 & \text{otherwise} \end{cases}.$$

Then $\delta_1(t), \ldots, \delta_n(t)$ constitute a set of n independent random variables from the Bernoulli distribution with parameter θ, where, $\theta = P\{\delta_i(t) = 1\} = P\{X_i \leq t\} = F_X(t)$.

As

$$S_n(x) = \sum_{i=1}^{n} \frac{\delta_i(x)}{n},$$

the random variable $nS_n(x)$ is the sum of n independent Bernoulli random variables which follows the binomial distribution with parameter $\theta = F_X(x)$.

Remarks

(i) $ES_n(x) = F_X(x)$, $\text{var}(S_n(x)) = \dfrac{F_X(x)[1 - F_X(x)]}{n}$

(ii) $S_n(x) \xrightarrow{p} F_X(x)$

Glivenko-Cantelli Theorem

$S_n(x)$ converge uniformly to $F_X(x)$, that is $\forall \in > 0$,

$$\lim_{n\to\infty} P\left\{ \sup_{-\infty < x < \infty} |S_n(x) - F_X(x)| > \in \right\} = 0.$$

Theorem 11.2

$$P\left\{ \frac{\sqrt{n}\,[S_n(x) - F_X(x)]}{\sqrt{F_X(x)[1 - F_X(x)]}} \le t \right\} = \phi(t),$$

where

$$\phi(t) = \int_{-\infty}^{t} \frac{e^{-\frac{x^2}{2}}}{\sqrt{2\pi}}\, dx.$$

Proof: From central limit theorem and previous theorem, the limiting distribution of

$$\frac{nS_n(x) - nF_X(x)}{\sqrt{nF_X(x)[1 - F_X(x)]}} = \frac{\sqrt{n}\,[S_n(x) - F_X(x)]}{\sqrt{F_X(x)[1 - F_X(x)]}}$$

is standard normal.

The Kolmogorov-Smirnov One Sample Statistic

A random sample $X_1, X_2, \ldots, X_n$ is taken from unknown distribution function $F_X(x)$. An empirical statistic $S_n(x)$ is the point estimate of $F_X(x)$. The deviation between true function and its image, $|S_n(x) - F_X(x)|$ should be small $\forall\ x$. Thus, the statistic $D_n = \sup |S_n(x) - F_X(x)|$ is a reasonable measure for our estimate. D_n is called the Kolmogorov-Smirnov one sample statistic, which is non-parametric as the distribution of D_n does not depend on $F_X(x)$.

One sided Kolmogorov-Smirnov statistics are

$$D_n^+ = \sup_x [S_n(x) - F_X(x)],\ D_n^- = \sup_x [F_X(x) - S_n(x)].$$

Theorem 11.3 The statistics D_n, D_n^+, D_n^- are distribution free for any continuous F_X.

Proof:

$$D_n = \sup_x |S_n(x) - F_X(x)| = \max_x (D_n^+, D_n^-).$$

Let $X_{(0)} = -\infty$, $X_{(n+1)} = \infty$. Then

$$D_n^+ = \sup_x [S_n(x) - F_X(x)]$$

$$= \max_{0 \le i \le n} \sup_{X_{(i)} < x < X_{(i+1)}} \left[\frac{i}{n} - F_X(x)\right]$$

$$= \max_{0 \le i \le n} \left[\frac{i}{n} - \left(\inf_{X_{(i)} < x < X_{(i+1)}} F_X(x)\right)\right]$$

$$= \max_{0 \le i \le n} \left[\frac{i}{n} - F_X(X_{(i)})\right]$$

$$= \max\left\{\max_{0 \le i \le n} \left[\frac{i}{n} - F_X(X_{(i)})\right], 0\right\}.$$

Similarly,

$$D_n^- = \max\left\{\max_{0 \le i \le n} \left[F_X(X_{(i)}) - \frac{i-1}{n}\right], 0\right\},$$

$$D_n = \max\left\{0, \max\left[\frac{i}{n} - F_X(X_{(i)})\right], \max\left[F_X(X_{(i)}) - \frac{i-1}{n}\right]\right\}.$$

Thus, D_n, D_n^+, D_n^- depend only on $F_X(X_{(i)})$ which has $u(0, 1)$ distribution and hence they all are distribution free.

For large samples, Kolmogorov gave the following approximation to the sampling distribution of D_n.

If F_X is any continuous distribution function, then for every $Z \geq 0$

$$\lim_{n\to\infty} P\left(D_n \leq \frac{Z}{\sqrt{n}}\right) = L(Z),$$

where

$$L(Z) = 1 - 2\Sigma(-t)^{i-1} e^{-2i^2 Z^2}$$

and $$\lim_{n\to\infty} P\left(D_n^+ < \frac{Z}{\sqrt{n}}\right) = 1 - e^{-2Z^2},$$

$$\lim_{n\to\infty} (4n(D_n^+)^2 < 4Z^2) = 1 - e^{-\frac{Z^2}{2}} \sim \chi_2^2$$

Application of the *K-S* Statistic

(i) To test $H: F_X(x) = F_0(x)\ \forall\ x$ vs $K: F_X(x) \neq F_0(x)$, reject H when $D_n > D_{n,\alpha}$ and reject H for $K: F_X(x) > F_0(x)$ when $D_n^+ > D_{n,\alpha}^+$.

(ii) To estimate F_X, one finds the minimum sample size n, such that $P\{D_n < \epsilon\} = 1 - \alpha, \epsilon = D_{n,\alpha}$ is given. Solve $Z_\alpha = \epsilon\sqrt{n}$.

(iii) To find confidence bounds, solve $P(D_n > D_{n,\alpha}) = \alpha$

$\Leftrightarrow P\{\sup |S_n(x) - F_X(x)| < D_{n,\alpha}\} = 1 - \alpha$

$\Leftrightarrow P\{S_n(x) - D_{n,\alpha} < F_X(x) < D_{n,\alpha} + S_n(x)\} = 1 - \alpha$

Comparison of χ^2 and *K-S* Test

Chi-square test requires grouped data whereas *K-S* test does not require grouped data, which is the advantage of *K-S* test. Chi-square test uses only partial data and it is affected by the number of classes and their widths whereas *K-S* test uses complete data. Chi-square test is sensitive to deviations between observed and expected frequencies whereas *K-S* test is sensitive to observed and expected cumulative frequencies.

The sampling distribution of chi-square test of g is approximated, so it is used for large n only. When classes are combined, chi-square is affected by the ways of combination whereas *K-S* test, for finite n, exact distribution of D_n is available and hence it is used for any sample size. Chi-square test provides only test whereas *K-S* test provides not only test but it also estimates sample size and confidence bounds.

In chi-square test the alternative is only two sided whereas in *K-S* test the alternative is flexible; two sided and one sided. Chi-square test is not easily applicable whereas *K-S* test is easily applicable in all cases.

For chi-square test discrete hypothesized distribution has no problem whereas in *K-S* test lack of continuity will have problem in application.

In chi-square test even when the hypothesized distribution is not completely specified, the parameters are estimated and test can be done whereas in *K-S* test substituting the estimates of unspecified parameters in H which will give statistic $\hat{D}_n$, whose distribution cannot be obtained.

The approximate power function of chi-square test is available whereas exact power function of *K-S* test is not available. Chi-square test is advisable for large n whereas *K-S* test is advisable for small n.

The *K-S* Two-Sample Test

Let $X_{(1)}, X_{(2)}, \ldots, X_{(m)}$ and $Y_{(1)}, Y_{(2)}, \ldots, Y_{(n)}$ be two observed samples from $F_X(x)$ and $F_Y(y)$. Then the empirical distribution function $S_m(x)$ and $T_n(x)$ are defined as

$$S_m(x) = \begin{cases} 0 & \text{if} \quad x < X_{(1)} \\ \dfrac{k}{m} & \text{if} \quad X_{(k)} < x < X_{(k+1)};\, k = 1, 2, \ldots, m-1 \\ 1 & \text{if} \quad x \geq X_{(m)} \end{cases}$$

$$T_n(x) = \begin{cases} 0 & \text{if} \quad x < Y_{(1)} \\ \dfrac{k}{n} & \text{if} \quad Y_{(k)} < x < Y_{(k+1)};\, k = 1, 2, \ldots, n-1. \\ 1 & \text{if} \quad x \geq Y_{(n)} \end{cases}$$

In the combined ordered arrangements of $m + n$, $S_m(x)$ and $T_n(x)$ are the respective proportions of X and Y which do not exceed x. If the null hypothesis

$$H: F_Y(x) = F_X(x) \ \forall \ x,$$

is true, we have two samples from the same population. The two-sided *K-S* statistic to test H against

$$K: F_Y(x) \neq F_X(x) \ \forall \ x, \text{ is}$$

$$D_{m,n} = \max_x |S_m(x) - T_n(x)|$$

and it rejects H when

$$D_{m,n} \geq c_\alpha$$

$D_{m,n}$ is completely distribution free as each one of $F_Y(x)$ and $F_X(x)$ is $u(0, 1)$. The exact distribution is not so easy but the asymptotic distribution is

$$\lim_{m,n\to\infty} P\left(\sqrt{\frac{mn}{m+n}} D_{m,n} \leq Z\right) = 1 - 2\Sigma(-1)^{i-1} e^{-\frac{2i^2}{Z^2}}.$$

Reject one-sided hypothesis $H^+: F_Y(x) = F_X(x)$ when $D^+_{m,n} \geq c_\alpha$ where

$$D^+_{m,n} = \max_x [S_m(x) - T_n(x)],$$

and

$$\lim_{m,n\to\infty} P\left(\sqrt{\frac{mn}{m+n}} D^+_{m,n} \leq Z\right) = 1 - e^{-2Z^2}.$$

PROBLEMS

11.1 Following 8 independent observations are taken from some distribution with median M. 4.2, 6.5, 2.9, 5.6, 6.7, 3.9, 7.1, 4.8. Test the hypothesis $H: M = 4.5$ vs $K: M > 4.5$.

11.2 The following information is on 16 rats for the increase in weight due to nutritive food and usual food.

Usual food: 50, 72, 49, 65, 39, 43, 35, 52

Nutritive food: 53, 46, 40, 70, 42, 35, 61, 74.

Using sign test show that the increase is same at 5% level. Also compare it with t-test.

11.3 Obtain 10 independent observations from two uniform distributions

$$f(x) = 1, 0 < x, 1$$

$$g(x) = 4, 0 < x < \frac{1}{4}$$

by using random number table and test the hypothesis $H: f(x) = g(x)$ with appropriate non-parametric test.

11.4 The following observations are taken from X and Y random variables having p.d.f. $f(x)$ and $g(x) = f(x - \Delta)$, respectively:

$$X = 40, 45, 43, 52, 53, 58$$

$$Y = 35, 42, 46, 48, 50, 54, 60, 62.$$

Test the hypothesis $H = \Delta = 0$ vs $\Delta > 0$ by using

(i) Median test (ii) Run test.

11.5 By using Mann-Whitney test at 5% level, test whether both samples have come from the same distribution

$$X: 2.3, 2.4, 2.4, 2.5, 2.7, 2.9, 2.9$$

$$Y: 2.6, 2.8, 3.0, 3.1, 3.1, 3.2, 3.3$$

11.6 The weights of 15 male cats and 12 female cats are given below.

Male: 13.7, 9.3, 15.6, 11.7, 9.1, 9.6, 12.8, 13.6, 11.2, 14.9, 10.7, 8.3, 8.0, 7.6, 9.4

Female: 9.5, 10.1, 10.2, 10.1, 7.4, 7.3, 9.0, 7.6, 9.5, 8.2, 8.7, 7.2

Use Wilcoxon sign test to test equality of weights and compare it with *t*-test.

11.7 To measure the effect of oxygen on nervous system 11 cats and 10 rabbits are selected. The observations are given below

Cat: 45, 43, 33, 25, 26, 28, 29, 24, 35, 36, 37

Rabbit: 23, 27, 30, 44, 38, 42, 41, 40, 39, 34.

By using Median test, Run test, Mann-Whitney test, at 5% level test whether both types of observations are coming from the same population. Compare them with Kolmogorov-Smirnov test.

11.8 Eight pairs of students are selected for efficiency of English teaching by two methods. The marks obtained in both methods are given below.

Method 1: 80, 66, 72, 69, 39, 60, 56, 71

Method 2: 62, 63, 76, 72, 33, 53, 43, 72.

Test whether both methods have same effect by appropriate non-parametric test. Test whether there is difference in efficiency by adopting different teaching methods using Wilcoxon sign rank test.

12

CHAPTER

UMP Tests for Truncated Distributions

12.1 INTRODUCTION

In this chapter we will discuss the testing of parameters for the class of truncated distributions (non-regular) which depends on parameter or parameters in the range, such as $U(0, \theta)$ – right truncated. Pareto distribution – left truncated. The UMP test for testing the parameter $H: \theta = 0$, which a is two-sided hypothesis and is available in many textbooks like Lehmann (2005) but the testing of the whole class of family of distributions has not been studied so far. So we will study it in this chapter. We derive the UMP test for testing two-sided hypothesis for the class of truncated one-parameter family of distributions and its power function in section 12.2 whereas its applications to some specific distributions like generalized uniform, uniform, power function, negative exponential, pareto distributions are dealt with in section 12.3.

12.2 UMP TESTS

We use the following set-up for one-parameter family of distributions.

Let X be continuous random variable with probability density function (p.d.f.) as

$$f_X(x;\theta_1) = \begin{cases} g(\theta_1)h(x), & a < x < \theta_1 < b \\ 0, & \text{otherwise} \end{cases} \quad \ldots(12.2.1)$$

or

$$f_X(x;\theta_2) = \begin{cases} g(\theta_2)h(x), & a < \theta_2 < x < b \\ 0, & \text{otherwise} \end{cases} \quad \ldots(12.2.2)$$

where a and b are known constants, $g(\theta_i), i=1,2$ is everywhere differentiable function, $h(x)$ is positive absolutely continuous function of x. The function $f_X(x;\theta_1)$ is called right-truncated function of x and $f_X(x;\theta_2)$ is called left-truncated function of x.

Let $X_1, X_2, \ldots, X_n$ be a random sample from X and $X_{(1)}, X_{(2)}, \ldots, X_{(n)}$ be its corresponding order statistics.

We shall use the following result for (12.2.1) and (12.2.2).

As
$$h(x)=\frac{d}{dx}\int_a^x h(u)du=\frac{d}{dx}[g(x)]^{-1},$$

$$h(x)=-g'(x)[g(x)]^{-2} \qquad \ldots(12.2.3)$$

Theorem 12.1: With the above set-up, given a random sample from right-truncated distribution, the UMP level-α test, $\phi_1(x)$ for testing $H_1:\theta_1=\theta_{10}$ vs $K_1:\theta_1\neq\theta_{10}$ for right-truncated distribution is

$$\phi_1(x)=\begin{cases}1, & \text{if } X_{(n)}<c_1 \text{ or } X_{(n)}>\theta_{10}\\ 0, & \text{otherwise}\end{cases} \qquad \ldots(12.2.4)$$

where c_1 is determined by solving the equation

$$\left[\frac{g(\theta_{10})}{g(c_1)}\right]^n=\alpha \qquad \ldots(12.2.5)$$

and UMP level-α test, $\phi_2(x)$ for testing $H_2:\theta_2=\theta_{20}$ vs $K_2:\theta_2\neq\theta_{20}$ for left-truncated distribution is

$$\phi_2(x)=\begin{cases}1, & \text{if } X_{(1)}>c_2 \text{ or } X_{(1)}<\theta_{20}\\ 0, & \text{otherwise}\end{cases} \qquad \ldots(12.2.6)$$

where c_2 is determined by solving the equation

$$\left[\frac{g(\theta_{20})}{g(c_2)}\right]^n=\alpha \qquad \ldots(12.2.7)$$

Further the power function $\beta_{\phi_1}(\theta_1)$ for the test (12.2.4) is

$$\beta_{\phi_1}(\theta_1)\begin{cases}\left[\dfrac{g(\theta_{10})}{g(\theta_{10})}\right]^n=1, & \text{if } \theta_1<c_1\\ \left[\dfrac{g(\theta_1)}{g(c_1)}\right]^n, & \text{if } c_1<\theta_1<\theta_{10}\\ 1-\left[\dfrac{g(\theta_1)}{g(\theta_{10})}\right]^n+\left[\dfrac{g(\theta_1)}{g(c_1)}\right]^n, & \text{if } \theta_1\geq\theta_{10}\end{cases} \quad \ldots(12.2.8)$$

and the power function, $\beta_{\phi_2}(\theta_2)$ for the test (2.6) is

$$\beta_{\phi_2}(\theta_2)\begin{cases}1-\left[\dfrac{g(\theta_2)}{g(\theta_{20})}\right]^n+\left[\dfrac{g(\theta_2)}{g(c_2)}\right]^n, & \text{if } \theta_2>c_2\\ \left[\dfrac{g(\theta_2)}{g(c_2)}\right]^n, & \text{if } \theta_{20}<\theta_2<c_2\\ \left[\dfrac{g(\theta_{20})}{g(\theta_{20})}\right]^n=1, & \text{if } \theta_2\leq\theta_{20}\end{cases} \quad \ldots(12.2.9)$$

Proof: By using $F_X(x)=\dfrac{g(\theta_1)}{g(x)}$ and (12.2.3) the p.d.f. of $X_{(n)}$ for right truncated distribution with p.d.f. (12.2.1) is

$$f_{X_{(n)}}(x)=\begin{cases}n\left[\dfrac{g(\theta_1)}{g(x)}\right]^{n-1}\dfrac{d}{dx}\left[\dfrac{g(\theta_1)}{g(x)}\right], & \text{if } x<\theta_1\\ 0, & \text{otherwise}\end{cases} \quad \ldots(12.2.10)$$

Since $X_{(n)}$ is complete sufficient statistic for θ_1, the UMP level-α test is a function of $X_{(n)}$ for testing the hypothesis $H_1:\theta_1=\theta_{10}$ vs $K:\theta_1\neq\theta_{10}$

The critical region of the UMP-test comprises two sets

$$A=\{x:X_{(n)}>\theta_{10}\} \quad \text{or} \quad B=\{x:X_{(n)}<c_1\}$$

As $P_{H_1}\{A\}=0$, to make the test of level-α we choose c_1 such that

$$P_{H_1}\{B\}=\alpha$$

$$\Rightarrow \qquad P_{H_1}\{X_{(n)}<c_1\}=\alpha$$

$$\Rightarrow \qquad \int_0^{c_1} n\left[\frac{g(\theta_{10})}{g(x)}\right]^{n-1}\frac{d}{dx}\left[\frac{g(\theta_{10})}{g(x)}\right]dx=\alpha \text{, by virtue of (12.2.10)}$$

$$\Rightarrow \qquad \int_0^{\frac{g(\theta_{10})}{g(c_1)}} ny^{n-1}dy=\alpha \text{, by keeping } y=\frac{g(\theta_{10})}{g(x)}$$

$$\Rightarrow \qquad \left[\frac{g(\theta_{10})}{g(c_1)}\right]^n=\alpha$$

Which is equation (12.2.5)

Thus, the UMP test for testing H_1 at level-α is given by (12.2.4).

By using $F_X(x)=1-\dfrac{g(\theta_2)}{g(x)}$ and (12.2.3), the p.d.f. of $X_{(1)}$ for left truncated distribution of X with p.d.f. (12.2.2) is

$$f_{X_{(1)}}(x)=\begin{cases} n\left[\dfrac{g(\theta_2)}{g(x)}\right]^{n-1}\dfrac{d}{dx}\left[\dfrac{g(\theta_2)}{g(x)}\right], & \text{if } \theta_2<x \\ 0, & \text{otherwise} \end{cases} \qquad \text{...(12.2.11)}$$

Similarly the critical region of UMP level-α test for testing $H_2:\theta_2=\theta_{20}$ vs $K:\theta_2\neq\theta_{20}$ will depend upon complete sufficient statistic $X_{(1)}$ and hence its critical region comprises of two sets

$$A_1=\{x:X_{(1)}<\theta_{20}\} \quad \text{or} \quad B_1=\{x:X_{(1)}>c_2\}$$

As $P_{H_2}\{A_1\}=0$, we have to select c_2 such that

$$P_{H_2}\{B_1\}=\alpha$$

$$\Rightarrow \qquad P_{H_2}\{X_{(1)} > c_2\} = \alpha$$

$$\Rightarrow \qquad \int_{c_1}^{\infty} n\left[\frac{g(\theta_{20})}{g(x)}\right]^{n-1} \frac{d}{dx}\left[\frac{g(\theta_{20})}{g(x)}\right] dx = \alpha,$$

which gives on integration equation (12.2.7) and the UMP level-α test is given by (12.2.6).

The power function $\beta_{\phi_1}(\theta_1)$ for testing $H_1:\theta_1 = \theta_{10}$ vs $K:\theta_1 \neq \theta_{10}$ is

$$\beta_{\phi_1}(\theta_1) = P_{\theta_1}\{X_{(n)} < c_1\} + P_{\theta_1}\{X_{(n)} > \theta_{10}\}$$

and for testing $H_2:\theta_2 = \theta_{20}$ vs $K:\theta_2 \neq \theta_{20}$ the power function $\beta_{\phi_2}(\theta_2)$ is

$$\beta_{\phi_2}(\theta_2) = P_{\phi_2}\{X_{(1)} > c_2\} + P_{\phi_2}\{X_{(1)} < \theta_{20}\}$$

where c_1 and c_2 are obtainable by solving equations (12.2.5) and (12.2.7), respectively.

Using the p.d.f. (12.2.10) of $X_{(n)}$, we get general form of the power function, $\beta_{\phi_1}(\theta_1)$ for three different ranges of θ_1 as under.

Case I Let $\theta_1 < c_1$

Using test function $\phi_1(x)$ given in (12.2.4), we have

$$\beta_{\phi_1}(\theta_1) = \int_0^{c_1} n\left[\frac{g(\theta_1)}{g(x_1)}\right]^{n-1} \frac{d}{dx_1}\left[\frac{g(\theta_1)}{g(x_1)}\right] + o$$

$$= \left[\frac{g(\theta_1)}{g(x_1)}\right]^n \Bigg|_0^{c_1}$$

$$= \left[\frac{g(\theta_1)}{g(c_1)}\right]^n \qquad \ldots(12.2.12)$$

Case II Let $\theta_1 > \theta_{10}$

$$\beta_{\phi_1}(\theta_1) = \int_{\theta_{10}}^{\theta} n\left[\frac{g(\theta_1)}{g(x_1)}\right]^{n-1} \frac{d}{dx_1}\left[\frac{g(\theta_1)}{g(x_1)}\right] + \int_{0}^{c_1} n\left[\frac{g(\theta_1)}{g(x_1)}\right]^{n-1} \frac{d}{dx_1}\left[\frac{g(\theta_1)}{g(x_1)}\right]$$

$$= 1 - \int_{0}^{\theta_{10}} n\left[\frac{g(\theta_1)}{g(x_1)}\right]^{n-1} \frac{d}{dx_1}\left[\frac{g(\theta_1)}{g(x_1)}\right] + \left[\frac{g(\theta_1)}{g(c_1)}\right]^{n}$$

$$= 1 - \left[\frac{g(\theta_1)}{g(\theta_{10})}\right]^{n} + \left[\frac{g(\theta_1)}{g(c_1)}\right]^{n} \qquad \ldots(12.2.13)$$

Case III Let $c_1 < \theta_1 < \theta_{10}$

$$\beta_{\phi_1}(\theta_1) = \int_{0}^{c_1} n\left[\frac{g(\theta_1)}{g(x_1)}\right]^{n-1} \frac{d}{dx_1}\left[\frac{g(\theta_1)}{g(x_1)}\right] + \int_{\theta_{10}}^{\theta_1} n\left[\frac{g(\theta_1)}{g(x_1)}\right]^{n-1} \frac{d}{dx_1}\left[\frac{g(\theta_1)}{g(x_1)}\right]$$

$$= \left[\frac{g(\theta_1)}{g(x_1)}\right]^{n}\Bigg|_{0}^{c_1}$$

$$= \left[\frac{g(\theta_1)}{g(c_1)}\right]^{n} \qquad \ldots(12.2.14)$$

Similarly, for testing $H_2 : \theta_2 = \theta_{20}$ vs $K : \theta_2 \neq \theta_{20}$ the power function $\beta_{\phi_2}(\theta_2)$ can be obtained by using the p.d.f. (12.2.11) of $X_{(1)}$ for three different ranges of θ_2 as under.

Case I Let $\theta_2 > c_2$

Using test function $\phi_2(x)$ given in (12.2.6), we have

$$\beta_{\phi_2}(\theta_2) = \int_{0}^{\theta_{20}} n\left[\frac{g(\theta_2)}{g(x)}\right]^{n-1} \frac{d}{dx}\left[\frac{g(\theta_2)}{g(x)}\right] + \int_{c_2}^{\infty} n\left[\frac{g(\theta_2)}{g(x)}\right]^{n-1} \frac{d}{dx}\left[\frac{g(\theta_2)}{g(x)}\right]$$

$$= 1 - \int_{\theta_{20}}^{\infty} n\left[\frac{g(\theta_2)}{g(x)}\right]^{n-1} \frac{d}{dx}\left[\frac{g(\theta_2)}{g(x)}\right] + \left[\frac{g(\theta_2)}{g(c_2)}\right]^{n}$$

$$= 1 - \left[\frac{g(\theta_2)}{g(\theta_{20})}\right]^{n} + \left[\frac{g(\theta_2)}{g(c_2)}\right]^{n} \qquad \ldots(12.2.15)$$

Case II Let $\theta_2 < \theta_{20}$

$$\beta_{\phi_2}(\theta_2) = \int_{\theta_{20}}^{\infty} n\left[\frac{g(\theta_2)}{g(x)}\right]^{n-1} \frac{d}{dx}\left[\frac{g(\theta_2)}{g(x)}\right] + \int_{c_2}^{\theta_2} n\left[\frac{g(\theta_2)}{g(x)}\right]^{n-1} \frac{d}{dx}\left[\frac{g(\theta_2)}{g(x)}\right]$$

$$= 1 + 0 \qquad \text{...(12.2.16)}$$

Case III Let $\theta_{20} < \theta_2 < c_2$

$$\beta_{\phi_2}(\theta_2) = \int_{c_2}^{\infty} n\left[\frac{g(\theta_2)}{g(x)}\right]^{n-1} \frac{d}{dx}\left[\frac{g(\theta_2)}{g(x)}\right] + \int_{c_2}^{\theta_2} n\left[\frac{g(\theta_2)}{g(x)}\right]^{n-1} \frac{d}{dx}\left[\frac{g(\theta_2)}{g(x)}\right]$$

$$= \left[\frac{g(\theta_2)}{g(x)}\right]^n \Bigg|_{c_2}^{\infty} + 0$$

$$= \left[\frac{g(\theta_2)}{g(c_2)}\right]^n \qquad \text{...(12.2.17)}$$

Thus power function, $\beta_{\phi_1}(\theta_1)$ for testing $H_1:\theta_1 = \theta_{10}$ vs $K_1:\theta_1 \neq \theta_{10}$ is given by (12.2.8) and $\beta_2(\theta_2)$ for testing $H_2:\theta_2 = \theta_{20}$ vs $K:\theta_2 \neq \theta_{20}$ is given by (12.2.9).

12.3 APPLICATIONS

In this section we have applied the results of general theory obtained in earlier section to some specific distributions such as generalized uniform, uniform, power function, negative exponential, Pareto distributions.

(i) Generalized Uniform Distribution

Let X have generalized uniform distribution with p.d.f. as

$$f_X(x;\theta) = \begin{cases} \dfrac{a+1}{\theta^{(a+1)}} x^a, & 0 < x < \theta, a > -1 \text{ (known)} \\ 0, & \text{otherwise} \end{cases}$$

$$= g(\theta)h(x)$$

where $$g(\theta)=\frac{a+1}{\theta^{(a+1)}},\ h(x)=x^a$$

so $$\frac{g(\theta)}{g(x)}=\left(\frac{x}{\theta}\right)^{(a+1)}$$

the UMP level-α test for testing the hypothesis $H:\theta=\theta_0$ vs $K:\theta\neq\theta_0$ is given by (12.2.4) where, c is obtained by solving (12.2.5), that is, by solving

$$\left[\frac{g(\theta_0)}{g(c)}\right]^n=\left[\left(\frac{c}{\theta_0}\right)^{(a+1)}\right]^n=\alpha$$

$$c=\theta_0\alpha^{\frac{1}{n(a+1)}}$$

and the test function is

$$\phi(x)=\begin{cases}1, & \text{if } X_{(n)}<\theta_0\alpha^{\frac{1}{n(a+1)}} \text{ or } X_n>\theta_0\\ 0, & \text{otherwise}\end{cases}$$

Hence (12.2.8) gives the power function $\beta_\phi(\theta)$ of the test as

$$\beta_\phi(\theta)=\begin{cases}1, & \text{if} \quad \theta<\theta_0\alpha^{-\frac{1}{n(a+1)}}\\ \alpha\left(\frac{\theta_0}{\theta}\right)^{n(a+1)}, & \text{if} \quad \theta_0\alpha^{-\frac{1}{n(a+1)}}<\theta<\theta_0\\ 1-(1-\alpha)\left(\frac{\theta_0}{\theta}\right)^{n(a+1)}, & \text{if} \quad \theta>\theta_0\end{cases}$$

(ii) Uniform Distribution

Let X have uniform distribution with p.d.f. as

$$f_X(x;\theta)=\begin{cases}\frac{1}{\theta}, & 0<x<\theta\\ 0, & \text{otherwise}\end{cases}$$

$$=g(\theta)h(x)$$

where $g(\theta)=\frac{1}{\theta}$, $h(x)=1$

For testing $H:\theta=\theta_0$ vs $K:\theta\neq\theta_0$ α-level UMP test is given by (12.2.4) where, c is obtained by solving (12.2.5), that is, by solving

$$\left[\frac{g(\theta_0)}{g(c)}\right]^n=\left(\frac{c}{\theta_0}\right)^n=\alpha$$

$$\Rightarrow \qquad c=\theta_0\alpha^{\frac{1}{n}}$$

and the test function is

$$\phi(x)=\begin{cases}1, & \text{if } X_{(n)}<\theta_0\alpha^{\frac{1}{n}} \text{ or } X_{(n)}>\theta_0\\ 0, & \text{otherwise}\end{cases}$$

Using $\frac{g(\theta)}{g(x)}=\frac{x}{\theta}$, (12.2.8) gives the power function $\beta_\phi(\theta)$ of the test as

$$\beta_\phi(\theta)=\begin{cases}1, & \text{if} \quad \theta<\theta_0\alpha^{\frac{1}{n}}\\ \alpha\left(\frac{\theta_0}{\theta}\right)^n, & \text{if} \quad \theta_0\alpha^{\frac{1}{n}}<\theta<\theta_0\\ 1-(1-\alpha)\left(\frac{\theta_0}{\theta}\right)^n, & \text{if} \quad \theta\geq\theta_0\end{cases}$$

(iii) Power Function Distribution

Let X have power function distribution with p.d.f. as

$$f_X(x;\theta)=\begin{cases}\frac{k}{\theta^k}x^{k-1}, & 0<x<\theta, k \text{ is known}\\ 0, & \text{otherwise}\end{cases}$$

$$=g(\theta)h(x)$$

where $$g(\theta)=\frac{k}{\theta^k},\ h(x)=x^{k-1}$$

so $$\frac{g(\theta)}{g(x)}=\left(\frac{x}{\theta}\right)^k$$

The UMP level-α test for testing $H:\theta=\theta_0$ vs $K:\theta\neq\theta_0$ is given by (12.2.4) where, c is obtained by solving (12.2.5), that is, by solving

$$\left[\frac{g(\theta_0)}{g(c)}\right]^n=\alpha$$

$$\left(\frac{c}{\theta_0}\right)^k=\alpha^{\frac{1}{n}}$$

$$c=\theta_0\alpha^{\frac{1}{nk}}$$

and the test function is

$$\phi(x)=\begin{cases}1, & \text{if } X_{(n)}<\theta_0\alpha^{\frac{1}{nk}} \text{ or } X_{(n)}>\theta_0\\ 0, & \text{otherwise}\end{cases}$$

Using $\frac{g(\theta)}{g(x)}=\left(\frac{x}{\theta}\right)^k$, (12.2.8) gives the power function $\beta_\phi(\theta)$ of the test as

$$\beta_\phi(\theta)=\begin{cases}1, & \text{if } \quad \theta<\theta_0\alpha^{\frac{1}{nk}}\\ \alpha\left(\frac{\theta_0}{\theta}\right)^{nk}, & \text{if } \quad \theta_0\alpha^{\frac{1}{nk}}<\theta<\theta_0\\ 1-(1-\alpha)\left(\frac{\theta_0}{\theta}\right)^{nk}, & \text{if } \quad \theta\geq\theta_0\end{cases}$$

(iv) Negative Exponential Distribution

Let X have negative exponential distribution with p.d.f. as

$$f_X(x;\theta)=\begin{cases} e^{-(x-\theta)}, & \theta<x<\infty \\ 0, & \text{otherwise} \end{cases}$$

$$= g(\theta)h(x)$$

where $$g(\theta)=e^{\theta},\ h(x)=e^{-x}$$

so that $$\frac{g(\theta)}{g(x)}=e^{-(x-\theta)}$$

The UMP level-α test for testing $H:\theta=\theta_0$ vs $K:\theta\neq\theta_0$ is given by (12.2.6) where, c is obtained by solving (12.2.7), that is, by solving

$$\left[\frac{g(\theta_0)}{g(c)}\right]^n=\alpha$$

$$\left[\frac{e^{\theta_0}}{e^c}\right]^n=\alpha$$

$$e^{-(c-\theta_0)}=\alpha^{\frac{1}{n}}$$

$$c=\theta_0-\frac{1}{n}\log\alpha$$

and so the test function is

$$\phi(x)=\begin{cases} 1, & \text{if } X_{(1)}>\theta_0-\frac{1}{n}\log\alpha \text{ or } X_{(1)}<\theta_0 \\ 0, & \text{otherwise} \end{cases}$$

By using (12.2.9) the power function $\beta_\phi(\theta)$ of the test is given by

$$\beta_\phi(\theta) = \begin{cases} 1-(1-\alpha)\left(\dfrac{\theta_0}{\theta}\right)e^{-n(\theta_0-\theta)}, & \text{if } \quad \theta > \theta_0 - \log \alpha/n \\ \alpha e^{-n(\theta_0-\theta)}, & \text{if } \quad \theta_0 < \theta < \theta_0 - \log \alpha/n \\ 1, & \text{if } \quad \theta < \theta_0 \end{cases}$$

(v) Pareto Distribution

Let X have Pareto distribution with p.d.f. as

$$f_X(x;\theta) = \begin{cases} a\theta^a x^{-(a+1)}, & \text{if } \theta < x < \infty, a > -1 \\ 0, & \text{otherwise} \end{cases}$$

$$= g(\theta)h(x)$$

where $$g(\theta) = a\theta^a, \; h(x) = x^{-(a+1)}$$

so that $$g(x) = ax^a \text{ and } \frac{g(\theta)}{g(x)} = \left(\frac{\theta}{x}\right)^a$$

The UMP level-α test for testing $H:\theta = \theta_0$ vs $K:\theta \neq \theta_0$ is given by (2.5) where, c is obtained by solving (2.7), that is, by solving

$$\left[\frac{g(\theta_0)}{g(c)}\right]^n = \alpha$$

$$\left(\frac{c}{\theta_0}\right)^{-an} = \alpha$$

$$c = \theta_0 \alpha^{-\frac{1}{an}}$$

and the test function is

$$\phi(x) = \begin{cases} 1, & \text{if } X_{(1)} > \theta_0\alpha^{-\frac{1}{an}} \text{ or } X_{(1)} < \theta_0 \\ 0, & \text{otherwise} \end{cases}$$

By using (2.9), the power function $\beta_\phi(\theta)$ of the test is

$$\beta_\phi(\theta) = \begin{cases} 1-(1-\alpha)\left(\dfrac{\theta_0}{\theta}\right)^{-an}, & \text{if} \quad \theta > \theta_0 \alpha^{-\frac{1}{an}} \\ \alpha\left(\dfrac{\theta_0}{\theta}\right)^{-an}, & \text{if} \quad \theta_0 < \theta < \theta_0 \alpha^{-\frac{1}{an}} \\ 1, & \text{if} \quad \theta < \theta_0 \end{cases}$$

13
CHAPTER

Asymptotic Tests for Several Exponential Family of Distributions

13.1 INTRODUCTION

Several books like Lehmann (2005) have discussed the testing of equality of means of k-normal distributions and have obtained uniformly most powerful invariant (UMPI) test for it. In this chapter we will consider the asymptotic testing of equality of parameters of several exponential family of distributions. Section 13.2 is devoted for asymptotic tests for testing equality of parameters of several exponential family of distributions and section 13.3 discusses the applications of it to some specific distributions like Binomial, Poisson, Exponential and Gamma distributions.

13.2 ASYMPTOTIC TESTS

Let $X_1, X_2, \ldots, X_k$ be K independently distributed exponential family with probability function (p.f.), $f_{X_i}(x_i; \theta_i)$ as

$$f_{X_i}(x_i; \theta_i) = a(\theta_i)b(x_i)e^{-\theta_i x_i}, -\infty < x_i < \infty, i = 1, 2, \ldots, k \qquad \ldots(13.2.1)$$

Let $X_{i1}, X_{i2}, \ldots, X_{in}$ be a random sample of size n from each X_i, $i = 1, 2, \ldots, k$ and let $w_i = \sum_{j=1}^{n} X_{ij}, i = 1, 2, \ldots, k$. Then the joint p.f. of $X_{i1}, X_{i2}, \ldots, X_{ik}$ can be written as

$$f_X(x;\theta) = q(\theta)h(x)e^{-\sum_{i=1}^{k}\theta_i w_i(x)}, \quad \text{where} \quad x = (x_1, x_2, \ldots, x_k),$$

$$\theta = (\theta_1, \theta_2, \ldots \theta_k), w_i(x) = wi, \text{ say } i = 1, 2, \ldots k$$

Let $$\underline{w} = (w_1, w_2, \ldots w_k)', \overline{w}_i = \frac{1}{n} wi, \underline{\overline{w}} = (\overline{w}_1, \overline{w}_2, \ldots, \overline{w}_k)'.$$

Theorem 13.1: With the above set-up of observations, the asymptotic UMPI level-α test for testing the hypothesis, $H:\theta_1 = \theta_2 = \cdots = \theta_k$ vs K: $\theta_i \neq \theta_j$ for at least one pair $i \neq j$, is to reject H whenever

$$\frac{n-k+1}{k-1}(C\underline{\overline{w}})'(C\Sigma C')^{-1}(C\underline{\overline{w}}) \geq F_{k-1, n-k+1;\alpha} \qquad \ldots(13.2.2)$$

where $F_{k-1, n-k+1;\alpha}$ is upper α% point of Snedekar F-distribution with $(k-1, n-k+1)$ d.f. and C is any $(k-1)\times k$ contrast matrix of constants and Σ is variance -covariance matrix of $w_1, w_2, \ldots w_k$.

Proof: We note that w_i is complete sufficient statistic for θ_i, $i = 1, 2, \ldots \theta_k$ and the test depends on function of w_i. Further, it is easy to show that

$$E(w_i) = q(\theta)\frac{\partial}{\partial\theta_i}[q(\theta)]^{-1} = q(\theta)di, i = 1, 2, \ldots k \qquad \ldots(13.2.3)$$

$$\text{Var}(w_i) = q(\theta)\frac{\partial^2}{\partial\theta_i^2}[q(\theta)]^{-1} - \left\{q(\theta)\frac{\partial}{\partial\theta_i}[q\theta)]^{-1}\right\}^2$$

$$= q(\theta)d_{ii} - (q(\theta)d_i)^2, i = 1, 2, \ldots k \qquad \ldots(13.2.4)$$

where $$d_i = \frac{\partial}{\partial\theta_i}[q(\theta)]^{-1}, d_{ii} = \frac{\partial^2}{\partial\theta_i^2}[q(\theta)]^{-1}, i = 1, 2, \ldots k.$$

Since $X_{ij}, i = 1, 2, \ldots k; j = 1, 2, \ldots n$ are all independently distributed. $\text{cov}(w_i, w_j) = 0$ for $i \neq j$.

Due to multivariate central limit theorem, $\underline{w} = (w_1, w_2, \ldots w_k)'$ is asymptotically distributed as k-variate normal distribution with mean vector

$$q(\theta)(d_1, d_2, \ldots, d_k)' = q(\theta)D_1'$$

and variance -covariance matrix, Σ as

$$\Sigma = q(\theta)D - [q(\theta)]^2 D_1' D_1$$

where $$D_1 = (d_1, d_2, \ldots, d_k), D = \text{diag}(d_{11}, d_{22}, \ldots, d_{kk})$$

Thus,

$$\underline{w} \xrightarrow{d} N_k(q(\theta)D_1', q(\theta)D - (q(\theta)^2 D_1' D_1)$$

and

$$C\underline{w} \xrightarrow{d} N_{k-1}(q(\theta)CD_1', C[q(\theta)D - [q(\theta)]^2 D_1' D_1]C')$$

where C is $(k-1) \times k$ contrast matrix and let it be

$$C = \begin{bmatrix} 1 & 0 & 0 & \cdots & 0 & -1 \\ 0 & 1 & 0 & \cdots & 0 & -1 \\ \vdots & \vdots & \vdots & \cdots & \vdots & \vdots \\ 0 & 0 & 0 & \cdots & 1 & -1 \end{bmatrix}$$

As

$$H: \theta_i = \theta_k \Leftrightarrow \frac{\partial}{\partial \theta_i}[q(\theta)]^{-1} = \frac{\partial}{\partial \theta_k}[q(\theta)]^{-1}, \; i = 1, 2, \ldots k-1$$

$$\Leftrightarrow \qquad q(\theta)\frac{\partial}{\partial \theta_i}[q(\theta)]^{-1} = q(\theta)\frac{\partial}{\partial \theta_k}[q(\theta)]^{-1}$$

$$\Leftrightarrow \qquad q(\theta)CD_1' = 0, \text{ say } H_1$$

Thus, $H: \theta_1 = \theta_2 = \ldots = \theta_k \Leftrightarrow H_1 : q(\theta)CD_1' = 0$ where $\theta = (\theta_1, \theta_2, \ldots \theta_k)$ and for testing H_1, we need the distribution of $C\underline{w}$ which is asymptotically $N_{k-1}(q(\theta)CD_1', C\Sigma C^1)$. By the usual multivariate theory of testing the mean vector of multivariate normal distribution equal to zero vector is to reject H_1, in this case, whenever

$$\frac{(n-1)-(k-1)+1}{k-1}(C\underline{\overline{w}})'(C\Sigma C')^{-1}(C\underline{\overline{w}}) \geq F_{k-1,\, n-k+1;\, \alpha}$$

where $F_{k-1,n-k+1;\alpha}$ is upper α% point of Snedekar F-distribution with $(k-1, n-k+1)$ d.f. Since the test depends on normal variates, the textbook theory of testing equality of location parameters of k normal variates, the test turns out as UMPI test and hence in this case also the test for testing equality of parameters of k exponential family is asymptotic UMPI test.

The computation of quadratic form $(C\underline{\overline{w}})'(C\Sigma C')^{-1}(C\underline{\overline{w}})$ can be simplified as under. When H is true we have

(i) $C\underline{\overline{w}} = (\overline{w}_1 - \overline{w}_k, \overline{w}_2 - \overline{w}_k, \ldots, \overline{w}_{k-1} - \overline{w}_k)$

(ii) $wC' = I_{k-1} + \underline{1}\underline{1}'$, where $\underline{1}$ is $(k-1)$ vector of unity

(iii) $(CwC')^{-1} = I_{k-1} + \frac{1}{k}\underline{1}1'$

Using the above three results in the quadratic form and simplifying one gets

$$(C\underline{\overline{w}})'(C\Sigma C')^{-1}(C\underline{\overline{w}}) = \sum_{i=1}^{k}\overline{w}_i^2 - \frac{1}{k}[(k-1)\overline{\overline{w}} + \overline{w}_k]^2, \text{ where } \overline{\overline{w}} = \frac{1}{k}\sum_{i=1}^{k}\overline{w}_i . \quad \ldots(13.2.5)$$

Particularly, when k = 2, c = (1 – 1), $C\Sigma C' = 2$,

$$C\underline{\overline{w}} = (\overline{w}_1 - \overline{w}_2)(C\underline{\overline{w}})'(C\Sigma C')^{-1}(C\underline{\overline{w}}) = \frac{(\overline{w}_1 - \overline{w}_2)^2}{2}.$$

As $\frac{\overline{w}_1 - \overline{w}_2}{\sqrt{2}} \xrightarrow{d} N(0, 1)$,

$$(C\underline{\overline{w}})'(C\Sigma C')^{-1}(C\underline{\overline{w}}) \xrightarrow{d} \chi_1^2 .$$

Thus, we reject H at level α when

$$(n-2)(C\underline{\overline{w}})'(C\Sigma C')^{-1}(C\underline{\overline{w}}) = \frac{(n-2)}{2}(\overline{w}_1 - \overline{w}_2)^2 \geq (n-2)\chi_{1;\alpha}^2$$

where $\chi_{1;\alpha}^2$ is upper α% point of χ^2 on one d.f.

13.3 APPLICATIONS TO SOME SPECIFIC DISTRIBUTIONS

(i) Binomial Distribution

Consider K independently distributed binomial distribution with p.f.

$$P(X_i = x_i; p_i) = \binom{n}{x_i} p_i^{x_i} (1-p_i)^{n-x_i}, \; i = 1, 2, \dots k, \; x_i = 1, 2, \dots n$$

$$= (1-p_i)^n \binom{n}{x_i} e^{x_i \log \frac{p_i}{1-p_i}}$$

$$= a(p_i) b(x_i) e^{x_i \log \frac{p_i}{1-p_i}},$$

where $\quad a(p_i) = (1-p_i)^n, \; b(x_i) = \binom{n}{x_i}$

Making the naturalization of the family by taking $\log \frac{p_i}{1-p_i} = \theta_i$,

that is, $p_i = \frac{e^{\theta_i}}{1+e^{\theta_i}}$, $a(p_i) = (1-p_i)^n = (1+e^{\theta_i})^{-n} = q(\theta_i)$, say and using the results (13.2.3) and (13.2.4) and simplifying them, we get

$$w_i = \sum_{i=1}^{n} x_{ij},$$

$$i = 1, 2, \dots k,$$

$$E(w_i) = np_i,$$

$$\text{Var}(w_i) = np_i(1-p_i),$$

$$i = 1, 2, \dots k$$

$$\text{cov}(w_i, w_j) = 0 \text{ for } i \neq j.$$

$$\underset{\sim}{w} \xrightarrow{d} N_k(\underset{\sim}{\mu}, \Sigma)$$

where
$$\underline{\mu} = (np_1, np_2, \ldots, np_k)',$$
$$\Sigma = \text{diag}(np_1(1-p_1),$$
$$np_2(1-p_2), \ldots, np_k(1-p_k))$$

and

$$C\underline{w} \xrightarrow{d} N_k(C\underline{\mu}, C\Sigma C'),$$

where C is any $(k-1)\times k$ constant matrix such that

$$C = \begin{bmatrix} 1 & 0 & 0 & \cdots & 0 & -1 \\ 0 & 1 & 0 & \cdots & 0 & -1 \\ \vdots & \vdots & \vdots & \cdots & \vdots & \vdots \\ 0 & 0 & 0 & \cdots & 1 & -1 \end{bmatrix}$$

The asymptotic UMPI level-α test for testing $H: p_1 = p_2 = \ldots = p_k$ is given by (13.2.2) and the quadratic form involved in it can be computed by using (13.2.5), in which

$$w_i = \sum_{i=1}^{n} x_{ij},\ i = 1, 2, \ldots k,\ \bar{w}_i = \frac{w_i}{n},\ i = 1, 2, \ldots k,\ \bar{\bar{w}} = \frac{1}{nk}\sum_{i=1}^{k}\sum_{j=1}^{n} x_{ij}.$$

(ii) Poisson Distribution

Let $X_1, X_2, \ldots, X_k$ be independently distributed as Poisson with parameters $\lambda_1, \lambda_2, \ldots, \lambda_k$. Then the joint p.f. is

$$P(X_1 = x_1, X_2 = x_2, \ldots, X_k = x_k; \lambda_1, \lambda_2, \ldots, \lambda_k) = \frac{e^{-\Sigma\lambda_i}\prod_{i=1}^{k}\lambda_i^{x_i}}{x_1!\, x_2! \ldots x_k!}$$

$$= e^{-\Sigma\lambda_i} \cdot \frac{1}{x_1!\, x_2! \ldots x_k!} \cdot e^{\Sigma x_i \log \lambda_i}$$

$$= g(\lambda)h(x)e^{\Sigma x_i \theta_i}$$

where $\lambda = (\lambda_1, \lambda_2, \ldots, \lambda_k),\ x = (x_1, x_2, \ldots, x_k)$

by naturalizing this family by taking $\theta_i = \log \lambda_i$, $i = 1, 2, \ldots, k$, the joint p.f. is rewritten as

$$P_W(w;\theta) = a(\theta)h(w_i)e^{\Sigma\theta_i w_i},\ \theta = (\theta_1, \theta_2, \ldots, \theta_k),\ w = (w_1, w_2, \ldots, w_k)$$

where $$a(\theta) = e^{-\sum_{i=1}^{k} e^{\theta_i}} = q(\theta) \text{ as } \lambda_i = e^{\theta_i},\ i = 1, 2, \ldots, k$$

and $\text{cov}(w_i, w_j) = 0$ for $i \neq j$, the variance covariance matrix, Σ of $w_1, w_2, \ldots, w_k$ will be

$$\Sigma = \text{diag}(\lambda_1, \lambda_2, \ldots, \lambda_k)$$

By multivariate central limit theorem,

$$\underline{w} \xrightarrow{d} N_k(\underline{\lambda}, \Sigma) \text{ where } \underline{\lambda} = (\lambda_1, \lambda_2, \ldots, \lambda_k), \Sigma = \text{diag}(\lambda_1, \lambda_2, \ldots, \lambda_k)$$

Using $(k-1)\times k$ constant matrix C as

$$C = \begin{bmatrix} 1 & 0 & 0 & \cdots & 0 & -1 \\ 0 & 1 & 0 & \cdots & 0 & -1 \\ \vdots & \vdots & \vdots & \cdots & \vdots & \vdots \\ 0 & 0 & 0 & \cdots & 1 & -1 \end{bmatrix}$$

$$C\underline{w} \xrightarrow{d} N_{k-1}(C\underline{\lambda}, C\Sigma C')$$

The asymptotic UMPI level-α test for testing $H: \lambda_1 = \lambda_2 = \ldots = \lambda_k$ is given by (13.2.2) and the quadratic form involved in it can be computed by using (2.5) in which

$$w_i = \sum_{i=1}^{n} x_{ij},\ i = 1, 2, \ldots, k,\ \bar{w}_i = \frac{w_i}{n},\ i = 1, 2, \ldots k,\ \bar{\bar{w}} = \frac{1}{nk}\sum_{i=1}^{k}\sum_{j=1}^{n} x_{ij}.$$

(iii) Exponential Distribution

Let $X_i (i = 1, 2, \ldots, k)$ be k independently distributed exponential distributions with p.d.f.

$$f_{X_i}(x_i, \lambda_i) = \lambda_i e^{-\lambda_i x_i}, \; x_i > 0, \; \lambda_i > 0, \; i = 1, 2, \ldots, k$$

$$= a(\lambda_i) h(x) e^{-\lambda_i x_i}$$

where $\qquad a(\lambda_i) = \lambda_i \, , h(x) = 1$

For a random sample $X_{i1}, X_{i2}, \ldots, X_{in} (i = 1, 2, \ldots k)$ on X_i

let $\qquad w_i = \sum_{j=1}^{n} x_{ij} \, , \; i = 1, 2, \ldots k \, .$

Using results (13.2.3) and (13.2.4);

$$E(w_i) = -\lambda_i \frac{\partial}{\partial \lambda_i}\left(\frac{1}{\lambda_i}\right) = \frac{1}{\lambda_i}, i = 1, 2, \ldots, k$$

$$\text{Var}(w_i) = \frac{1}{\lambda_i^2}, \; , i = 1, 2, \ldots, k \, , \; \text{cov}(w_i, w_j) = 0 \; \text{ for } \; i \neq j$$

$$\underline{w} = (w_1, w_2, \ldots, w_k) \xrightarrow{d} N_k(\underline{\lambda}, \Sigma)$$

where $\qquad \underline{\lambda} = \left(\frac{1}{\lambda_1}, \frac{1}{\lambda_2}, \ldots, \frac{1}{\lambda_k}\right),$

$$\Sigma = \text{diag}\left(\frac{1}{\lambda_1^2}, \frac{1}{\lambda_2^2}, \ldots, \frac{1}{\lambda_k^2}\right)$$

Under $\qquad H : \lambda_i = \lambda_2 = \ldots \lambda_k \, ,$

$$\underline{\lambda} = \frac{1}{\lambda_1}(1, 1, \ldots, 1)' = \frac{1}{\lambda_1} 1' \, ,$$

$$\Sigma = \frac{1}{\lambda_1^2} I_k$$

and

$$\underline{w} \xrightarrow{d} N_k\left(\frac{1}{\lambda_1}1, \frac{1}{\lambda_1^2} I_k\right)$$

Also under H,

$$C\underline{w} \xrightarrow{d} N_{k-1}\left(\frac{1}{\lambda_1}C\underline{1}, \frac{1}{\lambda_1^2}CC'\right)$$

The asymptotic UMPI level-α test is given by (13.2.2) and the quadratic form involved in it can be computed by using (13.2.5) in which

$$w_i = \sum_{i=1}^{n} x_{ij}\,,\; i=1,2,\ldots k\,,\; \overline{w}_i = \frac{w_i}{n}\,,\; i=1,2,\ldots k,\; \overline{\overline{w}} = \frac{1}{nk}\sum_{i=1}^{k}\sum_{j=1}^{n} x_{ij}\,.$$

(iv) Gamma Distribution

Let $X_i(i=1,2,\ldots,k)$ be k independently distributed as gamma with p.d.f. as

$$f_{X_i}(x_i;\theta_i) = \frac{1}{\Gamma(\theta_i+1)} x_i^{\theta_i} e^{-x_i},\; x_i>0,\; \theta_i>0\,,\; i=1,2,\ldots,k$$

$$= a(\theta_i)\cdot e^{-x_i}\cdot e^{\theta_i \log x_i},\text{ where, } a(\theta_i) = \frac{1}{\Gamma(\theta_i+1)}$$

For a random sample $X_{i1}, X_{i2}, \ldots, X_{in}(i=1,2,\ldots k)$ from X_i

let
$$w_i = \sum_{j=1}^{n} x_{ij},\; i=1,2,\ldots k\,.$$

Using results (13.2.3) and (13.2.4);

$$E(w_i) = \frac{1}{\Gamma(\theta_i+1)}\frac{\partial}{\partial\theta_i}(\Gamma(\theta_i+1)), i=1,2,\ldots,k$$

and

$$\text{Var}(w_i) = \frac{1}{\Gamma(\theta_i+1)} \frac{\partial^2}{\partial \theta_i^2}(\Gamma(\theta_i+1)) - \left(\frac{1}{\Gamma(\theta_i+1)} \frac{\partial}{\partial \theta_i}(\Gamma(\theta_i+1))\right)^2, i = 1, 2, \ldots, k,$$

$E(w_i) = d_i$ and $\text{Var}(w_i) = d_{ii} - (d_i)^2$, $i = 1, 2, \ldots, k$ by using

$$d_i = \frac{1}{\Gamma(\theta_i+1)} \frac{\partial}{\partial \theta_i}(\Gamma(\theta_i+1)),\ d_{ii} = \frac{1}{\Gamma(\theta_i+1)} \frac{\partial^2}{\partial \theta_i^2}(\Gamma(\theta_i+1))$$

Since $\text{cov}(w_i, w_j) = 0$ for $i \neq j$, the variance-covariance matrix, Σ of $w_1, w_2, \ldots, w_k$ will be

$$\Sigma = \text{diag}(d_{11}, d_{22}, \ldots, d_{kk}) - \text{diag}(d_1^2, d_2^2, \ldots, d_k^2)$$

$$= D - D_1 D_1',$$

where $\quad D = \text{diag}(d_{11}, d_{22}, \ldots, d_{kk}),\ D_1 = (d_1, d_2, \ldots, d_k)'$

under the hypothesis

$$H: \theta_1 = \theta_2 = \ldots = \theta_k \Leftrightarrow d_1 = d_2 = \ldots = d_k \Rightarrow CD_1' = 0$$

where

$$C = \begin{bmatrix} 1 & 0 & 0 & \cdots & 0 & -1 \\ 0 & 1 & 0 & \cdots & 0 & -1 \\ \vdots & \vdots & \vdots & \cdots & \vdots & \vdots \\ 0 & 0 & 0 & \cdots & 1 & -1 \end{bmatrix}$$

is $(k-1) \times k$ contrast matrix.

By multivariate central limit theorem, $(w_1, w_2, \ldots, w_k)$ is asymptotic distributed as k-variate normal with mean vector D_1 and variance-covariance matrix $\Sigma = D - D_1 D_1'$, that is,

$$\underline{w} \xrightarrow{d} N_k(D_1, \Sigma) \quad \text{and} \quad C\underline{w} \xrightarrow{d} N_{k-1}(CD_1, C\Sigma C')$$

when H is true,

$$C\underline{w} \xrightarrow{d} N_{k-1}(0, C\Sigma C')$$

The asymptotic UMPI level-α test is given by (13.2.2) and the quadratic form involved in it can be computed by using (13.2.5) in which

$$w_i = \sum_{i=1}^{n} x_{ij},\ i = 1, 2, \ldots k,\ \overline{w}_i = \frac{w_i}{n},\ i = 1, 2, \ldots k,\ \overline{\overline{w}} = \frac{1}{nk} \sum_{i=1}^{k} \sum_{j=1}^{n} x_{ij}.$$

14

CHAPTER

Asymptotic Tests for Several Truncated Distributions

14.1 INTRODUCTION

In this chapter we will obtain asymptotic test for testing equality of k truncated one parameter family of distributions. After obtaining asymptotic distribution of k truncated one-parameter family of distribution in section 14.2, asymptotic test for testing equality of parameters will be obtained in section 14.3. Further by taking $k = 2$, the necessary asymptotic conditional distribution is obtained in section 14.4 and the asymptotic test for testing equality of parameters has been discussed in section 14.5. The results of this test have been applied to some distributions like generalized uniform, uniform, truncated exponential, negative exponential and Pareto distributions in section 14.6.

14.2 ASYMPTOTIC CONDITIONAL DISTRIBUTIONS

In the following theorems we shall use the set-up of K one-truncation parameter family of independent distributions and their random samples with corresponding ordered samples.

Let $X_1, X_2, \ldots, X_k$ be independently continuous random variables with probability density function (p.d.f.) as

$$f_{X_i}(x_i, \theta_i) = g_i(\theta_i)h_i(x_i),\ a < x_i < \theta_i < b;\ i = 1, 2, \ldots, k \qquad \ldots(14.2.1)$$

or

$$f_{x_i}(x_i, \theta_i) = g_i(\theta_i)h_i(x_i),\ a < \theta_i < x_i < b;\ i = 1, 2, \ldots, k \qquad \ldots(14.2.2)$$

where a, b are known constants, $g_i(\theta_i); i=1,2,\ldots,k$ are everywhere differentiable functions, $h_i(x_i); i=1,2,\ldots,k$ are positive absolutely continuous functions. (14.2.1) is p.d.f. related to right truncated parameters family of distributions and (14.2.2) is that of left truncated parameter family of distributions.

Let $X_{i1}, X_{i2}, \ldots, X_{in_i}; i=1,2,\ldots,k$ be random samples either from (14.2.1) or (14.2.2) and let $X_{i(1)}, X_{i(2)}, \ldots, X_{i(n_i)}; i=1,2,\ldots,k$ be corresponding order statistics.

Also we shall use the following result for (14.2.1) or (14.2.2)

$$h_i(x_i) = -g_i'(x_i)[g_i(x_i)]^{-2}, i=1,2,\ldots,k \qquad \ldots(14.2.3)$$

as $$\frac{d}{dx_i}\int_a^{x_i} h_i(u_i)du_i = h_i(x_i) = \frac{d}{dx_i}[g_i(x_i)]^{-1}$$

Theorem 14.1: With the above set-up if

$$Y_i = n_i[1-g_i(\theta_i)[g_i(x_{i(n_i)})]^{-1}], i=1,2,\ldots,k$$

where $x_{i(n_i)}$ is the realised value of the largest order statistics $X_{i(n_i)}$ for the right truncation parameter family of distributions and if $Y_i = n_i[1-g_i(\theta_i)[g_i(x_{i(1)})]^{-1}], i=1,2,\ldots,k$ where, $x_{i(1)}$ is the realised value of smallest order statistics $X_{i(1)}$ for the left truncated parameter family of distributions then $Y_1, Y_2, \ldots, Y_k$ are asymptotically independently distributed as $E(Y_i; 0, 1), i=1,2,\ldots,k$ where, $E(T;\mu,\sigma)$ has p.d.f.

$$f_T(t;\mu,\sigma) = \sigma^{-1}e^{-\sigma^{-1}(t-\mu)}, t>\mu, \sigma>0$$

Proof: Considering ordered random sample, $X_{i(1)}, X_{i(2)}, \ldots, X_{i(n_i)}$ of right truncation parameter family of distributions with p.d.f. given in (1), the p.d.f. of the largest order sufficient statistics, $X_{i(n_i)}$ is

$$f_{X_{i(n_i)}}(x_{i(n_i)};\theta_i) = n_i[F_X(x_{i(n_i)})]^{n_i-1} f_X(x_{i(n_i)})$$

$$= n_i[g_i(\theta_i)[g_i(x_{i(n_i)})]^{-1}]^{n_i-1} g_i(\theta_i)[g_i'(x_{i(n_i)})[g_i(x_{i(n_i)})]^{-2}]dx_{i(n_i)}$$

$$\ldots(14.2.4)$$

Similarly, for left truncation parameters family of distributions with p.d.f. (14.2.2), the p.d.f. of smallest order sufficient statistic, $X_{i(1)}$ is

$$f_{X_{i(1)}}(x_{i(1)};\theta_i)=n_i[g_i(\theta_i)[g_i(x_{i(1)})]^{-1}]^{n_i-1}g_i(\theta_i)[g_i'(x_{i(1)})[g_i(x_{i(1)})]^{-2}]dx_{i(1)} \quad \ldots(14.2.5)$$

Making the transformation as:

$$Y_i=n_i[1-g_i(\theta_i)[g_i(x_{i(n_i)})]^{-1}],\ i=1,2,\ldots,k \text{ in (14.2.4)} \quad \ldots(14.2.6)$$

and

$$Y_i=n_i[1-g_i(\theta_i)[g_i(x_{i(1)})]^{-1}],\ i=1,2,\ldots,k \text{ in (14.2.5)} \quad \ldots(14.2.7)$$

using $g_i(\theta_i)[g_i'(x_{i(n_i)})[g_i(x_{i(n_i)})]^{-2}]dx_i=\dfrac{dy_i}{n_i}$ for (14.2.6)

and $g_i(\theta_i)[g_i'(x_{i(1)})[g_i(x_{i(1)})]^{-2}]dx_i=\dfrac{dy_i}{n_i}$

for (14.2.7) the p.d.f. of Y_i in both cases turn out as

$$f_{Y_i}(y_i)=n_i\left(1-\frac{y_i}{n_i}\right)^{n_i-1}\frac{dy_i}{n_i},\ i=1,2,\ldots,k$$

As $n_i\to\infty$, the asymptotic distribution of Y_i has p.d.f. as

$$f_{Y_i}(y_i)\xrightarrow{d}e^{-y_i},\ i=1,2,\ldots,k$$

that is, $Y_1,Y_2,\ldots,Y_k$ are all asymptotically independently distributed as $E(Y_i;0,1), i=1,2,\ldots,k$. Since Y_i depend on $X_i^{'}$s and since $X_i^{'}$s are independent, $Y_i(i=1,2,\ldots,k)$ are independently distributed.

14.3 ASYMPTOTIC TESTING FOR EQUALITY OF PARAMETERS OF *K* ONE-TRUNCATION PARAMETER FAMILY OF DISTRIBUTIONS

With the use of likelihood ratio test principle, testing equality of truncation parameter of K one-truncation parameter family of distribution is done in the following theorem. However, test turns out as asymptotic UMPU test.

Theorem 14.2: For K right truncated parameter family of distributions defined by (14.2.1), the asymptotic UMPU test for testing the hypothesis $H:\theta_1 = \theta_2 = \ldots = \theta_k$ is to reject at level α whenever

$$\frac{n-k}{k-1}(C\underline{\overline{w}})'(CC')^{-1}(C\underline{\overline{w}}) \geq F_{n-k,\,k-1};\alpha \qquad \ldots(14.3.1)$$

where $F_{n-k,\,k-1};\alpha$ is upper α% point of Snedekar F-distribution with $(n-k, k-1)$ d.f. and C is any $(k-1)\times k$ contrast matrix of constants,

$$\underline{w}_i = (w_{i1}, w_{i2}, \ldots, w_{in})', i = 1, 2, \ldots, k, \overline{w}_i = \frac{1}{n}\sum_{j=1}^{n} w_{ij}, j = 1, 2, \ldots, k,$$

$$\underline{\overline{w}} = (\overline{w_1}, \overline{w_2}, \ldots, \overline{w_k})', w_i = \lambda_i Y_i, \lambda_i = \frac{\theta_i}{n_i}, i = 1, 2, \ldots, k, Y_i\ i = 1, 2, \ldots, k$$

is defined either in (14.2.6) or (14.2.7).

Proof: According to Theorem 14.1, as $Y_1, Y_2, \ldots, Y_k$ are asymptotically independently distributed as $E(Y_i; 0, 1), i = 1, 2, \ldots, k$ then $w_1, w_2, \ldots, w_k$ are asymptotically independently distributed as $E(w_i; 0, \lambda_i), i = 1, 2, \ldots, k$. The p.d.f. of w_i will be

$$f_{W_i}(w_i, \lambda_i) = \frac{1}{\lambda_i} e^{-\frac{w_i}{\lambda_i}}, i = 1, 2, \ldots, k$$

Now testing

$$H:\theta_1 = \theta_2 = \ldots = \theta_k \Leftrightarrow H_1: \frac{\theta_1}{n_1} = \frac{\theta_2}{n_2} = \ldots = \frac{\theta_k}{n_k}$$

when $\qquad n_1 = n_2 = \ldots n_k \Leftrightarrow H_2: \lambda_1 = \lambda_2 = \ldots = \lambda_k.$

Let $(k-1)\times k$ contrast matrix C be

$$C = \begin{bmatrix} 1 & 0 & 0 & \cdots & 0 & -1 \\ 0 & 1 & 0 & \cdots & 0 & -1 \\ \vdots & \vdots & \vdots & \cdots & \vdots & \vdots \\ 0 & 0 & 0 & \cdots & 1 & -1 \end{bmatrix}$$

Then $H: C\underline{\theta} = 0 \Leftrightarrow H_2: C\underline{\lambda} = 0,$

where $\qquad \underline{\theta} = (\theta_1, \theta_2, \ldots, \theta_k)', \underline{\lambda} = (\lambda_1, \lambda_2, \ldots, \lambda_k)'$

As $w_1, w_2, \ldots, w_k$ are all independently asymptotically distributed as $E(w_i; 0, \lambda_i), i = 1, 2, \ldots, k$, the joint distribution of $w_1, w_2, \ldots, w_k$ is asymptotically distributed as k-variate normal distribution with mean vector, $\underline{\lambda} = (\lambda_1, \lambda_2, \ldots, \lambda_k)$ and variance covariance matrix as identity matrix of order k, I_k, that is,

$$\underline{w} = (w_1, w_2, \ldots, w_k)' \xrightarrow{d} N_k(\underline{\lambda}, I)$$

Let $\underline{w}_i = (w_{i1}, w_{i2}, \ldots, w_{in})', i = 1, 2, \ldots, k$ be k random samples each of size n from

$$w_i, i = 1, 2, \ldots, k,$$

$$\overline{w}_i = \frac{1}{n}\sum_{j=1}^{n} w_{ij}, \; j = 1, 2, \ldots, k,$$

$$\underline{\overline{w}} = (\overline{w_1}, \overline{w_2}, \ldots, \overline{w_k})'$$

and
$$\overline{\overline{w}} = \frac{1}{k}\sum_{i=1}^{k} \overline{w}_i.$$

For testing H equivalently testing $H_2 : C\underline{\lambda} = 0$ we need the distribution of $C\underline{W}$ which is $N_{k-1}(C\underline{\lambda}, CC')$ as $n \to \infty$. By the usual multivariate theory of testing the mean vector of multivariate normal distribution equal to zero vector is to reject H_2, in this case, whenever

$$\frac{n-(k-1)-1}{k-1}(c\underline{\overline{w}})'(cc')^{-1}(c\underline{\overline{w}}) \geq F_{n-k,\,k-1};\alpha$$

where $F_{n-k,\,k-1};\alpha$ is upper $\alpha\%$ point of Snedekar F-distribution with $(n-k, k-1)$ d.f., which is asymptotic UMPU test of level α. The computation of quadratic form, $(C\underline{\overline{w}})'(CC')^{-1}(C\underline{\overline{w}})$ can be simplified as under.

As $\quad \underline{C\overline{w}} = (\overline{w}_1 - \overline{w}_k, \overline{w}_2 - \overline{w}_k, \ldots, \overline{w}_{k-1} - \overline{w}_k),$

$CC' = I_{k-1} + \underline{1}\,\underline{1}'$ where $\underline{1}$ is $(k-1)$-vector of unity.

$$(CC')^{-1} = I_{k-1} - \frac{1}{k}\underline{1}\,\underline{1}'$$

Using these results of $C\underline{\overline{w}}$ and $(CC')^{-1}$ the quadratic form $(C\underline{\overline{w}})'(CC')^{-1}(C\underline{\overline{w}})$ simplifies to

$$(C\underline{\overline{w}})'(CC')^{-1}(C\underline{\overline{w}}) = \sum_{i=1}^{k} \overline{w}_i^2 - \frac{1}{k}[(k-1)\overline{\overline{w}} + \overline{w}_k]^2$$

where
$$\overline{w}_i = \frac{1}{n}\sum_{j=1}^{n} w_{ij},$$

and
$$\overline{\overline{w}} = \frac{1}{k}\sum_{i=1}^{k} \overline{w}_i.$$

Particularly, when $K = 2$, $C = (1-1)$, $CC' = 2$, $C\underline{\overline{w}} = (\overline{w}_1 - \overline{w}_2)$,

$$(C\underline{\overline{w}})'(CC')^{-1}(C\underline{\overline{w}}) = \frac{(\overline{w}_1 - \overline{w}_2)^2}{2}, \text{ since } \frac{\overline{w}_1 - \overline{w}_2}{\sqrt{2}} \xrightarrow{d} N(0,1)$$

Under null hypothesis $\theta_1 = \theta_2$,

$$(n-2)(C\underline{\overline{w}})'(CC')^{-1}(C\underline{\overline{w}}) \xrightarrow{d} (n-2)\chi_1^2$$

when H is true.

14.4 ASYMPTOTIC CONDITIONAL DISTRIBUTION OF TWO INDEPENDENT ONE-TRUNCATION PARAMETER FAMILY OF DISTRIBUTIONS

Let $K = 2$ in p.d.f. (14.2.1), that is,

$$f_{x_1}(x_1, \theta_i) = g_1(\theta_1)h_1(x_1),\ a < x_1 < \theta_1 < b \qquad \ldots(14.4.1)$$

$$f_{x_2}(x_2, \theta_2) = g_2(\theta_2)h_2(x_2),\ a < x_2 < \theta_2 < b \qquad \ldots(14.4.2)$$

be p.d.f. of two-independent right truncated parameter family of distributions.

For testing the equality of parameters, $\theta_1 = \theta_2$, conditional distribution is needed, which is obtained in the following theorem.

Theorem 14.3: For the independently distributed random variables X_1 and X_2 having p.d.f. (14.4.1) and (14.4.2) and setting $Y_i = n_i[1 - g_i(\theta_i)[g_i(x_{i(n)})]^{-1}], i = 1, 2$ for right truncation parameter family, $w_i = \lambda_i Y_i$, $\lambda_i = \frac{\theta_i}{n_i}, i = 1, 2$, $\lambda_i = \frac{\theta_i}{n_i}, i = 1, 2, T = w_1 + w_2$. Then the asymptotic conditional distribution of w_2 given $T = t$ has p.d.f. as

$$f_{w_2|T}(w_2 \mid T = t) = \frac{ve^{-vw_2}}{1 - e^{-vt}}, 0 < w_2 < t, v = \frac{1}{\lambda_2} - \frac{1}{\lambda_1}.$$

Proof: As shown in Theorem 14.1, $Y_1 \xrightarrow{d} E(0, 1)$ and $Y_2 \xrightarrow{d} E(0, 1)$ are independently distributed and hence, $w_1 = \lambda_1 Y_1 \xrightarrow{d} E(0, \lambda_1)$ and $w_2 = \xrightarrow{d} E(0, \lambda_2)$

The asymptotic joint p.d.f. of w_1 and w_2,

$$f_{w_1, w_2}(w_1, w_2) = \frac{1}{\lambda_1 \lambda_2} e^{-\frac{w_1}{\lambda_1} - \frac{w_2}{\lambda_2}}$$

transforms to the joint distribution of T_2 and T by the transformation $(w_1, w_2) \to (w_2, T), T = w_1 + w_2$ as

$$f_{w_2, T}(w_1, w_2) = \frac{1}{\lambda_1 \lambda_2} e^{-\frac{t}{\lambda_1} - w_2\left(\frac{1}{\lambda_2} - \frac{1}{\lambda_1}\right)}, 0 < w_2 < t \qquad \text{...(14.4.3)}$$

and integrating out with respect to T_2, the asymptotic marginal p.d.f. of T will be

$$f_T(t) = \frac{e^{-\frac{t}{\lambda_1}} - e^{-\frac{t}{\lambda_2}}}{\lambda_1 - \lambda_2} \qquad \text{...(14.4.4)}$$

Using (14.4.3) and (14.4.4) the asymptotic conditional distribution of w_2 given $T = t$ has p.d.f.

$$f_{w_2|T}(w_2 \mid T=t)=\frac{ve^{-vw_2}}{1-e^{-vt}},\ 0<w_2<t \qquad \text{...(14.4.5)}$$

where
$$v=\frac{1}{\lambda_2}-\frac{1}{\lambda_1}$$

14.5 ASYMPTOTIC TESTS FOR TESTING PARAMETERS OF TWO ONE-TRUNCATION PARAMETER FAMILY OF DISTRIBUTIONS

Many often situations arise where truncation point θ_i and population distribution with p.d.f. $f_{X_i}(x_i;\theta_i)$ changes during sampling due to known or unknown causes such as laboratory testing of blood samples changes due to different testers, due to different methods, due to different timings so that the parameter value θ_1, before laboratory testing and the values of parameter θ, such as $\theta_2, \theta_3, \ldots, \theta_k$due to the above said known or unknown causes. One would like to test the hypothesis of equality of parameters, $H: \theta_1=\theta_2=\ldots=\theta_k$. The following theorem helps to obtain reasonable tests.

Theorem 14.4: For the p.d.f.s (14.4.1) and (14.4.2) asymptotic UMPU level-α tests for testing

(i) $H_1: \theta_1-\theta_2>\theta_0$

(ii) $H_2: \theta_1=\theta_2, 0<\theta_2<\theta_1, \theta_0>0$ are (i) reject H_1 at level-α whenever $w_2 > c(t)$ where $c(t)$ is given by

$$c(t)=v_0\log[1-(1-\alpha)(1-e^{-t})]^{-1},\ v_0=\frac{n\theta_0}{\theta_1\theta_2} \qquad \text{...(14.5.1)}$$

and (ii) accept H_2 at level-α whenever $c_1(t)<w_2<c_2(t)$ for given t, and $c_1(t)$ and $c_2(t)$ are obtained by solving the following two equations

$$e^{-c_1(t)}-e^{-c_2(t)}=(1-\alpha)(1-e^{-t}) \qquad \text{...(14.5.2)}$$

$$c_2(t)e^{-c_2(t)}-c_1(t)e^{-c_1(t)}=(1-\alpha)te^{-t} \qquad \text{...(14.5.3)}$$

Proof: As $0<\theta_1<\theta_2, \theta_0>0$, the hypothesis

$$H_1:\theta_1-\theta_2>\theta_0 \text{ versus } K_1:\theta_1-\theta_2<\theta_0 \Leftrightarrow H_1':\nu>\nu_0>0 \text{ vs } K_1':0<\nu<\nu_0$$

where

$$\nu=\frac{1}{\lambda_2}-\frac{1}{\lambda_1},\ \lambda_i=\frac{\theta_i}{n_i}, i=1,2,\ n_1=n_2=n,\ \lambda_0=\frac{\theta_1\theta_2}{n\theta_0}, \nu_0=\frac{1}{\lambda_0}$$

and similarly when H_2 is true, that is $H_2:\theta_1-\theta_2=0 \Leftrightarrow H_2':\nu=0$, when $n_1=n_2$.

Thus, testing H_1 and H_2 is same as testing H_1' and H_2', respectively. In Theorem 14.3 it is shown that the asymptotic conditional distribution of w_2 given $w_1+w_2=T=t$ is truncated exponential distribution with known truncation point with p.d.f. given in (14.4.5). Since this conditional p.d.f.

$$f_{w_2|T}(w_2\,|t;\nu)=\frac{\nu e^{-\nu w_2}}{1-e^{-t}}, 0<w_2<t$$

has monotone likelihood ratio in $(-w_2)$, using textbook theory (Lehmann), the asymptotic UMPU test of level-α for testing $H_1':\nu>\nu_0>0$ is

$$\phi(w_2,T)=\begin{cases}1, & \text{if} \quad w_2>c(t), \text{ for given } t\\ 0, & \text{otherwise}\end{cases}$$

where $C(t)$ is determined by

$$E_{H_1'}[\phi(w_2,T)|\,t]=\alpha \Rightarrow P_{H_1}[w_2>c(t)\,|T=t]=\alpha$$

Which when simplified gives the value of $C(t)$ by (14.5.1). Further asymptotic UMPU test of level-α for testing $H_2':\nu=0$ is

$$\phi(w_2,T)=\begin{cases}1, & \text{if} \quad w_2<c_1(t) \text{ or } w_2>c_2(t), \quad \text{for given } t\\ 0, & \qquad\qquad\qquad\qquad \text{otherwise}\end{cases}$$

where $c_1(t)$ and $c_2(t)$ are determined by solving the equations (14.5.2) and (14.5.3)

$$E_{H_2'}[\phi(w_2,T)|t]=\alpha \Rightarrow P_{H_2'}[c_1(t)<w_2<c_2(t)|T=t]=1-\alpha$$

Which gives equation (14.5.2) and

$$EE_{H_2'}[w_2\phi(w_2, T)|t] = \alpha E_{H_2'}(w_2|t)$$

$$\Rightarrow \qquad E_{H_1'}[w_2\phi(w_2, T)|t] = \alpha E_{H_1'}(w_2|t)$$

$$\Rightarrow \qquad \int_{c_1(t)}^{c_2(t)} \frac{w_2 e^{-w_2}}{1-e^{-t}} dw_2 = (1-\alpha)\int_0^t \frac{w_2 e^{-w_2}}{1-e^{-t}} dw_2$$

which when evaluated gives equation (14.5.3). One may use incomplete gamma function to evaluate (14.5.2) and (14.5.3).

Remark 1

Similar results are true for two independent left truncated parameter family of distributions.

Theorem 14.5: For K one-truncation parameter family of distributions with truncation parameters $\theta_1, \theta_2, \ldots, \theta_k$ the asymptotic UMPU tests for testing (i) $\theta_i > \theta_j; i \neq j$ and (ii) $\theta_i = \theta_j$ are same as obtained in Theorem 14.4.

Proof: Analogues to Theorem 14.4 by observing that the necessary conditional distribution of the complete sufficient statistic for the parametric function to be tested by conditioning the distribution of those of nuisance parameters vanishes in the marginal distribution and this yield conditional distribution obtained in (14.4.5).

14.6 APPLICATION TO SPECIFIC ONE-TRUNCATION PARAMETERS FAMILY OF DISTRIBUTIONS

The results of Theorems (14.1), (14.2), (14.4) have been applied for testing the parameters of some specific distributions such as uniform, power function, generalized uniform, pareto, negative exponential, truncated exponential distributions.

(i) Generalised Uniform Distribution

Let $X_1, X_2, \ldots, X_k$ be independently distributed as generalised uniform distribution with p.d.f. as

$$f_{X_i}(x_i; \theta_i) = \frac{a+1}{\theta_i^{a+1}} x_i^a,\ a > -1,\ a < x_i < \theta_i,\ i = 1, 2, \ldots, k\ = g(\theta_i)h(x_i)$$

where $\quad (\theta_i) = \dfrac{a+1}{\theta_i^{a+1}},\ h(x_i) = x_i^a,\ i = 1, 2, \ldots, k$

so that $\dfrac{g(\theta_i)}{g(x_{i(n)})} = \left(\dfrac{x_{i(n)}}{\theta_i}\right)^{a+1}$, where, $x_{i(n)}$ is the realised value of nth order statistics $X_{i(n)}$, corresponding to the random samples $X_{i1}, X_{i2}, \ldots, X_{in}$, $i = 1, 2, \ldots, k$ on random variables $X_i, i = 1, 2, \ldots, k$. Due to Theorem (14.1), the asymptotic distributions of

$$Y_i = n\left(1 - \frac{g(\theta_i)}{g(x_{i(n)})}\right) = n\left(1 - \left(\frac{x_{i(n)}}{\theta_i}\right)^{a+1}\right), i = 1, 2, \ldots, k$$

are $E(Y_1, 0, 1), E(Y_2, 0, 1), \ldots, E(Y_k, 0, 1)$ which are independently distributed.

With the use of Theorem (14.2) the asymptotic UMPU level-α test is given by (14.3.1) for testing the hypothesis $H: \theta_1 = \theta_2 = \ldots = \theta_k$.

For two independent generalised uniform distributions of X_1 and X_2 being particular test of k generalised uniform distributions, $H = 2$ using the results of Theorem (14.4) the critical region of the asymptotic UMPU level-α test is given by (14.5.1) for testing $H_1: \theta_1 - \theta_2 > \theta_0$ and acceptance region of the asymptotic UMPU level-α test for testing $H_2: \theta_1 = \theta_2$ can be obtained by solving the equations (14.5.2) and (14.5.3) for $c_1(t)$ and $c_2(t)$.

(ii) Uniform Distribution

Uniform distribution being a particular case of generalised uniform distribution defined in 5(i) above with $a = 0$. The asymptotic UMPU tests for testing equality of parameters in k uniform distributions, that is, $H: \theta_1 = \theta_2 = \ldots = \theta_k$ is given by (14.3.1) and asymptotic UMPU tests for testing $H_1: \theta_1 - \theta_2 > \theta_0$ and $H_2: \theta_1 = \theta_2$ are given by (14.5.1), (14.5.2), and (14.5.3).

(iii) Power Function Distribution

Again power function distribution being a particular case of generalised uniform distribution defined in 5(i) above with $a = c - 1, c > 0$. The asymptotic UMPU tests for testing $H: \theta_1 = \theta_2 = \ldots = \theta_k$ is given by (14.3.1) and asymptotic UMPU tests for testing $H_1: \theta_1 - \theta_2 > \theta_0$ and $H_2: \theta_1 = \theta_2$ are given by (14.5.1) and (14.5.2) respectively.

(iv) Right Truncated Parameter Exponential Distribution

Let $X_1, X_2, \ldots, X_k$ be independently distributed as right truncated parameter exponential distributions with p.d.f. as

$$f_{X_i}(x_i; \beta, \theta_i) = \frac{\beta e^{-\beta x_i}}{1 - e^{-\beta \theta_i}};\ 0 < x_i < \theta_i, i = 1, 2, \ldots, k; \beta > 0$$

is known

$$= g(\theta_i)h(x_i);\ i = 1, 2, \ldots, k$$

where $\qquad g(\theta_i) = (1 - e^{-\beta\theta_i})^{-1}, h(x_i) = \beta e^{-\beta x_i}, i = 1, 2, \ldots, k$

so that

$$g(\theta_i)g(x_{i(n)})^{-1} = (1 - e^{-\beta x_{i(n)}})(1 - e^{-\beta\theta_i})^{-1}, i = 1, 2, \ldots, k$$

and making the transformation

$$Y_i = n_i(1 - g(\theta_i)g(x_{i(n)})^{-1})$$

The p.d.f. of nth order statistic $X_{i(n)}$

$$f_{X_{i(n)}}(x_{i(n)}) = n_i(g(\theta_i)g(x_{i(n)})^{-1})g(\theta_i)\frac{d}{dx_{i(n)}}g(x_{i(n)})^{-1}$$

Transfer to the asymptotic independent distributions of $Y_1, Y_2, \ldots, Y_k$ with p.d.f's as $E(Y_1, 0, 1), E(Y_2, 0, 1), \ldots, E(Y_k, 0, 1)$ due to Theorem 14.1. Using Theorem (14.2), the asymptotic UMPU level-α test for testing $H: \theta_1 = \theta_2 = \ldots = \theta_k$ is given by (14.3.1). For two independent right truncated parameter distributions, that is, for $K = 2$, the results of Theorem 14.4 give the critical region such as the critical region of asymptotic UMPU level-α test for testing $H_1: \theta_1 - \theta_2 > \theta_0$ is given by (14.5.1) and the acceptance region of asymptotic UMPU level-α test is given by solving the equations (14.5.2) and (14.5.3) for testing $H_2: \theta_1 = \theta_2$.

(v) Negative Exponential Distribution

Let $X_1, X_2, \ldots, X_k$ be independently distributed as negative exponential distribution with p.d.f. as

$$f_{X_i}(x_i, \theta_i) = e^{-(x_i - \theta_i)}; 0 < \theta_i < x_i < \infty, i = 1, 2, \ldots, k$$

$$= g(\theta_i)h(x_i);\ i = 1, 2, \ldots, k$$

where $$g(\theta_i)=e^{\theta_i} \quad \text{and} \quad h(x_i)=e^{x_i}, i=1,2,\ldots,k$$

so that $g(\theta_i)g(x_{i(1)})^{-1}=e^{-(x_{i(1)}-\theta_i)}$ $i=1,2,\ldots,k$ where, $x_{i(1)}$ is the realised value of first order statistic $X_{i(1)}$.

Due to Theorem 14.1, the asymptotic distributions of

$$Y_i=n_i(1-e^{-(x_{i(1)}-\theta_i)}), i=1,2,\ldots,k$$

are independently distributed as $E(0,1), E(0,1),\ldots, E(0,1)$. Using Theorem 14.2, the asymptotic UMPU level-α test for testing $H:\theta_1=\theta_2=\ldots=\theta_k$ is to reject it when (2) holds. For $K=2$, that is, for two independently distributed negative exponential distributions, applying the results of Theorem 14.4, the critical region of asymptotic UMPU level-α test for testing $H_1:\theta_1-\theta_2>\theta_0$ is given by (14) and the acceptance region of the asymptotic UMPU level-α test is given by the solution of equations (14.5.2) and (14.5.3) for testing $H_2:\theta_1=\theta_2$.

Remark 2

More general case of negative exponential distribution is for the p.d.f.

$$f_{X_i}(x_i,\theta_i)=ba^{-b(x_i-\theta_i)}; 0<\theta_i<x_i<\infty, i=1,2,\ldots,k$$

and a, b are known positive constants. If $a=e$ in the above distribution then the distribution turns out as left truncated exponential distribution with truncation point, θ_i.

(vi) Pareto Distribution

Let $X_1, X_2,\ldots, X_k$ be independently distributed as Pareto distribution with p.d.f's are

$$f_{X_i}(x_i,\theta_i)=a\cdot\theta_i^a\cdot x_i^{-(a+1)}; 0<\theta_i<x_i, i=1,2,\ldots,k$$

where a is a known positive constant.

$$f_{X_i}(x_i,\theta_i)=g(\theta_i)h(x_i); i=1,2,\ldots,k$$

where $$g(\theta_i)=a\cdot\theta_i^a, h(x_i)=x_i^{-(a+1)}, \quad i=1,2,\ldots,k$$

so that $g(\theta_i)g(x_{i(1)})^{-1} = \left(\dfrac{\theta_i}{x_{i(1)}}\right)^a$ $i = 1, 2, \ldots, k$ where, $x_{i(1)}$ is the realised value of first order statistic $X_{i(1)}$.

Due to Theorem 14.1, the asymptotic distributions of

$$Y_i = n_i\left(1 - \left(\frac{\theta_i}{x_{i(1)}}\right)^a\right), i = 1, 2, \ldots, k$$

are independently distributed as $E(0, 1), E(0, 1), \ldots, E(0, 1)$.

Due to Theorem 14.2, the asymptotic UMPU level-α test for testing $H: \theta_1 = \theta_2 = \ldots = \theta_k$ is to reject it when (2) holds. For two independently distributed Pareto distributions, applying the results of Theorem 14.4, the critical region of asymptotic UMPU level-α test for testing $H_1: \theta_1 - \theta_2 > \theta_0$ is given by (14.5.1) and the acceptance region of the asymptotic UMPU level-α test is given by the solution of equations (14.5.2) and (14.5.3) for testing $H_2: \theta_1 = \theta_2$.

Appendix

For usual probability distributions we recall here the density and the first two moments of the distributions which are used in this book. The reader may use the book by Johnson and Kotz (1969-1972). The densities are given with respect to the Lebesgue or the counting measure depending on the type of the distribution.

(1) Binomial Distribution, $b(n, p)$

$$f_X(x; n, p) = P(X = x) = \binom{n}{x} p^x (1-p)^{n-x},\ x = 0, 1, \ldots n,\ 0 \le p \le 1.$$

and $$E(X) = np,\ \mathrm{var}(X) = np(1-p)$$

(2) Hypergeometric Distribution, Hyp (N, n, M)

$$f_X(x; M) = \frac{\binom{M}{x}\binom{N-M}{n-x}}{\binom{N}{n}},\ x = 0, 1, 2 \ldots, n$$

$$E(X) = \frac{nM}{N},\ \mathrm{var}(X) = \frac{M(N-M)n(N-n)}{N^2(N-1)},\ n \le N \text{ and } M \le N.$$

(3) Poisson Distribution, $P(\lambda)$

$$f_X(x;\lambda)=\frac{e^{-\lambda^x}}{x!}, x=0,1,2,\ldots;\lambda>0$$

and $E(X)=\lambda$ and $\text{var}(X)=\lambda$

(4) Negative Binomial Distribution, Neg (n, p)

$$f_X(x;p)=\binom{n+x-1}{x}p^n(1-p)^x, x=0,1,2\ldots$$

$$0\le p\le 1$$

and

$$E(X)=\frac{n(1-p)}{p}, \text{var}(X)=\frac{n(1-p)}{p^2}$$

(5) Multinomial Distribution, $M_k(n; p_1, p_2, \ldots, p_k)$

$$f(x_1,x_2,\ldots,x_k;p_1,p_2,\ldots,p_k)=\binom{n}{x_1,x_2,\ldots,x_k}$$

$$\prod_{i=1}^{k}p_i^{xi}, x_i>0, p_i\ge 0, i=1,2,\ldots,k$$

$$\sum_{i=1}^{k}x_i=n, \sum_{i=1}^{k}p_i=1, E(X_i)=np_i, \text{var}(X_i)=np_i(1-p_i)$$

$$\text{cov}(X_i,X_j)=-np_ip_j(i\ne j), i=1,2,\ldots k\cdot j=1,2,\ldots k.$$

(6) Uniform Distribution, $u(a, b)$

$$f_X(x;\alpha,\beta)=\begin{cases}\dfrac{1}{\beta-\alpha}, & \text{if} \quad \alpha<x<\beta\\ 0, & \text{otherwise}\end{cases}$$

$$E(X)=\frac{\alpha+\beta}{2}, \text{var}(X)=\frac{(\beta-\alpha)^2}{12}$$

(7) Normal Distribution, $N(\mu, \sigma^2)$

$$f_X(x:\mu,\sigma)=\frac{1}{\sigma\sqrt{2\pi}}\exp\left[-\frac{(x-\mu)^2}{2\sigma^2}\right], -\infty<x<\infty, -\infty<\mu<\infty, \sigma>0$$

$$E(X)=\mu \text{ and } \operatorname{var}(X)=\sigma^2.$$

The p.d.f. of $N(0, 1)$ is denoted by $\phi(x)$ and distribution function of $N(0, 1)$ is denoted by $\Phi(x)$. Thus,

$$\phi(x)=\frac{1}{\sqrt{2\pi}}\exp\left[-\frac{(x)^2}{2\sigma^2}\right], -\infty<x<\infty$$

$$\Phi(x)=\int_{-\infty}^{x}\phi(t)dt, -\infty<x<\infty$$

If $\Phi(Z_\alpha)=1-\alpha$ then Z_α is called upper $\alpha\%$ point of $N(0,1)$ distribution.

(8) Gamma Distribution, $G(\propto, p)$

$$f_X(x;\alpha,p)=\frac{\alpha^p}{\Gamma(p)}e^{-\alpha x}x^{p-1}, 0<x<\infty, \alpha>0, p>0$$

$$E(X)=\frac{p}{\alpha}, \operatorname{var}(X)=\frac{p}{\alpha^2}$$

Particular Cases:

(i) $G\left(\frac{1}{2},\frac{n}{2}\right)$ is called chi-square distribution on n degree of freedom and it is denoted by χ_n^2.

(ii) $G(\alpha, 1)$ is called exponential distribution denoted by $\exp(\alpha)$ with p.d.f.

$$f_X(x;\alpha)=\alpha e^{-\alpha x}, x>0, \alpha>0$$

(iii) If $X - \theta$ has $G(1, 1)$ distribution, that is, if X has p.d.f.

$$f_X(x; \theta) = e^{-(x-\theta)}, \theta < x < \infty$$

then it is called negative exponential distribution which is a special case of $E(\mu, \sigma)$, the exponential distribution with p.d.f.

$$f(x; \mu, \sigma) = \begin{cases} \frac{1}{\sigma} e^{-(x-\mu)/\sigma, \ \mu<x<\infty} \\ 0, \quad \text{otherwise} \end{cases}$$

(9) Beta Distribution, Be(α, β)

$$f_X(x; \alpha, \beta) = \frac{x^{\alpha-1}(1-x)^{\beta-1}}{B(\alpha, \beta)}, 0 < x < 1; \alpha > 0; \beta > 0$$

where, $B(\alpha, \beta) = \frac{(\alpha)\Gamma(\beta)}{(\alpha+\beta)}$, $E(X) = \frac{\alpha}{\alpha+\beta}$, $\text{var}(X) = \frac{\alpha\beta}{(\alpha+\beta)^2(\alpha+\beta+1)}$

This is beta kind one distribution.
Second kind of beta distribution has p.d.f.

$$f_X(x; \alpha, \beta) = \frac{x^{\alpha-1}}{B(\alpha, \beta)(1+x)^{(\alpha+\beta)}}, 0 < x < \infty; \alpha > 0; \beta > 0$$

This is particular case of beta kind one distribution when $y = \frac{x}{1-x}$ and $x \sim B_e(\alpha, \beta)$ If $X_i \sim G(\alpha_i, \beta_i), i = 1, 2$ are independent then $\frac{X_1}{X_2}$ has beta type-2 distribution.

(10) Chi-square Distribution χ_n^2

$$f_X(x; n) = \frac{1}{2^{\frac{n}{2}}\Gamma\left(\frac{n}{2}\right)} e^{-\frac{x}{2}} x^{\frac{n}{2}-1}, x > 0$$

$E(X) = n$, $\text{var}(X) = 2n$.

This is also sampling distribution of $\sum_{i=1}^{n} X_i^2$ when $X_1, X_2, \ldots, X_n$ are independently distributed as $N(0, 1)$.

(11) Student's *t*-Distribution, t_n

$$f_T(t, n) = \frac{\Gamma\left(\frac{n+1}{2}\right)}{\sqrt{n\pi}\,\Gamma\left(\frac{n}{2}\right)}\left(1+\frac{t^2}{n}\right)^{-\frac{(n+1)}{2}}, -\infty < t < \infty.$$

$E(T) = 0$ if $n > 1$ and $\text{var}(T) = \frac{n}{n-2}$ if $n > 2$. This distribution is also sampling distribution of $T = \frac{X}{\sqrt{\frac{Y}{n}}}$

where $X \sim N(0,1)$ and Y has independent χ_n^2 distribution.

(12) Fisher's *F*-Distribution, $F_{m,n}$

$$f(x; m, n) = \frac{\Gamma\left(\frac{(m+n)}{2}\right) m^{\frac{n}{2}} n^{\frac{m}{2}}}{\Gamma\left(\frac{m}{2}\right)\Gamma\left(\frac{n}{2}\right)} \frac{x^{\frac{(m-2)}{2}}}{(m+nx)^{\frac{(m+n)}{2}}}, 0 < x < \infty.$$

$$E_{m,n}(X) = \frac{n}{n-2}, n > 2, \text{var}_{m,n}(X) = \frac{2n^2(m+n-2)}{m(n-4)(n-2)^2}, n > 4$$

(13) Non-Central Chi-square Distribution, $\chi_n^2(\lambda)$

$$f_X(x, \lambda) = \frac{1}{2}\left(\frac{x}{\lambda}\right)^{\frac{p-2}{2}} I_{\frac{p-2}{2}}(\sqrt{\lambda x})e^{-\frac{\lambda+x}{2}}$$

$E(X) = p + \lambda$, $\text{var}(X) = np + 4\lambda$.

This can be derived as the sampling distribution of $\sum_{i=1}^{n} X_i^2$ when $X_i \sim N(\mu_i, 1)$

$i = 1, 2, \ldots, n$ and $\sum_{i=1}^{n} \mu_i = \lambda$.

(14) Pareto Distribution, Pa(α, x_0)

$$f_X(x;\alpha,x_0)=\frac{\alpha x_0^{\alpha}}{x^{\alpha+1}},\ x_0<x<\infty$$

$$E(X)=\frac{\alpha\, x_0}{\alpha-1},\ \alpha>1,\ \text{var}(X)=\frac{\alpha x_0^2}{(\alpha-1)^2(\alpha-2)},\ \alpha>2$$

(15) Cauchy Distribution, $C(\theta)$

$$f_X(x;\theta)=\frac{1}{\pi[1+(x-\theta)^2]},\ -\infty<x<\infty$$

Mean and other moments do not exist. When $\theta = 0$, it reduces to t-distribution on 1 degree of freedom.

(16) Laplace Distribution, La(θ)

$$f_X(x;\theta)=\frac{1}{2}e^{-|x-\theta|},\ -\infty<x<\infty,\ \theta\in R'$$

$$E(X)=\theta,\ \text{var}(x)=2$$

(17) Multivariate Normal distribution, $N_p(\mu, \Sigma)$

$$f_X(x;\mu,\Sigma)=|\Sigma|^{-\frac{1}{2}}(2\pi)^{-\frac{p}{2}}\exp\left[-\frac{1}{2}(x-\mu)'\Sigma^{-1}(x-\mu)\right]$$

$\mu\in R^p$ and Σ is pXp positive definite matrix

$$E(X)=\mu \text{ and } E[(x-\mu)(x-\mu)']=\Sigma$$

When Σ is not positive definite, the $N_p(\mu,\Sigma)$ has no density w.r.t. the Lebesgue measure R^p. For $p = 1$, it is univariate normal $X\sim N(\mu,\sigma^2)$ distribution and $y = \log x$ has log-normal distribution with y p.d.f. as

$$f_X(x:\mu,\sigma)=\frac{1}{\sqrt{2\pi}\sigma x}\exp\left[-\frac{(\log x-\mu)^2}{2\sigma^2}\right], x>0, -\infty<\mu<\infty, \sigma>0$$

$$E(X)=\exp\left(\mu+\frac{\sigma^2}{2}\right)$$

$$\text{var}(x)=[\exp(\sigma^2)-1][\exp(2\mu+\sigma^2)]$$

(18) Distribution of Order Statistics

If x_1, x_2, ..., x_n is a random sample from a random variable X with p.d.f. $f(x)$ and if $(x_1, x_2, \ldots x_n)$ is realized value of (x_1, x_2, ..., x_n) then by arranging $x_1, x_2, \ldots x_n$ in ascending order of magnitude the random variable associated with r^{th} observation, taking any value of confuragation of 1, 2, ..., n is called r^{th} order statistic denoted by $X_{(r)}$.

Thus, $x_{(1)}$, $x_{(2)}$, ..., $x_{(n)}$ are the corresponding order statistics of x_1, x_2, ..., x_n. The p.d.f. or r^{th} order statistic $X_{(r)}$ is

$$f_{X_{(r)}}(x)=\frac{n!}{(r-1)!(n-r)!}[F_X(x)]^{r-1}[1-F_X(x)]^{n-r}f_X(x), x\in R$$

where, $F_X(x)$ is the distribution function of x. Particularly, when $r = n$, the distribution of maximum order statistic, $X_{(n)}$ is

$$F_{X(n)}(x)=n[F_X(x)]^{n-1}f_X(x), x\in R$$

and when $r = 1$, it reduces to distribution of minimum order statistic, $X_{(1)}$ is

$$f_{X_{(1)}}(x)=n[1-f_X(x)]^{n-1}f_X(x), x\in R$$

Also the joint distribution of $x_{(1)}$, $x_{(2)}$, ..., $x_{(n)}$ is

$$f_{X_{(1)}, X_{(2)}, \ldots, X_{(n)}}(x_1, x_2, \ldots x_n)=n!\, f_{X_1}(x_1), f_{X_2}(x_2), \ldots, f_{X_n}(x_n)$$

The joint distribution of $X_{(r)}, X_{(s)}$, $r < s$ is

$$f_{X_{(r)}, X_{(s)}}(x, y) = \frac{n![F_X(x)]^{r-1}[F_X(y) - F_X(x)]^{s-r-1}[1 - F_X(y)]^{n-s}}{(r-1)!(s-r-1)!(n-s)!} f_X(x) f_X(y)$$

Particularly, when $r = 1$, $s = n$, the joint distribution of $x_{(1)}, x_{(n)}$ will be

$$f_{X_{(1)}, X_{(n)}}(x, y) = n(n-1)[F_X(y) - F_X(x)]^{n-2} f_X(x) f_X(y)$$

From this one can get the distribution of the range, $R = X_{(n)} - X_{(1)}$.

References

1. Basu (1955) : On Statistics independent of a complete sufficient statistics, Sankhya, 15, 377-380 and Sankhya, 20, (1958), 223-226.
2. Ferguson (1967) : Mathematical Statistics, A decision theoretic approach, Academic Press.
3. Gupta, R.C. (1974) : Distribution of the sum of the independent decapitated generalized Poisson Variables, Sankhya B, 36, 212-214.
4. Lehmann, E.L. (1947) : On optimum test of composite hypotheses with constraints, Ann. Math. Stat. Vol. 18, 473-494.
5. Lehmann, E.L. (2005) : Testing Statistical Hypotheses, Springer, New York.
6. Lehmann, E.L. (2003) : Theory of Point Estimation, Springer, New York.
7. Patel, S.R. (1978 a) : Minimum variance unbiased estimation of multivariate modified power series distribution, Metrika, 25, 155-161.
8. Patel, S.R. (1978 b) : Minimum variance unbiased estimation of left truncated multivariate power series distribution, Metrika, 25, 209-219
9. Patel, S.R. (1979 a) : Minimum variance unbiased estimation of left truncated multivariate modified power series distribution, Metrika, 26, 87-94
10. Patel, S.R. (1979 b) : Multivariate generalized negative binomial distribution, Metron XXXVII, 59-69.

11. Patel, S.R. (1981) : On multivariate generalized logarithmic series distribution, Metron XXXIX, 53-63.

12. Patel, S.R. (1990) : Unbiased estimation in Type-II censored samples from two-truncation parameter family, Common Stat. Theoy and Math., 19(10), 3809-3832.

13. Patel, S.R. (1991 a) : On unbiased estimation in two-truncation parameter density, Metron, XLIX-N1-4, 31-XII

14. Patel, S.R. (1991 b) : Unbiased estimation in doubly type-II censored samples from two-truncation parameter families, Common Stat. Theoy and Math., 20(7), 2219-2148.

15. Patel, S.R. (1991 c) : Unbiased estimation in sampling from two-truncation parameter families when both samples are type-II censored, Proceedings of the symposium on distribution theory, Kochi, Kerala, India, 79-92.

16. Patel, S.R. (1992 a) : Unbiased estimation of parametric function in sampling from one and two-truncation parameter families, Aust. J. Stat., 34(1), 47-54.

17. Patel, S.R. (1992 b) : Unbiased estimation of parametric function insampling from two-truncation parameter families, Common Stat. Theoy and Math., 21(3), 843-860.

18. Patel, S.R (1992 c) : Mixed effect exponential linear model, Common Stat. Theoy and Math., 21(9), 2721-2740.

19. Patel, S.R. (1994) : Unbiased estimation in doubly type-II censored samples from truncation parameter families in presence of an outlier, Sankhya series B, Vol. 56 part-2, 165-172.

20. Patel, S.R. (1999) : Asymptotic distribution for the order statistics of a truncated one parameter family of distribution, Communication in Statistics: Theory and Method, Vol. 28(8), 1823-1856.

21. Patel, S.R (2001) : On asymptotic distribution for two-truncation parameter family, Journal of Statistical Planning and Inference, 92(2001), 39-46.

22. Patel, S.R. (2010) : Confidence limits for the C.V data of field experiments on mustard crop. Agri. Sci. Digest 30(3), 230-231.

23. Patel, S.R. (2013) : Confidence limits for the CV data of the field experiments on wheat crop; Pearson Type-IV distribution; International Journal of Science and Nature 4(2), 338-340.

24. Patel, S.R. (2018) : Asymptotic test for truncation parameters of several one-truncation parameter family of distribution, Research and Reviews: Journal of Statistics, 7(2), 35-41.

25. Patel, S.R. (2018) : On asymptotic testing of equality of parameters of *k* exponential family of distributions, International Journal of Scientific Research in Mathematical and Statistical Sciences, 5(6), 177-182.

26. Patel, S.R. (2019) : On uniformly most powerful tests for truncated parameters of several one-parameter-truncated family of distributions, Research and Reviews: Journal of Statistics, 8(2), 1-9.

27. Rao, C.R. (1945) : Information and accuracy attainable in the estimation of statistical parameters, Bull, Cal. Math. Soc., 37, 81-91.

28. Rao, C.R. (1973) : Linear Statistical Inference, John Wiley & Sons.

29. Rohatgi, V.K. (1989) : Unbiased estimation of parametric function in sampling from two one-truncation parameter families, Aust. J. of Stat. 31, 327-332.

30. Roy & Mitra S.K. (1957) : Unbiased minimum variance estimation in a class of discrete distribution. Sankhya 18, 371-378.

31. Wald, A. (1947) : Sequential Analysis, Wiley, New York.

32. Zacks, S. (1971) : The Theory of Statistical Inference, John Wiley, New York.

Index

G

H

I

K

L

M